Recent Results in Cancer Research

Volume 193

For further volumes:
http://www.springer.com/series/392

Mei Hwei Chang · Kuan-Teh Jeang

Editors

Viruses and Human Cancer

From Basic Science to Clinical Prevention

 Springer

Editors

Mei Hwei Chang
Department of Pediatrics
National Taiwan University Hospital
Taipei
Taiwan, R.O.C.

Kuan-Teh Jeang
Molecular Virology Section, Laboratory of
  Molecular Microbiology
National Institute of Allergy
  and Infectious Disease
Bethesda, MD
USA

ISSN 0080-0015
ISBN 978-3-642-38964-1   ISBN 978-3-642-38965-8 (eBook)
DOI 10.1007/978-3-642-38965-8
Springer Heidelberg New York Dordrecht London

Library of Congress Control Number: 2013944531

Printed on acid-free paper

Springer is part of Springer Science+Business Media (www.springer.com)

Obituary

This special book "Viruses and Human Cancer: From Basic Science to Clinical Prevention", is dedicated to the memory of Dr. Kuan-Teh Jeang. He was the Chief of the Molecular Microbiology, National Institute of Allergy and Infectious Diseases (NIAID), National Institutes of Health (NIH), Bethesda, USA. It is with deep sadness to learn that Dr. Kuan-Teh Jeang, the co-editor of this book, passed away on January 27, 2013. His great devotion and contribution in research and the propagation for the knowledge will be missed forever in the scientific community.

Dr. Jeang is a highly respected world scientific leader, particularly in the field of molecular mechanisms of HIV and HTLV-1 pathogenesis and oncogenesis. When I was invited by Professor Hans-Joerg Senn and Professor Peter Schlag to editor this book, Dr. Jeang was the most suitable co-editor for this very important book in my mind. I felt truly honored that Dr. Jeang accepted my invitation to co-edit this book. He was such an inspiring and supportive co-editor of this book, I have been very grateful to have worked with him for this book.

Research on oncogenic viruses and related human cancers has advanced rapidly in the past decade. Most updated articles are focused mainly on a specific oncogenic virus and human cancer. A comprehensive updated monograph with integrated depth of knowledge in this regard is needed. In this book, leading researchers in the world delivered their most updated knowledge from basic science to clinical prevention to the readers. The book covers the virology, virus-induced inflammation and tissue injuries, oncogenic mechanisms, epidemiology, clinical and preventive or therapeutic strategies for the main human oncogenic viruses, including hepatitis B virus, hepatitis C virus, papilloma viruses, EB virus, Human Adult T-cell Leukemia-1 (HTLV-1), Human Herpes Virus-8 (HHV-8), and their related human cancers. The high risk hosts such as transplantation recipients are discussed. Future prospects, particularly the preventive strategies such as new vaccine development are also discussed.

Dr. Jeang had edited at least six books previous to this book. This book, "Viruses and Human Cancer: From Basics to Clinical Prevention" is an important book with his devotion not only as an editor but also an author for a chapter entitled "HTLV-1 and Leukemogenesis: Virus-cell Interactions in the Development of Adult T-cell Leukemia".

Dr. Jeang was born in Taiwan. At age 5, he moved with his family to Libya where he stayed until age 12, when he arrived in the United States. He was accepted to the Massachusetts Institute of Technology at age 16, and at age 18, he started his MD and Ph.D. training at the Johns Hopkins University of Medicine (Baltimore, USA); graduating with both degrees in 1984. He joined the NIH in 1985, had been a tenured investigator at the NIH for 27 years. His work has tremendous contribution to insights into HIV-1 gene expression and HTLV-1 transformation of cells, Mouse model of diseases, and Non-coding RNAs. He published more than 290 paper on peer reviewed journals, in the field of molecular virology, particularly on HIV and HTLV-I, with very high total citations >14,900 and H-index 65, ISI Web of Science.

In 2004, he and his colleagues founded the open access journal Retrovirology and served as the Editor-in-Chief of Retrovirology since 2004. His dedication had driven the excellent performance of the journal to one of the top ranking journals in virology. He also served as the Editor of Journal of Biomedical Science (1994–2004) and the Editor of Cell and Bioscience (2011–2013), an Associate Editor for Cancer Research (2003- present), and editor of Human Retrovirus and AIDS database (1994–1998). He was also on the editorial board of many other important scientific journals, such as the Journal of Biological Chemistry, and the Journal of Virology. He was the immediate Past-President of the Society of Chinese Bioscientists in America.

He was a gifted scientist and a remarkable and caring person. He was so passionate to link molecular virology to cancer. He also devoted a substantial amount of his intellectual effort to the promotion of scientific publication. His enthusiasm for the editorial work was showed by his extraordinary effort and dedication. While working on this book, he was very kind and liked to help when it is in need. He responded kindly and quickly in our e-mail communication. He leaves behind a unique heritage of ideas and spirit for others to pursue. Our condolence is with his wife Diane, and his children David, John, and Diana.

On behalf of the authors, I sincerely hope the readers like this special book, which is dedicated to the memory of Dr. Kuan-Teh Jeang.

Mei-Hwei Chang

Contents

Virus Infection and Human Cancer: An Overview

John T. Schiller and Douglas R. Lowy

Abstract

It is now estimated that approximately 10 % of worldwide cancers are attributable to viral infection, with the vast majority (>85 %) occurring in the developing world. Oncogenic viruses include various classes of DNA and RNA viruses and induce cancer by a variety of mechanisms. A unifying theme is that cancer develops in a minority of infected individuals and only after chronic infection of many years duration. The viruses associated with the greatest number of cancer cases are the human papillomaviruses (HPVs), which cause cervical cancer and several other epithelial malignancies, and the hepatitis viruses HBV and HCV, which are responsible for the majority of hepatocellular cancer. Other oncoviruses include Epstein–Barr virus (EBV), Kaposi's sarcoma-associated herpes virus (KSHV), human T-cell leukemia virus (HTLV-I), and Merkel cell polyomavirus (MCPyV). Identification of the infectious cause has led to several interventions that may reduce the risk of developing these tumors. These include preventive vaccines against HBV and HPV, HPV-based testing for cervical cancer screening, anti-virals for the treatment of chronic HBV and HCV infection, and screening the blood supply for the presence of HBV and HCV. Successful efforts to identify additional oncogenic viruses in human cancer may lead to further insight into etiology and pathogenesis as well as to new approaches for therapeutic and prophylactic intervention.

J. T. Schiller (✉) · D. R. Lowy
Laboratory of Cellular Oncology, Center for Cancer Research, National Cancer Institute, Bethesda, MD, USA
e-mail: schillej@mail.nih.gov

D. R. Lowy
e-mail: dl60z@nih.gov

M. H. Chang and K.-T. Jeang (eds.), *Viruses and Human Cancer*,
Recent Results in Cancer Research 193, DOI: 10.1007/978-3-642-38965-8_1,
© Springer-Verlag Berlin Heidelberg 2014

Contents

One of the most notable achievements in cancer etiology is the establishment that infection by a specific subset of human viruses is the primary cause of a substantial fraction of human malignancies. It was recently estimated that more than 10 % of the almost 13 million yearly cases of human cancer worldwide are caused by one of the following human viruses: Human papillomaviruses (HPV), hepatitis B virus (HBV), hepatitis C virus (HCV), Epstein–Barr virus (EBV), Kaposi's sarcoma-associated herpes virus (KSHV) (also called Human Herpes Virus 8), human T-cell leukemia virus (HTLV-1), and Merkel cell polyomavirus (MCPyV) (de Martel et al. 2012). An additional 5 % of worldwide cancers, mostly gastric cancer, are attributed to infection by the bacterium *Helicobacter pylori*. This conclusion is based on the cumulative results of a large number of laboratory experiments and epidemiological studies over the last five decades. As discussed in more detail in Chap. 2, the number of worldwide incident cases associated with each virus varies widely, from 600,000 for HPV to 2,100 for HTLV (Table 1). More than 85 % of the burden of virus-induced cancers is borne by individuals in the developing regions of the world (Table 1), with profound implications for translating the knowledge of virus-induced cancers into public health interventions.

Table 1 Number of new cancer cases attributable to specific viral infections by development status (de Martel et al. 2012)

Virus	World	Less developed	More developed
HPV	600,000	520,000	80,000
HBV	380,000	330,000	44,000
HCV	220,000	190,000	37,000
EBV	110,000	96,000	16,000
KSHV	43,000	39,000	4,000
HTLV	2,100	660	1,500

1   Viruses that Cause Human Cancer

The types of cancers induced by the various viruses and the fraction of these cancers attributed to the virus infection also varies widely (de Martel et al. 2012) (Table 2). In addition, viral prevalence in cases can vary depending on geographical location. HPVs, which normally infect stratified squamous epithelium, are causally associated with a number of anogenital cancers, from almost 100 % of cervical cancers to less than 50 % of vulvar cancers. They have more recently been implicated in oropharyngeal cancers, with prevalence estimates varying quite widely by region (Gillison 2008). Rates of HPV-associated oropharyngeal cancers appear to be substantially increasing, with current prevalences of over 50 % in the U.S. and several other (Arora et al. 2012) industrialized countries (Chaturvedi

Table 2 Prevalence of viruses in virus-associated cancers (de Martel et al. 2012)

Virus	Cancer	Geographical area	Prevalence in cases (%)
HPV	Cervix	World	100
HPV	Penile	World	50
HPV	Anal	World	88
HPV	Vulvar	World	43
HPV	Vaginal	World	70
HPV	Oropharynx	North America	56
HPV	Oropharynx	Southern Europe	17
HPV	Oropharynx	Japan	52
HBV	Liver	Developing	59
HBV	Liver	Developed	23
HCV	Liver	Developing	33
HCV	Liver	Developed	20
EBV	Hodgkin's lymphoma	Developing-children	90
EBV	Hodgkin's lymphoma	Developing-adults	60
EBV	Hodgkin's lymphoma	Developed	40
EBV	Burkitt's lymphoma	Sub-saharan Africa	100
EBV	Burkitt's lymphoma	Other regions	20–30
EBV	Nasopharyngeal carcinoma	High-incidence areas	100
EBV	Nasopharyngeal carcinoma	Low-incidence areas	80
KSHV	Kaposi's sarcoma	World	100
HTLV-1	Adult T-cell leukaemia and lymphoma	World	100
MCPyV	Merkel cell carcinoma	World	74

et al. 2011; Chaturvedi 2012) (Table 2). HBV and HCV have a strict tropism for hepatocytes and together are the major cause of liver cancer (El-Serag 2012). EBV normally infects epithelial cells and lymphocytes, especially B cells, and is the cause of most cases of Hodgkin's and Burkitt's lymphomas (Saha and Robertson 2011). It is also an etiological agent in most cases of an epithelial cancer, naso-pharyngeal carcinoma (Kutok and Wang 2006). KSHV is detected in virtually all Kaposi's sarcomas and is also strongly associated with two relatively rare B-cell neoplasias, multicentric Castleman's disease and primary effusion lymphoma (Gantt and Casper 2011). HTLV-1 also targets lymphocytes and is a primary cause of adult T-cell leukemia and lymphoma (Gallo 2011). MCPyV appears to be part of the normal flora of the skin and is causally related to approximately three-quarters of a relatively rare skin cancer, Merkel cell carcinoma (Arora et al. 2012).

Human tumor viruses encompass several distinct viral groups, including those with small DNA genomes (HPV, HBV, and MCPyV), large DNA genomes (EBV and KSHV), positive sense RNA genomes (HCV), and retroviruses (HTLV-1) (Table 3) (Butel and Fan 2012). Their specific mechanisms of carcinogenesis also vary widely. However, a common feature of human tumor viruses is that onco-genesis is an aberration of their normal viral life cycle and an uncommon outcome of infection. With some viruses, e.g. HPV and MCPyV, the viral genomes in cancer cells are usually altered by mutation and/or insertion into the host DNA, such that they can no longer produce infectious virions (Vinokurova et al. 2008; Arora et al. 2012). Virally associated cancers almost always arise as monoclonal events from chronic infections, usually after an interval of many years, indicating that the infections are just one component in a multi-step process of carcinogen-esis. A notable exception is KSHV-induced Kaposi's sarcoma, which can arise as a

Table 3 Basic features of human oncoviruses

Virus	Genome	Virion structure	Normal tropism	Year isolated (reference)
HPV16	Circular 7.9 kb DS DNA	55 nm naked Icosahedron	Stratified squamous epithelium	1983 (Dürst et al. 1983)
HBV	Circular 3.2 kb partial DS DNA	42 nm enveloped	Hepatocytes	1970 (Dane et al. 1970)
HCV	Linear 9.6 k nt positive sense RNA	Enveloped	Hepatocytes	1989 (Choo et al. 1989)
EBV	Linear 172 kb DS DNA	Enveloped	Epithelium and B cells	1964 (Epstein et al. 1964)
KSHV	Linear 165 kb DS DNA	Enveloped	Oropharyngeal epithelium	1994 (Chang et al. 1994)
HTLV-1	Linear 9.0 k nt positive sense RNA	Enveloped	T and B cells	1980 (Poiesz et al. 1980)
MCPyV	Circular 5.4 kb DS DNA	40 nm naked icosahedron	Skin	2008 (Feng et al. 2008)

polyclonal tumor within months of infection in immunosuppressed individuals (Mesri et al. 2010) (also see Chap. 13).

2   Oncogenic Mechanisms

As discussed in detail in later chapters of this book, the oncogenic mechanisms of most tumor viruses involve the continued expression of specific viral gene (oncogene) products that regulate proliferative or anti-apoptotic activities through an interaction with cellular gene products. Examples of oncoproteins include E6 and E7 of HPVs, LMP1 of EBV, and Tax of HTLV-1 (Chaps. 8, 10 and 11, respectively). Virally encoded microRNAs, for instance those of EBV, may also play a role in carcinogenesis by decreasing the expression of negative regulators of cell growth (Raab-Traub 2012). KSHV may act primarily by altering complex cytokine/chemokine networks (Mesri et al. 2010) (Chap. 13). In contrast, some tumor viruses, such as HCV and HBV, may induce cancer more indirectly, as a result of continued tissue injury and regeneration and the chronic inflammatory response of the host to persistent infection (Alison et al. 2011) (Chaps. 3, 5 and 6).

Some viruses, particularly retroviruses, can induce cancers by insertional mutagenesis in animal models (Fan and Johnson 2011). However, this mechanism has not been convincingly documented in humans, except in a few patients in experimental gene transfer trials involving delivery of high doses of recombinant retroviral vectors (Romano et al. 2009). HIV could also be considered a tumor virus in that HIV infection is a strong risk factor for several cancers, including most cancers that are associated with infections by other viruses (Parkin 2006). However, the effect of HIV infection on oncogenesis is thought to be indirect, by inhibiting normal host immune functions that would otherwise control or eliminate oncovirus infections and/or provide immunosurveillance of nascent tumors (Clifford and Franceschi 2009). Consistent with this conjecture, increases in many of the same cancers are seen in patients with other forms of immunosuppression (Rama and Grinyo 2010).

2.1   Causal Association of Viral Infection and Cancer

The causal associations between the seven viruses and specific cancers noted above are well established. They fulfill most, if not all, of the causality criteria proposed by Sir A. Bradford Hill in the early 1970s (Hill 1971). The strength and consistency of association between infection and cancer are high based upon multiple epidemiological studies in varying settings. For instance, the relative risk of HPV and KSHV infection for the development of cervical carcinoma and Kaposi's sarcoma, respectively, is over 100 in most studies. In some instances, establishing a strong association required identification of especially oncogenic types, e.g., HPV16 and 18 among mucosotropic HPVs, and a specific subset tumors, e.g., oropharyngeal among head and neck cancers. Temporality was established by demonstrating that infection proceeded cancer, usually by many years. In some cases, the viruses are consistently detected in well-established cancer precursor lesions, as is the case for HPV and high-grade cervical intraepithelial neoplasia (Chap. 8). Dose–response relationships were established by demonstrating that, for the most part, populations with higher prevalences of virus infection also had higher incidences of the associated tumor, e.g., HBV and liver cancer (El-Serag 2012) (Chap. 5). However, these associations were sometimes confounded by high prevalence of the oncovirus in the general population and variability in the prevalence of additional risk factors. An example is the high frequency of EBV infection in the general population and the strongly associated cofactor of malaria infection in the induction of EBV-positive Burkitt's lymphoma (Magrath 2012) (Chap. 10). Biological plausibility as oncogenic agents was established in numerous laboratory studies that identified the interaction of viral proteins with key regulators of proliferation and apoptosis, their immortalizing and transforming activity in vitro, and their oncogenic activity in animal models (Chaps. 4, 6, 8, 10, 11, and 13). These studies also support the criterion that the associations be in agreement with current understanding of disease pathogenesis, in this case, the process of tumorigenic progression. The last criterion, that removing the exposure prevents the disease, has been most convincing demonstrated for HBV, as discussed below and in Chap. 5.

2.2 Basic and Clinical Implications of Identifying a Viral Etiology

The identification of a virus as a central cause of a specific cancer can be an important discovery for several reasons. First, it can provide basic insights into the carcinogenic process, which in turn can identify potential cellular targets for interventions that are often relevant for both virus-associated and virus-independent tumors. For example, the tumor suppressors p53 and pRb were first identified as binding partners of the small DNA tumor viruses in experimental systems and later shown to be targets for human oncoviruses. They are also among the most frequently mutated genes in non-virally induced cancers (Howley and Livingston 2009) (Chap. 8).

Second, the presence of the virus can be used in cancer diagnosis or risk assessment. This aspect is well illustrated by the increasing use of HPV DNA testing to screen for cervix cancer risk (Schiffman et al. 2011). HPV DNA tests are more sensitive for the detection of high-grade premalignant lesions and cervical cancer than is the standard Pap test, so intervals between tests can be increased in women who test negative for high-risk HPV DNA in their cervix (Saslow et al. 2012). Another example is HCV screening to identify individuals at high risk of progression to liver cirrhosis and cancer. HCV screening was recently recommended in the U.S. for all individuals born between 1945 and 1965 (Smith et al. 2012).

Third, viral gene products provide potential targets for treatment for cancers, precancerous lesions, or chronically infected patients at high risk of cancer by therapeutic drugs or therapeutic vaccines. There has been substantial research activity in this potentially fruitful area, which, although it has not lead to viral-based treatment of malignancy or premalignant lesions, has had some clinical success in the development of anti-virals to treat chronic HBV and HCV infection (Chap. 14). Pegylated interferon alpha or a nucleoside/nucleotide analogue is currently being used to suppress HBV replication (Chap. 5) and thereby liver cirrhosis and risk of hepatocellular carcinoma. HCV infection can be similarly treated. However, sustained virologic responses are limited. Fortunately, several promising drug candidates that directly act on viral gene products are currently under development and appear to be more active and better tolerated (Poordad and Dieterich 2012) (Chap. 7). Similarly, treatment for KSHV infection in patients with AIDs decreases risk of developing Kaposi's sarcoma (Uldrick and Whitby 2011). In addition, Kaposi's sarcoma lesion often regress in HIV-infected individuals after initiation of HAART, but this is primarily an indirect effect related to the reconstitution of the immune system rather than direct activity of the drugs against KSHV (Uldrick and Whitby 2011). In addition, HAART treatment for HIV-infected individuals has been associated with risk reduction for the development of some, but not all, virally induced tumors (Shiels et al. 2011).

Fourth, the knowledge of a viral etiology can serve as the basis of cancer prevention measures. One approach involves behavioral interventions to reduce susceptibility to infection, e.g., limiting exposure to blood products in the case of HBV and HCV (Chaps. 5 and 7), limiting number of sexual partners in the case of HPV, or preventing HTLV-1 transmission by discouraging breast-feeding by infected mothers (Ruff 1994) (Chap. 12). Alternatively, the identification of human oncoviruses can be used to develop effective vaccines to prevent oncovirus infection. This approach has been successful implemented for HBV and HPV. HBV prophylactic vaccines were introduced more than 20 years ago (reviewed in Chap. 5). A dramatic reduction in childhood liver cancers of greater than two-thirds has been documented in Taiwan, a high-incidence region (Chang et al. 2009). A substantial reduction in adult liver cancer is expected in the near future as individuals who would have otherwise contracted HBV as infant reach the age of peak cancer incidence. HPV vaccines targeting HPV16 and 18 have been licensed for only 6 years (Schiller and Lowy 2012) (see Chap. 9). While substantial reductions in the incidences of HPV-associated cancer are not expected for at least another decade, there has already been a significant reduction in premalignant cervical disease and evidence of herd immunity developing in Australia, a country with high vaccination coverage and a screening program that includes relatively young women (Brotherton et al. 2011). There are considerable efforts underway to develop prophylactic and/or therapeutic vaccines against EBV and HCV (Wu et al. 2010; Feinstone et al. 2012) (Chap. 10). However, specific characteristics of their biology have made the development of effective vaccines challenging. These include viral latency and genetic instability in the case of EBV and HCV, respectively. Vaccines against these viruses would likely be commercially viable products. There has been less effort devoted to developing KSHV and HTLV-1 vaccines because the numbers of worldwide cancers they induce are lower, and they do not appear to be a frequent cause of medically important non-malignant disease, in contrast to EBV and HCV (Schiller and Lowy 2010). They also have proven susceptible to other interventions, specifically reduction in KSHV-induced Kaposi's sarcoma by treating HIV infection and reduction in transmission of HTLV-1 by discouraging infant breast-feeding by infected mothers (Chap. 12).

2.3 The Search for Additional Oncoviruses

Are there other human oncoviruses waiting to be discovered? The technologies of high through-put nucleic acid sequencing of entire cellular genomes, as now applied to a wide variety of human tumors, provide an unprecedented wealth of raw data for the hunt (Lizardi et al. 2011). The discovery of MCPyV illustrates how this technology can be employed to identify novel human oncoviruses (Feng et al. 2008). However, this type of search will miss oncoviruses with RNA genomes. In addition, identification of a viral nucleic acid sequence in a tumor is only the first step. It takes many additional laboratory, clinical, and epidemiological studies to establish that a viral infection is causally related to the development of a cancer, as opposed to being a passive parasite of the tumor. Establishment of causality can be particularly difficult in situations where the implicated virus is a common infection in the general population. It will also be difficult to establish causality if a virus is involved in the initiation of a tumor but not in its maintenance, a plausible but as yet unproven mechanism for human cancers, commonly referred to as a hit and run (Schiller and Buck 2011). There are suggestions that viral infections may be associated with several other cancers. For instance, the incidence of non-melanoma skin cancer increases dramatically after immunosuppression (Schiller and Buck 2011), and some epidemiological studies have linked the risk of prostate cancer with sexual activity variables, suggesting involvement of a sexually transmitted infectious agent (Sutcliffe 2010). Interesting arguments have also been made that colorectal carcinomas and childhood leukemia's may be caused by as yet unidentified viral infections (zur Hausen 2009, 2012).

There is no doubt that oncovirus studies have been at the forefront of biomedical research over the last several decades. They have provided important insights into basic cell biology and disease mechanisms and have generated important public health interventions for the control of major human cancers. This monograph provides an outstanding summary of the current state of the art of this dynamic field. We expect that continued research in this area will generate new and excites insights into the genesis of cancer and novel interventions to prevent and treat it.

References

Alison MR, Nicholson LJ, Lin WR (2011) Chronic inflammation and hepatocellular carcinoma. Recent Results Cancer Res 185:135–148
Arora R, Chang Y, More PS (2012) MCV and Merkel cell carcinoma: a molecular success story. Curr Opin Virol 2(4):489–498

Brotherton JM, Fridman M, May CL et al (2011) Early effect of the HPV vaccination programme on cervical abnormalities in Victoria, Australia: an ecological study. Lancet 377(9783):2085–2092

Butel JS, Fan H (2012) The diversity of human cancer viruses. Curr Opin Virol 2(4):449–452

Chang MH, You SL, Chen CJ et al (2009) Decreased incidence of hepatocellular carcinoma in hepatitis B vaccinees: a 20-year follow-up study. J Natl Cancer Inst 101(19):1348–1355

Chang Y, Cesarman E, Pessin MS et al (1994) Identification of herpesvirus-like DNA sequences in AIDS-associated Kaposi's sarcoma. Science 266(5192):1865–1869

Chaturvedi AK (2012) Epidemiology and clinical aspects of HPV in head and neck cancers. Head Neck Pathol 6(Suppl 1):S16–S24

Chaturvedi AK, Engels EA, Pfeiffer RM et al (2011) Human papillomavirus and rising oropharyngeal cancer incidence in the United States. J Clin Oncol 29(32):4294–4301

Choo QL, Kuo G, Weiner AJ et al (1989) Isolation of a cDNA clone derived from a blood-borne non-A, non-B viral hepatitis genome. Science 244(4902):359–362

Clifford GM, Franceschi S (2009) Cancer risk in HIV-infected persons: influence of CD4(+) count. Future Oncol 5(5):669–678

Dane DS, Cameron CH, Briggs M et al (1970) Virus-like particles in serum of patients with Australia-antigen-associated hepatitis. Lancet 1(7649):695–698

de Martel C, Ferlay J, Franceschi S et al (2012) Global burden of cancers attributable to infections in 2008: a review and synthetic analysis. Lancet Oncol 13(6):607–615

Dürst M, Gissmann L, Ikenberg H et al (1983) A papillomavirus DNA from a cervical carcinoma and its prevalence in cancer biopsy samples from different geographic regions. Proc Natl Acad Sci USA 80:3812–3815

El-Serag HB (2012) Epidemiology of viral hepatitis and hepatocellular carcinoma. Gastroenterology 142(6):1264–1273 e1261

Epstein MA, Achong BG, Barr YM et al (1964) Virus particles in cultured lymphoblasts from Burkitt's lymphoma. Lancet 1(7335):702–703

Fan H, Johnson C (2011) Insertional oncogenesis by non-acute retroviruses: implications for gene therapy. Viruses 3(4):398–422

Feinstone SM, Hu DJ, Major ME (2012) Prospects for prophylactic and therapeutic vaccines against hepatitis C virus. Clin Infect Dis 55(Suppl 1):S25–S32

Feng H, Shuda M, Chang Y et al (2008) Clonal integration of a polyomavirus in human Merkel cell carcinoma. Science 319(5866):1096–1100

Gallo RC (2011) Research and discovery of the first human cancer virus, HTLV-1. Best Pract Res Clin Haematol 24(4):559–565

Gantt S, Casper C (2011) Human herpesvirus 8-associated neoplasms: the roles of viral replication and antiviral treatment. Curr Opin Infect Dis 24(4):295–301

Gillison ML (2008) Human papillomavirus-related diseases: oropharynx cancers and potential implications for adolescent HPV vaccination. J Adolesc Health 43(4 Suppl):S52–S60

Hill A (1971) Statistical evidence and inference. Principles of medical statistics, 9th edn. Oxford University Press, New York, pp 309–323

Howley PM, Livingston DM (2009) Small DNA tumor viruses: large contributors to biomedical sciences. Virology 384(2):256–259

Kutok JL, Wang F (2006) Spectrum of Epstein-Barr virus-associated diseases. Annu Rev Pathol 1:375–404

Lizardi PM, Forloni M, Wajapeyee N (2011) Genome-wide approaches for cancer gene discovery. Trends Biotechnol 29(11):558–568

Magrath I (2012) Epidemiology: clues to the pathogenesis of Burkitt lymphoma. Br J Haematol 156(6):744–756

Mesri E, Cesarman E, Boshoff C (2010) Kaposi's sarcoma and its associated herpesvirus. Nat Rev Cancer 10(10):707–719

Parkin DM (2006) The global health burden of infection-associated cancers in the year 2002. Int J Cancer 118(12):3030–3044

Poiesz BJ, Ruscetti FW, Mier JW et al (1980) Detection and isolation of type C retrovirus particles from fresh and cultured lymphocytes of a patient with cutaneous T-cell lymphoma. Proc Natl Acad Sci U S A 77(12):7415–7419

Poordad F, Dieterich D (2012) Treating hepatitis C: current standard of care and emerging direct-acting antiviral agents. J Viral Hepat 19(7):449–464

Raab-Traub N (2012) Novel mechanisms of EBV-induced oncogenesis. Curr Opin Virol 2(4):453–458

Rama I, Grinyo JM (2010) Malignancy after renal transplantation: the role of immunosuppression. Nat Rev Nephrol 6(9):511–519

Romano G, Marino IR, Pentimalli F et al (2009) Insertional mutagenesis and development of malignancies induced by integrating gene delivery systems: implications for the design of safer gene-based interventions in patients. Drug News Perspect 22(4):185–196

Ruff AJ (1994) Breastmilk, breastfeeding, and transmission of viruses to the neonate. Semin Perinatol 18(6):510–516

Saha A, Robertson ES (2011) Epstein-Barr virus-associated B-cell lymphomas: pathogenesis and clinical outcomes. Clin Cancer Res 17(10):3056–3063

Saslow D, Solomon D, Lawson HW et al (2012) American Cancer Society, American Society for Colposcopy and Cervical Pathology, and American Society for Clinical Pathology screening guidelines for the prevention and early detection of cervical cancer. CA Cancer J Clin 62(3):147–172

Schiffman M, Wentzensen N, Wacholder S et al (2011) Human papillomavirus testing in the prevention of cervical cancer. J Natl Cancer Inst 103(5):368–383

Schiller JT, Buck CB (2011) Cutaneous squamous cell carcinoma: a smoking gun but still no suspects. J Invest Dermatol 131(8):1595–1596

Schiller JT, Lowy DR (2010) Vaccines to prevent infections by oncoviruses. Annu Rev Microbiol 64:23–41

Schiller JT, Lowy DR (2012) Understanding and learning from the success of prophylactic human papillomavirus vaccines. Nat Rev Microbiol 10(10):681–692

Shiels MS, Pfeiffer RM, Gail MH et al (2011) Cancer burden in the HIV-infected population in the United States.". J Natl Cancer Inst 103(9):753–762

Smith BD, Jorgensen C, Zibbell JE et al (2012) Centers for Disease Control and Prevention initiatives to prevent hepatitis C virus infection: a selective update. Clin Infect Dis 55(Suppl 1):S49–S53

Sutcliffe S (2010) Sexually transmitted infections and risk of prostate cancer: review of historical and emerging hypotheses. Future Oncol 6(8):1289–1311

Uldrick TS, Whitby D (2011) Update on KSHV epidemiology, Kaposi Sarcoma pathogenesis, and treatment of Kaposi Sarcoma. Cancer Lett 305(2):150–162

Vinokurova S, Wentzensen N, Kraus I et al (2008) Type-dependent integration frequency of human papillomavirus genomes in cervical lesions. Cancer Res 68(1):307–313

Wu TT, Blackman MA, Sun R (2010) Prospects of a novel vaccination strategy for human gamma-herpesviruses. Immunol Res 48(1–3):122–146

zur Hausen H (2009) Childhood leukemia's and other hematopoietic malignancies: interdependence between an infectious event and chromosomal modifications. Int J Cancer 125(8):1764–1770

zur Hausen H (2012) Red meat consumption and cancer: reasons to suspect involvement of bovine infectious factors in colorectal cancer. Int J Cancer 130(11):2475–2483

Epidemiology of Virus Infection and Human Cancer

Chien-Jen Chen, Wan-Lun Hsu, Hwai-I Yang, Mei-Hsuan Lee, Hui-Chi Chen, Yin-Chu Chien and San-Lin You

Abstract

The International Agency for Research on Cancer (IARC) has comprehensively assessed the human carcinogenicity of biological agents. Seven viruses including Epstein–Barr virus (EBV), hepatitis B virus (HBV), hepatitis C virus (HCV), Kaposi's sarcoma herpes virus (KSHV), human immunodeficiency virus, type-1 (HIV-1), human T cell lymphotrophic virus, type-1 (HTLV-1), and human papillomavirus (HPV) have been classified as Group 1 human carcinogens by IARC. The conclusions are based on the findings of epidemiological and mechanistic studies. EBV, HPV, HTLV-1, and KSHV are direct carcinogens; HBV and HCV are indirect carcinogens through chronic inflammation; HIV-1 is an indirect carcinogen through immune suppression. Some viruses may cause more than one cancer, while some cancers may be caused by more than one virus. However, only a proportion of persons infected by these

C.-J. Chen (✉) · W.-L. Hsu · H.-C. Chen · S.-L. You
Genomics Research Center, Academia Sinica, 128 Academia Road, Section 2,
Taipei 115, Taiwan
e-mail: chencj@gate.sinica.edu.tw

C.-J. Chen
Graduate Institute of Epidemiology and Preventative Medicine, National Taiwan University,
Taipei, Taiwan

H.-I. Yang · Y.-C. Chien
Molecular and Genomic Epidemiology Center, China Medical University Hospital
and Graduate Institute of Clinical Medical Science, China Medical University,
Taichung, Taiwan

M.-H. Lee
Institute of Clinical Medicine, National Yang-Ming University, Taipei, Taiwan

M. H. Chang and K.-T. Jeang (eds.), *Viruses and Human Cancer*,
Recent Results in Cancer Research 193, DOI: 10.1007/978-3-642-38965-8_2,
© Springer-Verlag Berlin Heidelberg 2014

oncogenic viruses will develop specific cancers. A series of studies have been carried out to assess the viral, host, and environmental cofactors of EBV-associated nasopharyngeal carcinoma, HBV/HCV-associated hepatocellular carcinoma, and HPV-associated cervical carcinoma. Persistent infection and high viral load are important risk predictors of these virus-caused cancers. Risk calculators incorporating host and viral factors have also been developed for the prediction of long-term risk of hepatocellular carcinoma. These risk calculators are useful for the triage and clinical management of infected patients. Both clinical trials and national programs of immunization or antiviral therapy have demonstrated a significant reduction in the incidence of cancers caused by HBV, HCV, and HPV. Future researches on gene–gene and gene–environment interaction of oncogenic viruses and human host are in urgent need.

Keywords

Cancer · EBV · Epidemiology · HBV · HCV · HIV · HPV · HTLV-I · KSHV

Contents

1   Introduction

The International Agency for Research on Cancer has comprehensively assessed the carcinogenicity of the biological agents to humans based on epidemiological and mechanistic evidence (IARC 2009). Seven viruses including Epstein–Barr virus (EBV), hepatitis B virus (HBV), hepatitis C virus (HCV), Kaposi's sarcoma herpes virus (KSHV), human immunodeficiency virus, type-1 (HIV-1), human T cell lymphotrophic virus, type-1 (HTLV-1), and several types of human papillomavirus (HPV) have been classified as Group 1 human carcinogen as shown in the Table 1.

There is sufficient evidence to conclude that EBV causes nasopharyngeal carcinoma, Burkitt's lymphoma, immune suppression-related non-Hodgkin lymphoma, extranodal NK/T cell lymphoma (nasal type), and Hodgkin's lymphoma in humans. The evidence for EBV-caused gastric carcinoma and lympho-epithelioma-like carcinoma is limited. HBV and HCV cause hepatocellular carcinoma

Table 1 Cancers caused by Group 1 oncogenic viruses with sufficient and limited evidence according to the IARC criteria

Virus	Cancer sites with sufficient evidence	Cancer sites with limited evidence
Epstein–Barr virus (EBV)	Nasopharyngeal carcinoma, Burkitt's lymphoma, immune suppression-related non-Hodgkin lymphoma, extranodal NK/T cell lymphoma (nasal type), Hodgkin's lymphoma	Gastric carcinoma, lympho-epithelioma-like carcinoma
Hepatitis B virus (HBV)	Hepatocellular carcinoma	Cholangiocarcinoma, non-Hodgkin lymphoma
Hepatitis C virus (HCV)	Hepatocellular carcinoma, non-Hodgkin lymphoma	Cholangiocarcinoma
Human immunodeficiency virus, type 1 (HIV-1)	Kaposi's sarcoma, non-Hodgkin lymphoma, Hodgkin's lymphoma, cancers of the cervix, anus, and conjunctiva	Cancers of the vulva, vagina and penis, non-melanoma skin cancer, hepatocellular carcinoma
Human papillomavirus type 16 (HPV-16)	Cancers of the cervix, vulva, vagina, penis, anus, oral cavity, oropharynx, and tonsil	Cancer of the larynx
Human papillomavirus type 18, 31, 33,35, 39, 45, 51, 52, 56, 58, 59 (HPV-18, 31, 33,35, 39, 45, 51, 52, 56, 58, and 59)	Cancer of the cervix	
Human papillomavirus type 26, 30, 34, 53, 66, 67, 68, 69, 70, 73, 82, 85, 97 (HPV- 26, 30, 34, 53, 66, 67, 68, 69, 70, 73, 82, 85, and 97)		Cancer of the cervix
Human T cell lymphotrophic virus, type-1 (HTLV-1)	Adult T cell leukemia and lymphoma	
Kaposi's sarcoma herpes virus (KSHV)	Kaposi's sarcoma, primary effusion lymphoma	Multicentric Castleman's disease

with sufficient evidence. The evidence for HCV-caused non-Hodgkin lymphoma, especially B-cell lymphoma, is sufficient, while the evidence for HBV-caused non-Hodgkin lymphoma is limited. There is also limited evidence to conclude that HBV and HCV cause cholangiocarcinoma. The evidence to conclude that HIV-1 causes Kaposi's sarcoma, non-Hodgkin lymphoma, Hodgkin's lymphoma, and cancers of the cervix, anus, and conjunctiva is sufficient. But the evidence for HIV-1 to cause cancers of the vulva, vagina, penis, non-melanoma skin cancer, and hepatocellular carcinoma is limited.

There is sufficient evidence to conclude that HPV-16 causes cancers of the cervix, vulva, vagina, penis, anus, oral cavity, oropharynx, and tonsil; but the evidence for HPV-16 to cause cancer of the larynx is limited. Cervical cancer is caused by several types of HPV including HPV-18, 31, 33, 35, 39, 45, 51, 52, 56, 58, and 59. The evidence for HPV-26, 30, 34, 53, 66, 67, 68, 69, 70, 73, 82, 85, and 97 to cause cervical cancer is limited. HTLV-1 causes adult T cell leukemia and lymphoma with sufficient evidence. There is sufficient evidence to conclude KSHV causes Kaposi's sarcoma and primary effusion lymphoma, but the evidence for KSHV to cause multicentric Castleman's disease is limited.

The proportion of cancers caused by infectious agents was recently estimated to be more than 20 % (IARC 2009). The identification of new cancer sites attributed to these agents means that more cancers are potentially preventable. This chapter will review mainly the epidemiology of oncogenic viruses and their associated cancers.

2   Prevalence of Oncogenic Virus Infection in the World

EBV is highly prevalent throughout the world with more than 90 % adults infected with EBV even in the remote populations (IARC 2009). The estimated number of persons infected with EBV is more than 5.5 billion. The age at primary infection of EBV varies significantly in the world. People live in overcrowded conditions with poor sanitation have a younger age at primary infection than those live in better environments. Two major types of EBV have been identified and differ in geographical distribution with EBV-2 more common in Africa and homosexual men. The role of specific EBV types in the development of difference cancers remains to be elucidated. As EBV infection is ubiquitous, the specific geographical distribution of EBV-related malignancies including endemic Burkitt lymphoma and nasopharyngeal carcinoma is more likely attributable to the variation in the distributions of other cofactors which may activate EBV replication.

Figure 1 shows the geographical variation in the prevalence of oncogenic viruses in the world. HBV infects more than 2.0 billion people in the world and more than 300 million of them are chronic HBV carriers (IARC 2009). There is a wide variation of chronic HBV infection in the world[1] as shown in Fig. 1a. Approximately 45 %, 43 %, and 12 % of the world population live in areas where the endemicity of chronic HBV infection is high (seroprevalence of hepatitis B surface antigen >8 %), medium (2–7 %), and low (<2 %). The prevalence is highest in sub-Saharan Africa, the Amazon Basin, China, Korea, Taiwan, and several countries in Southeast Asia. In areas of high endemicity, the lifetime risk of HBV infection is more than 60 % with most infections acquired from perinatal and child-to-child transmission, when the risk of becoming chronic infection is greatest. Perinatal (vertical) transmission is predominant in China, Korea, and

---

[1] CDC http://wwwnc.cdc.gov/travel/yellowbook/2012/chapter-3-infectious-diseases-related-to-travel/hepatitis-b.htm

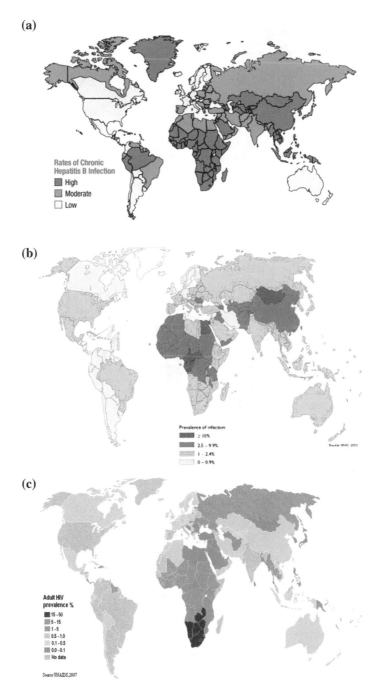

Fig. 1 Estimated prevalence (per 100) of Group 1 oncogenic viruses in the world. **a** HBV, **b** HCV, **c** HIV-1, **d** HPV, **e** HTLV-1, and **f** KSHV

(d)

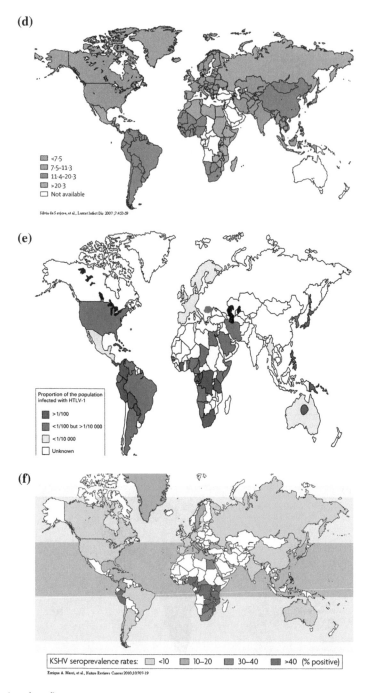

(e)

(f)

Fig. 1 (continued)

Taiwan where the seroprevalence of HBeAg in pregnant women is high, while child-to-child (horizontal) transmission is common in sub-Saharan Africa where HBeAg seroprevalence is low in mothers. In areas of medium endemicity, mixed HBV transmission patterns occur in infancy, early childhood, adolescence, and adulthood. In the low endemicity areas, most HBV infections occur in adolescents and young adults through injection drug use, male homosexuality, health care practice, and regular transfusion or hemodialysis.

In addition to the striking geographical variation in seroprevalence of HBsAg in the world, the distribution of eight genotypes of HBV varies significantly in different countries (IARC 2009). Genotype A is prevalent in Europe, Africa, and North America; genotypes B and C are prevalent in East and Southeast Asia; Genotype D is predominant in South Asia, Middle East, and Mediterranean areas; genotype E is limited to West Africa; genotypes F and G are found in Central and South America; and genotype H is observed in Central America.

HCV infects around 150 million people in the world showing an estimated prevalence of 2.2 % (IARC 2009) with a wide variation in different regions as shown in Fig. 1b. The estimates of HCV infection (seroprevalence of antibodies against HCV) range from <0.1 % in the United Kingdom and Scandinavia to 15–20 % in Egypt (Alter 2007). The high prevalence of HCV infection was observed in Mongolia, northern Africa, Pakistan, China, southern Italy, and some areas in Japan. There are at least six major genotypes of HCV have been identified. There is a wide variation in geographical distribution of HCV genotype in the world. The response to antiviral therapy also varies by HCV genotype. It is better in patients infected with genotype 2 or 3 than those with genotype 1 or 4.

HCV have two major transmission routes including injection drug use and iatrogenic exposures through transfusion, transplantation, and unsafe therapeutic injection. While iatrogenic transmission of HCV has been reduced after 1990 in developed countries such as Japan and Italy, it remains frequent in low-resource countries where disposable needles tend to be reused. Injection drug use is the most important transmission route for newly acquired HCV infection in developed countries. Transmission of HCV through perinatal, sexual, and accidental needle-stick exposures is less efficiently than iatrogenic exposure and injection drug use.

HIV infects estimated 34 million people in the world at the end of 2010 (IARC 2009).[2] An estimated 0.8 % of adults aged 15–49 years worldwide are living with HIV, and the burden varies considerably between countries and regions as shown in Fig. 1c. Sub-Saharan Africa remains most severely affected with a prevalence of 4.9 %. Although the prevalence of HIV infection is nearly 25 times higher in sub-Saharan Africa than in Asia, almost 5 million people are living with HIV in South, Southeast, and East Asia combined. After sub-Saharan Africa, regions most heavily affected are the Caribbean and Eastern Europe and Central Asia, where 1.0 % of adults were living with HIV in 2011. There were 2.5 million people including 0.39 million children were newly infected with HIV in 2011. Since

---

[2] UNAIDS http://data.unaids.org/pub/epislides/2012/2012_epiupdate_en.pdf

2001, annual HIV incidence has fallen in 33 countries, 22 of them in sub-Saharan Africa. However, incidence is accelerating again in Eastern Europe and Central Asia after having slowed in the early 2000s, and new infections are on the rise in the Middle East and North Africa.

HIV-1 infection is transmitted through three major routes: sexual intercourse, blood contact, and mother-to-child transmission. The HIV-1 infectivity is determined by the interaction of three factors of agent, host, and environments. The probability of HIV-1 transmission is highest for blood transfusion, followed by mother-to-child transmission, needle sharing, man-to-man sexual transmission, and lowest for woman-to-man sexual transmission.

HPV infection is very prevalent in most sexually active individuals will acquire at least one genotype of anogenital HPV infection during their lifetime (IARC 2009). The estimated oncogenic HPV DNA point prevalence has been reported as high as 10 % in a meta-analysis of 157,879 women with normal cytology, giving an estimate of 600 million people being infected (de Sanjose et al. 2007). The point prevalence was highest (20–30 %) in Africa, East Europe, and Latin America; and lowest (6–7 %) in southern and western Europe and Southeast Asia demonstrating a striking geographical variation as shown in Fig. 1d. The estimated point prevalence is highly dynamic because both incidence and clearance rates are high.

Among 13 oncogenic HPV types, the most prevalent types include 16, 18, 31, 33, 35, 45, 52, and 58. HPV 16 is the most common type in all regions with prevalence ranging 2.3–3.5 %. HPV infections are transmitted through direct skin-to-skin or skin-to-mucosa contact. Anogenital HPV types spread mainly through sexual transmission in teenagers and young adults. Non-sexual routes including perinatal and iatrogenic transmissions account for a minority of HPV infections.

HTLV-1 infects estimated 15–20 million people in the world (IARC 2009). HTLV-1 infection is characterized by the micro-epidemic hotspots surrounded by low prevalence areas as shown Fig. 1e (Proietti et al. 2005). The HTLV-1 infection prevalence ranges from <0.1 % in China, Korea, and Taiwan to 20 % in Kyushu and Okinawa of Japan. The regions of high endemicity include southwestern Japan, parts of sub-Saharan Africa, the Caribbean Islands, and South Africa. HTLV-1 has three major transmission routes: vertical transmission, sexual transmission, and parenteral transmission. Vertical transmission through breastfeeding has a high efficiency to result in mother-to-child infection. However, in utero infectivity is low due to limited trafficking of HTLV-1-infected lymphocytes across placenta. The efficiency of sexual transmission of HTLV-1 depends on the proviral load and use of condom. Parenteral transmission through transfusion is significantly reducing due to the sensitive serological examination of blood products. Needle sharing associated with injection drug use is another parenteral route for HTLV-1 transmission.

Infection prevalence of KSHV determined by serological tests varies significantly in the world (Dukers and Rezza 2003) as shown in Fig. 1f. It ranges from 2–3 % in northern Europe to 82 % in Congo (IARC 2009). The prevalence is generally low (<10 %) in northern Europe, the USA, and Asia, elevated in

Mediterranean region (10–30 %) and high in sub-Saharan Africa (>50 %). The KSHV is primarily transmitted via saliva. In the countries where KSHV prevalence is high, the infection occurs during childhood and increases with age. The transmission of KSHV among homosexual men is also via saliva. KSHV may also be transmitted with a low efficiency through prolonged injection drug use, blood transfusion, and organ transplantation.

3   Incidence of Some Virus-caused Cancers in the World

The world maps of age-adjusted incidence rates of some oncogenic virus-related cancers are shown in Fig. 2. The age-adjusted incidence rates of nasopharyngeal cancer range from <0.1 to 8.05 per 100,000 as shown in Fig. 2a. The highest incidence was observed in southern China, Southeast Asia, and sub-Saharan Africa, and the lowest incidence in Europe, western Africa, and Central America. Chinese ethnicity in different cancer registries has the highest incidence of nasopharyngeal cancer. As EBV infection is ubiquitous in humans, the uniquely high incidence of nasopharyngeal carcinoma suggesting Chinese lifestyles or genetic susceptibility may play an important role in the development of nasopharyngeal cancer.

The age-adjusted incidence rates of Burkitt lymphoma are shown in Fig. 2b. Central Africa, equatorial South America, Papua New Guinea, and Caribbean countries are endemic for Burkitt lymphoma, but the incidence rate of Burkitt Lymphoma is relatively low in other countries. As EBV infection is ubiquitous in humans, the extraordinarily high endemicity of Burkitt lymphoma in Africa suggesting local environments or genetic susceptibility may play an important role in the development of endemic Burkitt lymphoma.

The age-adjusted incidence rates of liver cancer range from 0.70 to 94.4 per 100,000 as shown in Fig. 2c. The highest incidence was observed in East Asia, Southeast Asia, Egypt, and sub-Saharan Africa, and the lowest incidence in Europe, Middle East, Australia, New Zealand, and Canada. The geographical variation in liver cancer incidence is consistent with that of seroprevalence of HBV and HCV.

The age-adjusted incidence rates of cervical cancer range from 2.14 to 56.29 per 100,000 as shown in Fig. 2d. The highest incidence was observed in Latin America, South Asia, and sub-Saharan Africa, and the lowest incidence in Europe, North America, Australia, New Zealand, and Middle East. The geographical variation in cervical cancer incidence is consistent with that of seroprevalence of oncogenic HPV.

The age-adjusted incidence rates of Kaposi's sarcoma range from <1.0 to 30 per 100,000 as shown in Fig. 2e. The highest incidence was observed in sub-Saharan Africa and the lowest incidence in Europe, Australia, North America, and East Asia. The geographical variation in Kaposi's sarcoma incidence is consistent with that of seroprevalence of KSHV.

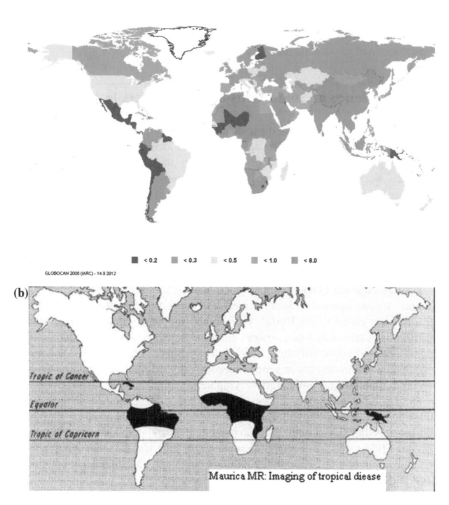

Fig. 2 Age-standardized incidence rate (per 100,000) of virus-caused cancers in the world.
a Nasopharynx, **b** Burkitt lymphoma **c** Liver, **d** Cervix uteri, and **e** Kaposi's sarcoma

4   Carcinogenic Mechanisms of Oncogenic Viruses

There are three major mechanisms of carcinogenesis for seven Group 1 oncogenic
viruses as shown in Table 2. They are defined as direct, indirect through chronic
inflammation, and indirect through immune suppression (IARC 2009). The direct
carcinogens include EBV, HPV, HTLV-1 and KSHV; the indirect carcinogens

(c) Estimated age-standardised incidence rate per 100,000
Liver: both sexes, all ages

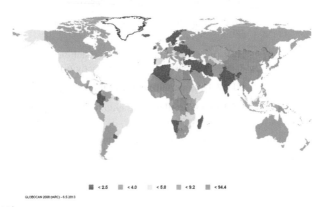

■ < 2.5   ■ < 4.0   ■ < 5.8   ■ < 9.2   ■ < 94.4

GLOBOCAN 2008 (IARC) - 6.5.2013

(d) Estimated age-standardised incidence rate per 100,000
Cervix uteri, all ages

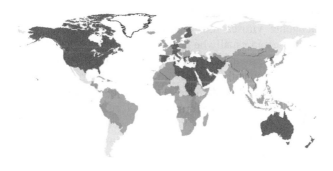

■ < 7.0   ■ < 12.9   ■ < 20.2   ■ < 29.6   ■ < 56.3

GLOBOCAN 2008 (IARC) - 6.5.2013

(e)

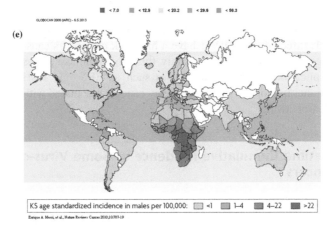

KS age standardized incidence in males per 100,000:   ☐ <1   ☐ 1–4   ■ 4–22   ■ >22

Enrique A. Mesri, et al., Nature Reviews Cancer 2010;10:707-19

Fig. 2 (continued)

Table 2 Established carcinogenic mechanisms of oncogenic viruses

Mechanism	Group 1 virus (carcinogenic properties)
Direct	EBV (cell proliferation, inhibition of apoptosis, genomic instability, cell migration)
	HPV (immortalization, genomic instability, inhibition of DNA damage response, anti-apoptotic activity)
	HTLV-1 (immortalization and transformation of T cells)
	KSHV (cell proliferation, inhibition of apoptosis, genomic instability, cell migration)
Indirect through chronic inflammation	HBV (inflammation, liver cirrhosis, chronic hepatitis)
	HCV (inflammation, liver cirrhosis, liver fibrosis)
Indirect through immune suppression	HIV-1 (immunosuppression)

through chronic inflammation include HBV and HCV; and the indirect carcinogen through immune suppression is HIV-1.

Direct oncogenic viruses have following characteristics: (1) The entire or partial viral genome can usually be detected in each cancer cell. (2) The virus can immortalize after the growth of target cells in vitro. (3) The virus expresses several oncogenes that interact with cellular proteins to disrupt cell-cycle checkpoints, inhibit apoptosis, and DNA damage response, cause genomic instability, and induce cell immortalization, transformation, and migration.

Both HBV and HCV cause hepatocellular carcinoma through chronic inflammation, which leads to the production of chemokines, cytokines, and prostaglandins secreted by infected cells and/or inflammatory cells. The chronic inflammation also leads to the production of reactive oxidative species with direct mutagenic effects to deregulate the immune system and promote angiogenesis, which is essential for the neovascularization and survival of tumors.

Individuals infected with HIV-1 have a high risk of cancers caused by another infectious agent. HIV-1 infection, mainly through immunosuppression, leads to increased replication of oncogenic viruses such as EBV and KSHV. Although antiretroviral therapy lowers the risk of many cancers associated with HIV-1, risks remain high worldwide.

5   Lifetime Cumulative Incidence of Some Virus-caused Cancers

Some viruses may cause more than one cancer, while some cancers may be caused by more than one virus. However, only a proportion of persons infected by these oncogenic viruses will develop specific cancers. Table 3 shows the lifetime cumulative incidence of some virus-caused cancers. The cumulative lifetime

Table 3 Lifetime cumulative incidence and risk cofactors of virus-caused cancers

Virus (cancer)	Lifetime incidence	Viral factors	Host factors	Environmental factor
EBV (nasopharyngeal carcinoma)	Men, 2.0 %.	Elevated serotiter of antibodies against EBV, EBV viral load	Male gender, family history, genetic polymorphisms (xenobiotic metabolism, DNA repair, human leukocyte antigen)	Cantonese salted fish, Dietary nitrosamine, wood dust, formaldehyde, tobacco
HBV (hepatocellular carcinoma)	Men, 27.4 %; women, 8.0 %.	Persistent infection, viral load, genotype, mutant, serum HBsAg level	Elder age, male gender, obesity, diabetes, serum androgen and ALT level, family history, genetic polymorphisms (DNA repair, human leukocyte antigen, androgen, and xenobiotic metabolism)	Aflatoxins, alcohol, tobacco, carotenoids, selenium, HCV infection
HCV (hepatocellular carcinoma)	Men, 23.7 %; women, 16.7 %	Persistent infection, viral load, genotype, mutant	Elder age, male gender, obesity, diabetes, serum ALT level, family history, genetic polymorphisms	Alcohol, tobacco, betel, HBV or HTLV-1 infection, radiation
HPV (cervical carcinoma)	HPV-16, 34.3 %; HPV-52, 23.3 %; HPV-58, 33.4 %; Any oncogenic HPV, 20.3 %.	Persistent infection, viral load, genotype	Elder age, number of pregnancies, family history, serum estrogen level, genetic polymorphisms (DNA repair, human leukocyte antigen)	Tobacco, immunosuppression, HIV-1 infection, contraceptives, nutrients

(30–75 years old) risk of developing nasopharyngeal carcinoma was 2.2 % for men seropositive for IgA antibodies against EBV VCA or antibodies against EBV DNase and 0.48 % for those seronegative for both antibodies.

Only around one-quarter of patients with patients chronically infected with HBV will develop hepatocellular carcinoma showing a striking gender difference of 27.4 % for men and 8.0 % for women (Huang et al. 2011). The development of hepatocellular carcinoma caused by HBV has been considered as a multistage hepatocarcinogenesis with multifactorial etiology, which involved the interaction of HBV, chemical carcinogens, host characteristics, and genetic susceptibility (Chen et al. 1997; Chen and Chen 2002; Chen and Yang 2011; IARC 2009).

Around one-fifth of patients seropositive for antibodies against HCV (anti-HCV) will develop hepatocellular carcinoma showing a less significant gender difference of 23.7 % for men and 16.7 % for women (Huang et al. 2011). The lifetime cumulative risk of HCC for anti-HCV seropositives with and without detectable serum HCV RNA level was 3.53 % and 24.2 %, respectively. Many cofactors have been involved in the development of hepatocellular carcinoma in anti-HCV seropositives (Lee et al. 2010; IARC 2009).

The cumulative lifetime (30–75 years old) risk of cervical cancer for women who were infected by HPV16, HPV 52, HPV 58, and any Group 1 oncogenic HPV was 34.3 %, 23.3 %, 33.4 %, and 20.3 %, respectively. Women with persistent oncogenic HPV infection have a much higher cumulative risk of cervical cancer than those with the transient infection (Chen et al. 2011b).

6 Cofactors of Some Virus-caused Cancers

Only a proportion of persons infected by oncogenic viruses will develop specific cancers. The fact strongly suggests the involvement of cofactors in the oncogenic process. Carcinogenesis would result from the interaction of multiple risk factors including viral factors, host factors, and environmental factors as shown in Table 3. The viral factors include various infection markers such as viral load, genotypes, variants, mutants, and serotiter of antibodies. The host factors include age, gender, race, anthropometric characteristics, immune status, hormonal level, personal disease history, and family cancer history. The environmental factors include chemical carcinogens, nutrients, ionizing radiation, immunosuppressive drugs, and coinfections of other infectious agents. The contribution of several additional factors to the development of virus-associated cancers seems to be substantial, but has not yet been elucidated in detail.

Several cofactors for nasopharyngeal carcinoma have been reviewed previously (Chien and Chen 2003). The viral factors associated with EBV-caused nasopharyngeal carcinoma include the elevated serotiter of antibodies against EBV including anti-EBV VCA IgA, anti-EBV DNase, anti-EBNA1 (Chien et al. 2001; Hsu et al. 2009), and the elevated serum EBV DNA level (viral load). Host factors include male gender, family history of nasopharyngeal carcinoma (Hsu et al. 2011), and genetic polymorphisms of xenobiotic metabolism enzymes (Hildesheim et al. 1997), DNA repair enzymes (Cho et al. 2003), and human leukocyte antigen (Hildesheim et al. 2002; Hsu et al. 2012b). Environmental factors include consumption of Cantonese salted fish, high dietary intake of nitrite and nitrosamine (Ward et al. 2000), occupational exposure to wood dust and formaldehyde (Hildesheim et al. 2001), long-term tobacco smoking (Hsu et al. 2009), and low intake of plant vitamin, fresh fish, green tea, and coffee (Hsu et al. 2012a).

The viral factors associated with HBV-caused hepatocellular carcinoma include the elevated serum level HBeAg serostatus (Yang et al. 2002), serum HBV DNA level (Yang et al. 2002; Chen et al. 2006; Chen et al. 2009), HBV genotype and mutant types (Yang et al. 2008), and elevated serum HBsAg level (Lee et al. 2013). Host factors include elder age, male gender, elevated serum alanine aminotransferase (ALT) level, family history of hepatocellular carcinoma (Chen et al. 1991; Yu et al. 2000a; Yang et al. 2010), disease status of obesity and diabetes (Chen et al. 2008), elevated serum level of androgen and androgen-related genetic polymorphisms (Yu and Chen 1993; Yu et al. 2000b), and genetic polymorphisms of xenobiotic metabolism enzymes and DNA repair enzyme (Chen et al. 1996a; Yu et al. 1995a, 1999a, 2003). Environmental factors include aflatoxin exposure (Chen et al. 1996b; Wang et al. 1996), habits of alcohol consumption and tobacco smoking (Chen et al. 1991; Wang et al. 2003), inadequate intake of carotenoids and selenium (Yu et al. 1995b, 1999a, b), and coinfection with HCV (Huang et al. 2011).

The viral factors associated with HCV-caused hepatocellular carcinoma include the elevated serum level of HCV RNA and HCV genotype 1 (Lee et al. 2010; Huang et al. 2011). Host factors include elder age, male gender, obesity, diabetes, elevated serum ALT level, family history of hepatocellular carcinoma, and genetic polymorphisms (Sun et al. 2003; Chen et al. 2008; Lee et al. 2010; IARC 2009). Environmental factors include alcohol consumption, tobacco smoking, betel hewing, radiation exposure, and coinfection with HBV or HTLV-1 (Sun et al. 2003; Huang et al. 2011; IARC 2009).

The viral factors associated with oncogenic HPV-caused cervical cancer include the persistent infection, elevated viral load, HPV genotypes, and variants (Chen et al. 2011a, b; Chang et al. 2011; IARC 2009). Host factors include elder age, number of pregnancies, family history of cervical cancer, serum estrogen level, genetic polymorphisms of DNA repair enzymes, and human leukocyte antigen (Chen et al. 2011a; Chuang et al. 2012; IARC 2009). Environmental factors include tobacco smoking, immunosuppression, HIV-1 coinfection, use of oral contraceptives, and inadequate intake of micronutrients (IARC 2009).

7   Risk Calculators of HBV-caused Hepatocellular Carcinoma

As there are many risk predictors for each virus-caused cancer, it is useful to incorporate all these factors to develop a risk model or risk calculator for the prediction of cumulative cancer incidence. Such risk calculators may provide clinicians important information for the triage of patients who need intensive treatment from those who need only routine follow-up. Several risk models/calculators have been developed to predict the incidence of hepatocellular carcinoma of chronic hepatitis B patients (Yang et al. 2010), and only REACH-B score was externally validated (Yang et al. 2011). The REACH-B score has recently been used to examine the efficacy of antiviral therapy to reduce liver cancer risk of chronic hepatitis B patients.

Table 4 Scores assigned to risk predictors of HBV-caused hepatocellular carcinoma

Risk predictor	Risk score
Age	
30–34	0
35–39	1
40–44	2
45–49	3
50–54	4
55–59	5
60–64	6
Sex	
Female	0
Male	2
Family history of hepatocellular carcinoma	
No	0
Yes	2
Serum ALT levels (IU/L)	
<15	0
15–44	1
≥45	2
HBeAg/HBV DNA (copies/mL)/HBsAg (IU/mL)/Genotype	
Negative/<10^4/<100/any type	0
Negative/<10^4/100–999/any type	2
Negative/<10^4/≥1000/any type	2
Negative/10^4–10^6/<100/any type	3
Negative/10^4–10^6/100-999/any type	3
Negative/10^4–10^6/≥ 1000/any type	4
Negative/≥10^6/any level/B or B + C	5
Negative/≥10^6/any level/C	7
Positive/any level/any level/B or B+C	6
Positive/any level/any level/C	7

Table 4 shows a most recent HCC risk calculator for chronic hepatitis B patients (Lee et al. 2013). Predictors included in the risk calculator are age, gender, family history of hepatocellular carcinoma, serum ALT level, HBeAg serostatus, serum levels of HBV DNA and HBsAg, and HBV genotype. Risk scores are

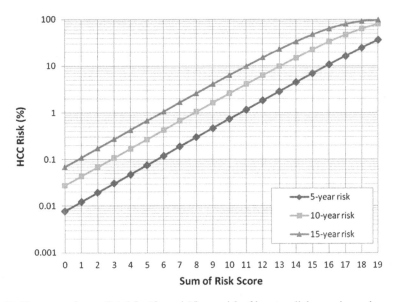

Fig. 3 Nomogram for predicted 5-, 10-, and 15-year risk of hepatocellular carcinoma by sum of risk score

assigned for various categories of risk predictors. For example, a 60-year-old (risk score = 6) male (risk score = 2) chronic hepatitis B patient who had a family history of hepatocellular carcinoma (risk score = 2), a serum ALT level of 90 IU/L (risk score = 2), a HBeAg-seropositive serostatus, a serum HBV DNA level of 10^7 copies/mL, a serum HBsAg level of 104 IU/mL, and a HBV genotype C infection (risk score = 7) has a sum of risk score of 19. The 5-, 10-, and 15-year cumulative risks of hepatocellular carcinoma by the sum of risk score are shown in the nomogram of Fig. 3. For the patient with a sum of risk score as high as 19, his 5- and 10- and 15-year risk of hepatocellular carcinoma will be 35 %, 80 %, and 90 %. In contrast, a 34-year-old women with no family history of hepatocellular carcinoma, a serum ALT level of 10 IU/L, a HBeAg-negative serostatus, a serum HBV DNA level of 10^3 copies/mL, and a serum HBsAg level of 50 IU/mL (sum of risk score = 0) has 5-, 10-, and 15-year risk of hepatocellular carcinoma of 0.0075 %, 0.025 %, and 0.065 %.

The risk calculators for other virus-caused cancers such as nasopharyngeal carcinoma and cervical cancer may also be helpful to improve the triage and clinical management of patients infected with other oncogenic viruses. The development of the risk calculators needs large-scale prospective cohorts, which have been followed for a long period of time with accurate measurements of risk predictors. Demographical characteristics, viral infection biomarkers, family history, and polymorphisms of genetic susceptibility may be incorporated to develop valid and useful cancer risk calculators.

Table 5 Evidence showing the reduction of incidence of virus-caused cancers through the preventive strategy of vaccination and antiviral therapy

Virus	Cancer	Preventive strategy
HBV	Hepatocellular carcinoma	Vaccination and antiviral therapy
HCV	Hepatocellular carcinoma	Antiviral therapy
HIV-1	Kaposi's sarcoma	Antiviral therapy
HPV	Cervical cancer	Vaccination

8   Cancer Incidence Reduction through Vaccination and Antiviral Therapy

The most effective strategy to prevent virus-caused cancers is through the vaccination to prevent viral infection or the antiviral therapy to eliminate oncogenic viruses in human host. Table 5 shows currently available vaccines or antivirals to prevent or treat patients with oncogenic viral infection. Vaccines are available for the prevention of HBV-caused hepatocellular carcinoma and HPV-caused cervical cancer, while antiviral therapies are available for the treatment of chronic infection of HBV, HCV, and HIV.

Many clinical trials have demonstrated the efficacy of the HPV vaccination to prevent cervical neoplasia, the precursor lesions of cervical cancer, and the efficacy of antiviral therapy to prevent hepatocellular carcinoma in cirrhotic patients (Liaw et al. 2004). The national HBV immunization program in Taiwan implemented in 1984 has successful reduced the incidence of hepatocellular carcinoma at ages 6–19 years in vaccinated birth cohorts (Chang et al. 1997, 2009; Chien et al. 2006). The HPV immunization program in Australia has effectively lowered the incidence of cervical neoplasia in vaccinated adolescent cohorts. A national antiviral therapy program was implemented in 2003 to control chronic hepatitis B or C in Taiwan. It is expected to reduce the incidence and mortality of hepatocellular carcinoma in treated adult patients. However, its efficacy to prevent hepatocellular carcinoma remains to be assessed.

9   Future Perspectives

Along with the advancement in proteomic and genomic medicine, more and more biomarkers associated with the development of virus-caused cancers have been identified. They may be applied for the risk prediction or early detection of the cancers. For example, multiple micro RNAs have been combined for the diagnosis of hepatocellular carcinoma. However, its efficacy and cost-effectiveness for early diagnosis of HCC should be further assessed and compared with those of other methods including abdominal ultrasonography (Chen and Lee 2011). More importantly, repeated measurements of biomarkers may further improve the risk prediction or early detection of virus-caused cancers (Chen 2005). For example, the trajectory of serum HBV DNA levels has been found to predict long-term risk of hepatocellular carcinoma effectively (Chen et al. 2011c). More longitudinal studies with regular follow-up examinations of various biomarkers are in urgent need to identify good molecular targets for the development of preventives, diagnostics, or therapeutics of various virus-caused cancers. The health economic assessment of these biopharmaceuticals may help the clinical application of them.

References

Alter MJ (2007) Epidemiology of hepatitis C virus infection. World J Gastroenterol 13:2436–2441

Chang MH, Chen CJ, Lai MS, Kong MS, Wu TC, Liang DC, Hsu HM, Shau WY, Chen DS (1997) Taiwan Childhood Hepatoma Study Group. Universal hepatitis B vaccination in Taiwan and the incidence of hepatocellular carcinoma in children. New Engl J Med 336:1855–1859

Chang MH, You SL, Chen CJ, Liu CJ, Lee CM, Lin SM, Chu HC, Wu TC, Yang SS, Kuo HS, Chen DS (2009) The Taiwan hepatoma study group. decreased incidence of hepatocellular carcinoma in hepatitis B vaccinees: a 20-year follow-up study. J Natl Cancer Inst 101:1348–1355

Chang YJ, Chen HC, Lee BH, You SL, Lin CY, Pan MH, Chou YC, Hsieh CY, Chen YM, Chen YJ, Chen CJ (2011) Unique variants of human papillomavirus genotypes 52 and 58 and risk of cervical neoplasia. Int J Cancer 129:965–973

Chen CJ (2005) Time-dependent events in natural history of occult hepatitis B virus infection: Importance of population-based long-term follow-up study with repeated measurements (Editorial). J Hepatology 42:438–440

Chen CJ, Chen DS (2002) Interaction of hepatitis B virus, chemical carcinogen and genetic susceptibility: multistage hepatocarcinogenesis with multifactorial etiology (editorial). Hepatology 36:1046–1049

Chen CJ, Lee MH (2011) Early diagnosis of hepatocellular carcinoma by multiple microRNAs: validity, efficacy and cost-effectiveness (editorial). J Clin Oncol 29:4745–4747

Chen CJ, Yang HI (2011) Natural history of chronic hepatitis B REVEALed. J Gastroenterol Hepatol 26:628–638

Chen CJ, Liang KY, Chang AS, Chang YC, Lu SN, Liaw YF, Chang WY, Sheen MC, Lin TM (1991) Effects of hepatitis B virus, alcohol drinking, cigarette smoking and familial tendency on hepatocellular carcinoma. Hepatology 13:398–406

Chen CJ, Yu MW, Liaw YF, Wang LW, Chiamprasert S, Matin F, Hirvonen A, Bell AB, Santella RM (1996a) Chronic hepatitis B carriers with null genotypes of glutathione S-transferase M1 and T1 polymorphisms who are exposed to aflatoxins are at increased risk of hepatocellular carcinoma. Am J Hum Genet, 59:128–134

Chen CJ, Wang LY, Lu SN, Wu MH, You SL, Li HP, Zhang YJ, Wang LW, Santella RM (1996b) Elevated aflatoxin exposure and increased risk of hepatocellular carcinoma. Hepatology 24:38–42

Chen CJ, Yu MW, Liaw YF (1997) Epidemiological characteristics and risk factors of hepatocellular carcinoma. J Gastroen Hepatol 12:S294–S308

Chen CJ, Yang HI, Su J, Jen CL, You SL, Lu SN, Huang GT, Iloeje UH (2006) Risk of hepatocellular carcinoma across a biological gradient of serum hepatitis B virus DNA level. JAMA 2006(295):65–73

Chen CL, Yang HI, Yang WS, Liu CJ, Chen PJ, You SL, Wang LY, Sun CA, Lu SN, Chen DS,
    Chen CJ (2008) Metabolic factors and risk of hepatocellular carcinoma by chronic hepatitis B/
    C virus infection: a follow-up study in Taiwan. Gastroenterology 135:111–121
Chen CJ, Yang HI, Iloeje UH (2009) REVEAL-HBV Study Group. Hepatitis B virus DNA levels
    and outcomes in chronic hepatitis B. Hepatology 49:S72–S84
Chen CF, Lee WC, Yang HI, Chang HC, Jen CL, Iloeje UH, Su J, Hsiao KC, Wang LY, You SL,
    Lu SN, Chen CJ (2011a) Changes in serum levels of HBV DNA and alanine aminotransferase
    determine risk for hepatocellular carcinoma risk. Gastroenterology 141:1240–1248
Chen HC, You SL, Hsieh CY, Lin CY, Pan MH, Chou YC, Liaw KL, Schiffman M, Hsing AW,
    Chen CJ (2011b) For CBCSP-HPV Study Group. Prevalence of genotype- specific human
    papillomavirus infection and cervical neoplasia in Taiwan: a community-based survey of
    10,602 women. Int J Cancer 128:1192–1203
Chen HC, Schiffman M, Lin CY, Pan MH, You SL, Chuang LC, Hsieh CY, Liaw KL, Hsing AW,
    Chen CJ (2011c) CBCSP-HPV study group. persistence of type-specific human papilloma-
    virus infection and increased long-term risk of cervical cancer. J Natl Cancer Inst
    103:1387–1396
Chien YC, Chen CJ (2003) Epidemiology and etiology of nasopharyngeal carcinoma: gene-
    environment interaction. Cancer Rev Asia-Pacific 1:1–19
Chien YC, Chen CJ, Liu MY, Yang HI, Hsu MM, Chen CJ, Yang CS (2001) Serologic markers
    of Epstein-Barr virus infection and nasopharyngeal carcinoma in Taiwanese men. New Engl J
    Med 345:1877–1882
Chien YC, Jan CF, Kuo HS, Chen CJ (2006) Nationwide hepatitis B vaccination program in
    Taiwan: Effectiveness in 20 years after it was launched. Epidemiol Rev 28:126–135
Cho EY, Hildesheim A, Chen CJ, Hsu MM, Chen IH, Mittl BF, Levine PH., Liu MY, Chen JY,
    Brinton LA, Cheng YJ, Yang CS (2003) Nasopharyngeal carcinoma and genetic polymor-
    phisms of DNA repair enzymes XRCC1 and hOGG1. Cancer Epidemiol Biomarker Prev,
    12:1100–1104
Chuang LC, Hu CY, Chen HC, Lin PJ, Lee B, Lin CY, Pan MH, Chou YC, You SL, Hsieh CY,
    Chen CJ (2012) Associations of human leukocyte antigen class II genotypes with human
    papillomavirus 18 infection and cervical intraepithelial neoplasia risk. Cancer 118:223–231
de Sanjose S, Diaz M, Castellsague X, Clifford G, Bruni L, Munoz N, Bosch FX (2007)
    Worldwide prevalence and genotype distribution of cervical human papillomavirus DNA in
    women with normal cytology: a meta-analysis. Lancet Infect Dis 7:453–459
Dukers NH, Rezza G (2003) Human herpesvirus 8 epidemiology: what we do and do not know.
    AIDS 17:1717–1730
Hildesheim A, Anderson LM, Chen CJ, Cheng YJ, Brinton LA, Daly AK, Reed CD, Chen IH,
    Caporaso NE, Hsu MM, Chen JY, Idle JR, Hoover RN, Yang CS, Chabra SK (1997) CYP2E1
    genetic polymorphisms and risk of nasopharyngeal carcinoma in Taiwan. J Natl Cancer Inst
    89:1207–1212
Hildesheim A, Dosemeci M, Chan CC, Chen CJ, Cheng YC, Hsu MM, Chen IH, Mittl BF, Sun B,
    Levine PH, Chen JY, Brinton LA, Yang CS (2001) Occupational exposure to wood,
    formaldehyde, and solvents and risk of nasopharyngeal carcinoma. Cancer Epidemiol
    Biomarker Prev 2001(10):1145–1153
Hildesheim A, Apple RJ, Chen CJ, Wang SS, Cheng YC, Klitz W, Mack SJ, Chen IH, Hsu MM,
    Yang CS, Brinton LA, Levine PH, Erlich HA (2002) Association of HLA class I and II alleles
    and extended haplotypes with nasopharyngeal carcinoma in Taiwan. J Natl Cancer Inst
    94:1780–1789
Hsu WL, Chen JY, Chien YC, Liu MY, You SL, Hsu MM, Yang CS, Chen CJ (2009)
    Independent effects of EBV and cigarette smoking on nasopharyngeal carcinoma: a 20-year
    follow-up study on 9,622 males without family history in Taiwan. Cancer Epidemiol
    Biomarkers Prev 18:1218–1226
Hsu WL, Yu KJ, Chien YC, Chiang JY, Cheng YJ, Chen JY, Liu MY, Chou SP, You SL, Hsu
    MM, Lou PJ, Wang CP, Hong JH, Leu YS, Tsai MH, Su MC, Tsai ST, Chao WY, Ger LP,

Chen PR, Yang CS, Hildesheim A, Diehl SR, Chen CJ (2011) Familial tendency and risk of nasopharyngeal carcinoma in Taiwan: effects of covariates on risk. Am J Epidemiol 173:292–299

Hsu WL, Pan WH, Chien YC, Yu KJ, Cheng YJ, Chen JY, Liu MY, Hsu MM, Lou PJ, Chen IH, Yang CS, Hildesheim A, Chen CJ (2012a) Lowered risk of nasopharyngeal carcinoma and intake of plant vitamin, fresh fish, green tea and coffee: a case-control study in Taiwan. PLoS ONE 7:e41779

Hsu WL, Tse KP, Liang S, Chien YC, Su WH, Yu KJ, Cheng YJ, Tsang NM, Hsu MM, Chang KP, Chen IH, Chen TI, Yang CS, Golstein AM, Chen CJ, Chang YS, Hildesheim A (2012b) Evaluation of human leukocyte antigen-A (HLA-A), other non-HLA markers on chromosome 6p21 and risk of nasopharyngeal carcinoma. PLoS ONE 7:e42767

Huang YT, Jen CL, Yang HI, Lee MH, Lu SN, Iloeje UH, Chen CJ (2011) REVEAL-HBV/HCV Study Group. Lifetime risk and gender difference of hepatocellular carcinoma among patients affected with chronic hepatitis B and C. J Clin Oncol 29:3643–3650

IARC (2009)(International Agency for Research on Cancer) A review of human carcinogens. Part B: biological agent. IARC, Lyon, IARC

Lee MH, Yang HI, Lu SN, Jen CL, Yeh SH, Liu CJ, Chen PJ, You SL, Wang LY, Chen WJ, Chen CJ (2010) Hepatitis C virus seromarkers and subsequent risk of hepatocellular carcinoma: long-term predictors from a community-based cohort study. J Clin Oncol 28:4587–4593

Lee MH, Yang HI, Liu J, Batrla-Utermann R, Jen CL, Iloeje UH, Jun J, Lu SN, You SL, Wang LY, Chen CJ (2013) For the R.E.V.E.A.L.-HBV Study Group. Models predicting long-term risk of liver cirrhosis and hepatocellular carcinoma in chronic hepatitis B patients: risk scores integrating characteristics of host and hepatitis B virus. Hepatology (in press)

Liaw YF, Sung JJ, Chow WC, Farrell G, Lee CZ, Yuen H, Tanwandee T, Tao QM, Shue K, Keene ON, Dixon JS, Gray DF, Sabbat J (2004) Lamivudine for patients with chronic hepatitis B and advanced liver disease. N Engl J Med 351:1521–31

Proietti FA, Carneiro-Proietti AB, Catalan-Soares BC, Murphy EL (2005) Global epidemiology of HTLV-1 infection and associated diseases. Oncogene 24:6058–6068

Sun CA, Wu DM, Lin CC, Lu SN, You SL, Wang LY, Wu MH, Chen CJ (2003) Incidence and co-factors of hepatitis C virus-related hepatocellular carcinoma: a prospective study of 12,008 men in Taiwan. Am J Epidemiol 157:674–682

Wang LY, Hatch M, Chen CJ, Levin B, You SL, Lu SN, Wu MH, Wu WP, Wang LW, Wang Q, Huang GT, Yang PM, Lee HS, Santella RM (1996) Aflatoxin exposure and the risk of hepatocellular carcinoma in Taiwan. Int J Cancer 67:620–625

Wang LY, You SL, Lu SN, Ho HC, Wu MH, Sun CA, Yang HI, Chen CJ (2003) Risk of hepatocellular carcinoma and habits of alcohol drinking, betel quid chewing, and cigarette smoking: a cohort of 2416 HBsAg-seropositive and 9421 HBsAg-seronegative male residents in Taiwan. Cancer Cause Control 14:241–250

Ward MH, Pan WH, Cheng YJ, Li FH, Brinton LA, Chen CJ, Hsu MM, Chen IH, Levine PH, Yang CS, Hildesheim A (2000) Dietary exposure to nitrite and nitrosamines and risk of nasopharyngeal cancer in Taiwan. Int J Cancer 2000(86):603–609

GLOBOCAN http://globocan.iarc.fr/factsheets/populations/factsheet.asp?uno=900

Yang HI, Lu SN, You SL, Sun CA, Wang LY, Hsiao K, Chen PJ, Chen DS, Liaw YF, Chen CJ (2002) Hepatitis B e antigen and the risk of hepatocellular carcinoma. New Engl J Med 347:168–174

Yang HI, Yeh SH, Chen PJ, Iloeje UH, Jen CL, Wang LY, Lu SN, You SL, Chen DS, Liaw YF, Chen CJ, REVEAL-HBV Group (2008) Association between Hepatitis B virus genotype and mutants and the risk of hepatocellular carcinoma. J Natl Cancer Inst, 100: 1134–1143

Yang HI, Sherman M, Su J, Chen PJ, Liaw YF, Iloeje UH, Chen CJ (2010) Nomograms for risk of hepatocellular carcinoma in patients with chronic hepatitis B virus infection. J Clin Oncol 28:2437–2444

Yang HI, Yuen MF, Chan HL, Han KH, Chen PJ, Kim DY, Ahn SH, Chen CJ, Wong VW, Seto WK (2011) Risk estimation for hepatocellular carcinoma in chronic hepatitis B (REACH-B): development and validation of a predictive score. Lancet Oncol 12:568–574

Yu MW, Chen CJ (1993) Elevated serum testosterone levels and risk of hepatocellular carcinoma. Cancer Res 53:790–794

Yu MW, Gladek-Yarborough A, Chiamprasert S, Santella RM, Liaw YF, Chen CJ (1995a) Cytochrome P-450 2E1 and glutathione S-transferase M1 polymorphisms and susceptibility to hepatocellular carcinoma. Gastroenterology 109:1266–1273

Yu MW, Hsieh HH, Pan WH, Yang CS, Chen CJ (1995b) Vegetable consumption, serum retinol level and risk of hepatocellular carcinoma. Cancer Res 55:1301–1305

Yu MW, Chiu YH, Chiang YC, Chen CH, Lee TH, Santella RM, Chen HD, Liaw YF, Chen CJ (1999a) Plasma carotenoids, glutathione S-transferases M1 and T1 genetic polymorphisms, and risk of hepatocellular carcinoma: Independent and interactive effects. Am J Epidemiol 149:621–629

Yu MW, Horng IS, Hsu KH, Chiang YC, Liaw YF, Chen CJ (1999b) Plasma selenium levels and risk of hepatocellular carcinoma among men with chronic hepatitis B virus infection. Am J Epidemiol 150:367–374

Yu MW, Chang HC, Liaw YF, Lin SM, Lee SD, Chen PJ, Hsiao TJ, Lee PH, Chen CJ (2000a) Familial risk of hepatocellular carcinoma among chronic hepatitis B carriers and their relatives. J Natl Cancer Inst 92:1159–1164

Yu MW, Cheng SW, Lin MW, Yang SY, Liaw YF, Chang HC, Hsiao TJ, Lin SM, Lee SD, Chen PJ, Liu CJ, Chen CJ (2000b) Androgen-receptor CAG repeat, plasma testosterone levels, and risk of hepatitis B-related hepatocellular carcinoma. J Natl Cancer Inst 92:2023–2028

Yu MW, Yang SY, Pan IJ, Lin CL, Liu CJ, Liaw YF, Lin SM, Chen PJ, Lee SD, Chen CJ (2003) Polymorphisms in XRCC1 and glutathione S-transferase genes and hepatitis B-related hepatocellular carcinoma. J Natl Cancer Inst 95:1485–1488

Virus Infection, Inflammation and Prevention of Cancer

Norman Woller and Florian Kühnel

Abstract

Our molecular understanding of cancer biology has made substantial progress during the last two decades. During recent years it became evident that inflammation is a major driving force in tumor development since chronic virus infection and carcinogenesis are closely correlated. These insights refined our view on the decisive role of persistent virus infection and chronic inflammation in tumor onset, growth, and metastatic progression. Explanations have been delivered how tumor cells interact and correspond with neighbouring epithelia and infiltrating immune cells for shaping the so-called 'tumor-microenvironment' and establishing tumor-specific tolerance. This extended view on malignant diseases should now allow for rational design of interventions targeting inflammation and underlying pathways for prevention and therapy of inflammation-associated cancer. This chapter outlines the role of virus-mediated inflammations in tumorigenesis thereby shedding light on the mechanisms of cancer-related inflammation and on characteristic features of the tumor-microenvironment, which has been recently identified to play a key role in maintenance and progression of tumors. Finally, the chapter discusses latest aspects in prevention of inflammation-related cancer and provides a short outlook on the future prospects of cancer immunotherapy.

N. Woller · F. Kühnel (✉)
Clinic for Gastroenterology, Hepatology, and Endocrinology, Hannover Medical School,
Carl-Neuberg-Str. 1, 30625, Hannover, Germany
e-mail: Kuehnel.Florian@mh-hannover.de

M. H. Chang and K.-T. Jeang (eds.), *Viruses and Human Cancer*,
Recent Results in Cancer Research 193, DOI: 10.1007/978-3-642-38965-8_3,
© Springer-Verlag Berlin Heidelberg 2014

Contents

In the 19th century, Rudolph Virchow hypothesized that tissue injury precedes tumor development at the same locus. Observations of leukocyte infiltrations in tumor tissue let him speculate that inflammatory processes of wound healing may be involved in tumor development (Balkwill and Mantovani 2001). Picking up these ideas, tumors have been later described as a result of 'possible overhealing' (Haddow 1972) or as 'wounds that do not heal' (Dvorak 1986) thereby referring to obvious similarities between tumor growth and wound healing such as fibroblast activation, attraction of leukocytes, cell proliferation and angiogenesis. During the last two decades, this notion was supported by molecular evidence confirming that inflammatory processes are essentially involved in the development of tumor and in emergence of metastases (Karin 2006; Guerra et al. 2007). With regard to etiology of cancer, germline mutations play a minor role in the development of cancer whereas up to 90 % of all cancers are associated with acquired somatic mutations, which are mainly acquired by environmental and life style factors. 30 % of global malignancies can be attributed to tobacco smoking and 35 % are due to dietary factors including obesity (Aggarwal et al. 2009). An estimated 20-25 % of the global cancer burden is associated with pathogen inflammations, the vast majority due to unresolved, persistent virus infections that drive inflammation (Hussain and Harris 2007; Parkin 2006). Moreover, an inflammatory component is also present in the microenvironment of tumors that are epidemiologically not related to classical pathogen infections but to other environmental risk factors such as tobacco smoking or inhalation of silica fibres thus illustrating the general role of inflammation in carcinogenesis. In this review, we summarize the key aspects of the role of virus-associated infections and inflammations for tumor development, and discuss applicable measures to prevent infection-associated cancers. Hereby, we will focus on infection-associated inflammatory processes involved in carcinogenesis that may offer attractive molecular targets for pharmacologic means in cancer prevention and therapy. In the following section, the clinically most relevant virus types that cause chronic tissue inflammation and cancer are briefly introduced.

1   Chronic Virus Infection as a Cause of Cancer

Virus infections belong to the most important causes of cancer (de Martel and Franceschi 2009). Among those tumors that are related to virus infections, solid tumors in the liver and cervix uteri are the most relevant entities in terms of global health burden. These tumors are mainly attributable to hepatitis B and C virus (HBV; HCV) and human papilloma virus (HPV), respectively. HBV is a small DNA virus from the hepadnavirus family which can be transmitted between humans mainly via blood contact or sexual intercourse. Liver infections by HBV frequently lead to severe complications such as acute hepatitis including the risk of liver failure and furthermore to cirrhosis and hepatocellular carcinoma if the infection persists. The causal relationship between HBV and HCC development is well established (Chen et al. 2006) and conservatively estimated 54 % of liver cancer cases are due to HBV infection. Though an effective vaccination is available since more than 20 years, HBV remains a major health problem since approximately 360 million chronic HBV carriers exist worldwide (Shepard et al. 2006; Custer et al. 2004). The risk to develop HCC as a late complication of liver infection is closely linked to the duration of the virus-mediated inflammation. Several pathogenic mechanisms are involved including oncogenic effects mediated by HBV proteins (such as HBx), by mutagenic insertion of parts of the viral DNA into the host cell genome, and by T cell-dependent autoimmunity (Farazi and DePinho 2006).

A second virus that is significantly involved in the development of liver cancer is HCV, a single stranded RNA virus from the flavivirus family. HCV infection is a blood borne disease that chronically infects 3 % of the world's population (Shepard et al. 2005). HCV-positive individuals have a highly significant risk to develop an HCC (Donato et al. 1998) and about 31 % of liver cancer cases are due to chronic HCV infections. The underlying pathogenetic mechanisms are not fully understood. As in case of HBV, HCV can establish persistent infections and malignant transformation results from a longer-lasting process of coincident inflammation, cell death and tissue renewal. Molecular studies have shown that several HCV proteins such as NS3, NS4B, NS5A and core actually have oncogenic potential (Farazi and DePinho 2006; Barth et al. 2008).

A further virus involved in development of infection- and inflammation-associated solid tumors is human papilloma virus (HPV) (Parkin and Bray 2006). HPV is a non-enveloped DNA virus that can induce benign and malignant lesions of skin and mucosa. The course of infection is usually asymptomatic and local immunity leads to virus clearance in most individuals (Plummer et al. 2007). In about 10 % of infected women HPV establishes a persistent infection, which elevates the risk for cervix cancer development. Among 100 different genotypes, several HPV strains are regarded as high risk strains due to their causal correlation with development of cervix carcinoma. In this context, the most virulent HPV-16 and HPV-18 genotypes account for the majority of cervical neoplasias (Smith et al. 2007). The pathogenic mechanisms underlying HPV-mediated oncogenesis are relatively well studied. Chronic infection with HPV is accompanied by integration of viral DNA sequences into the DNA of the host cell. The expression of early viral genes such as E6/E7 from viral integrates interferes with the activity of the tumor suppressors p53 and Rb, thus activating the cell cycle and inhibiting intrinsic apoptosis. Additional immunological or mutagenic influence factors can further promote cell transformation and tumor growth.

A further example for a clinically relevant oncogenic pathogen is the Epstein-Barr virus (EBV) which has been associated with several hematological malignancies such as Burkitt's lymphoma, Hodgkin- and non-Hodgkin-lymphoma. Further tumor-associated viruses are the human herpes virus 8 (HHV-8), the causal pathogen for Kaposi Sarcoma, the human T cell leukemia virus type 1 (HTLV-1), and Merkel cell polyoma virus.

2   Inflammation and Cancer

2.1   The Interconnection Between Inflammation and Tumor Development

Under steady-state conditions, maintenance of tissue-integrity and cell renewal is tightly regulated by the host. Upon disturbance of tissue-integrity by pathogen infection or tissue damage, an acute inflammation is elicited, a locally and timely limited immune mechanism that enables both effective elimination of the pathogen and wound healing responses. Though the preferential goal of an acute inflammation is the pathogen-free reconstitution of tissue integrity, infection is eventually not completely cleared by acute inflammatory mechanisms. This consequently is leading over to a chronic inflammatory response, which is in coincidental balance with pathogen persistence. Two events appear to be fundamental to establish chronic virus-mediated inflammations. First, tumor-associated viruses must find ways to subvert the host's antiviral immune defense or to retreat into immuno-privileged niches to persist in a latent state. Second, the rigor of antiviral immune responses must be adjusted to levels that outbalance sufficient control of viremia and prevention of severe immunopathology in the infected organ. These persistent infections lead to a mild, but chronic inflammation. In persistent infections, cells are not only subject of transformation by viral oncogenes. Accumulation of tumor-prone genetic alterations is also accelerated by enhanced cell turnover during chronic inflammations. Both features, chronic inflammation and viral transformation of cells, are important events of tumorigenesis (Rakoff-Nahoum and Medzhitov 2009). Moreover, inflammation has been identified to be of utmost importance to promote virtually every step of tumorigenesis (Karin 2006). Even the involvement of inflammatory processes in initial cell transformation is a matter of discussion. It has been shown that inflammation may increase mutation frequency in tissues even in the absence of defined extrinsic triggers (Sato et al. 2006; Bielas et al. 2006). Inflammatory mechanisms are involved in the process of 'immunosurveillance' which effectively limits tumor development at a very early stage. Triggered by aberrant oncogene activation, induction of cellular senescence in liver cells consequently leads to the elimination of senescent cells, orchestrated by CD4 T cells and M1 macrophages (Kang et al. 2011). On the other hand, immunosuppression or overcoming and escape from senescence further promote tumorigenesis at an early stage by inflammatory mechanisms such as senescence-associated cytokine secretion (Rodier et al. 2009).

Once an early neoplastic nodule is established, several kinds of leukocytes and mesenchymal cells are attracted, which establish the tumor-microenvironment and provide cancer cells with supplemental growth stimuli and cytokines. This smoldering inflammatory tumor-microenvironment is vital for tumor development and has been generally accepted as another hallmark of cancer (Hanahan and Weinberg 2011). These observations also attracted much attention towards the contribution of non-malignant cells to the tumor-microenvironment, their interactions, the released signalling molecules and central molecular pathways.

2.2 Cytokine Signalling and Molecular Pathways in Tumor-Associated Inflammation

Tumor-associated inflammation and tumorigenesis are closely interconnected mechanisms with the inflammatory process as a predominant driver of the malignant disease. Crucial endogenous factors of tumor-associated inflammation have already been identified in numerous studies such as the signal transducer activator of transcription-3 (STAT-3) and the transcription factor NFκB. These transcriptional activators are essentially involved in signal transduction and/or expression of inflammatory cytokines playing a role in tumor-associated inflammation, such as TNF-α, IL-6, IL-1β, IL-11, and IL-23 (Grivennikov et al. 2009; Fukuda et al. 2011; Lesina et al. 2011; Voronov et al. 2003; Langowski et al. 2006). NFκB is a well known key orchestrator in innate immune and inflammatory responses and is activated in both tumor cells as well as in immune cells involved in tumor-associated inflammation. NFκB translates a panoply of extrinsic and intrinsic danger signals into specific gene activation. NFκB lies downstream of toll-like-receptor (TLR)-MyD88 dependent pathways and processes signals from receptors of inflammatory cytokines such as TNF-α and IL-1β. Additionally, NFκB can be upregulated by endogenous genetic alterations in cancer cells. Upon upstream signalling, NFκB activates expression of proinflammatory factors such as cytokines, adhesion molecules, NO-synthase, stimulators of angiogenesis, and COX2, a crucial enzyme in prostaglandin synthesis. NFκB as well as STAT3 promote cell survival by expression of the antiapoptotic mediators Bcl-2, Bcl-XL,

Mcl-1, c-Flip, survivin, and IAPs thereby conferring resistance against antitumoral immune responses in cancer surveillance. Additionally, links between NFκB activation and the hypoxic response have been reported (Rius et al. 2008). Both NFκB and STAT3 also interfere with p53 functions in genome surveillance, DNA damage response and intrinsic apoptosis (Colotta et al. 2009; Ryan et al. 2000).

2.3 The Inflammatory Tumor-Microenvironment

Most prominent cells of the immune system recruited to the tumor-microenvironment are tumor-associated macrophages (TAMs). This cell type reflects a subtype of the monocyte-macrophage lineage and may be obligatory for invasion, metastasis, and angiogenesis (Condeelis and Pollard 2006). Due to the plasticity and diversity of these cells, two fundamental phenotypes are distinguished. The classical M1 activation is mediated by TLR ligands and IFN-γ, whereas alternative M2 activation is stimulated by IL-4/IL-13. The polarization of M1-M2 mirrors the polarization of Th1-Th2 T cells. M1 macrophages promote Th1 responses and release high levels of proinflammatory cytokines, reactive oxygen species and reactive nitrogen intermediates, whereas M2 polarized macrophages display immunoregulatory functions, promote tissue remodeling and display high expression of scavenger receptors (Gordon and Martinez 2010; Biswas and Mantovani 2010; Mantovani et al. 2002). TAMs found in advanced stages during tumor-development generally display an M2-like phenotype, which is characterized by a tumor-promoting activity of tissue-remodelling and angiogenesis together with a high IL-10 expression. These cells are polarized by tumor-resident lymphocytes of different origin, depending on the organ (Biswas and Mantovani 2010). In breast carcinogenesis, plasma cells promote TAM-polarization (DeNardo et al. 2009; Pedroza-Gonzalez et al. 2011) whereas in development of skin tumors IL-4 secreting Th2 cells induce the M2-like phenotype (Schioppa et al. 2011; Andreu et al. 2010). Additionally, tumor cells have the ability to directly influence macrophages towards a cancer-promoting mode by secretion of different components. These include components of the extracellular matrix, the cytokines M-CSF and IL-10, and chemokines like CCL2, CCL17, CCL18, and CXCL4 (Mantovani et al. 2008; Erler et al. 2009; Kim et al. 2009; Roca et al. 2009). It has been shown, that a high TAM infiltration correlates with a poor prognosis (Murdoch et al. 2008).

Besides the central role of TAMs, several other innate and adaptive immune cells are found in the tumor-microenvironment, and for almost all of them a protumorigenic role has been demonstrated. These include T and B cells, dendritic cells, mast cells, myeloid-derived suppressor cells, and neutrophils. Only NK cells lack an established tumor-supporting function so far. Mature T cells present in solid tumors are classically defined as cytotoxic CD8 T cells (CTLs) and CD4 helper cells (Th). The latter are further classified in Th1, Th2, Th17, and regulatory T (Treg) cells. The polarization of these cells dictate, whether T cell subsets (including CD8 T cells) promote tumor development and metastasis (Roberts et al.

2007). Additionally, polarization correlates with survival in some cancers, due to invasion of tumor-specific CTLs and Th1 cells, as shown in melanoma, pancreatic cancer, colon cancer, and multiple myeloma (Galon et al. 2006; Laghi et al. 2009; Swann and Smyth 2007). Moreover, presence of tumor-infiltrating lymphocytes (TILs) with high ratios of CD8/CD4 and Th1/Th2 indicates an improved prognosis in breast cancer (Kohrt et al. 2005). Treg cells are mainly suspected to mediate tumor tolerance by suppression of antitumor immune responses (Gallimore and Simon 2008). However, Erdman and colleagues could demonstrate that Tregs can also inhibit cancer-associated inflammation thereby playing an antitumorigenic role in malignant diseases (Erdman et al. 2005). Myeloid-derived suppressor cells (MDSCs) are often recruited and activated by the inflammatory tumor-microenvironment by multiple factors, such as IL-6, IL-1ß, and VEGF (Gabrilovich and Nagaraj 2009). Activated MDSCs in turn release pro-inflammatory factors which results in a positive feedback loop promoting cancer-associated inflammation. Furthermore, MDSCs are not only capable to suppress adaptive immune responses, but also influence the cytokine production of macrophages (Sinha et al. 2007). Tumor-associated neutrophils (TAN) have been identified to play a role in cancer-related inflammation as well (Cassatella et al. 2009; Mantovani 2009). Their previously unnoticed plasticity can exert tumor-promoting and tumoricidal functions, depending on the polarization by TGF-ß (Fridlender et al. 2009).

Cancer-associated fibroblasts (CAFs) are a major component of the tumor stroma and an integral part of the cancer-related inflammatory environment. They are activated by IL-1 and share many characteristics with activated fibroblasts in wound healing processes. CAFs have a proinflammatory signature and promote tumor growth and angiogenesis (Erez et al. 2010).

During tumorigenesis, malignant cells frequently interact with immune cells resulting in their polarization into a tumor-supportive phenotype. The tumor-microenvironment is to a large extent borne by feedback of these tumor-promoting immune cells finally mediating tolerance by adaptation and manipulation of infiltrating lymphocytes within the tumor-microenvironment.

Many, if not all innate and adaptive immune cells that are involved in tumor development, exhibit a more or less significant plasticity, which can even be reverted under certain conditions (Sharma et al. 2010; Fridlender et al. 2009). Therefore, it can be assumed that cancer-associated inflammation and antitumor immunity are able to coexist simultaneously and thereby mutually affect each other during tumor progression. However, a growing tumor mass might finally quench antitumor immunity, thus achieving the tumor's complete immune escape in advanced stages of tumor development (Koebel et al. 2007).

2.4   Inflammation, Hypoxia and Angiogenesis in Tumor-Growth

Hypoxic conditions generally appear in solid malignancies, when the demand for oxygen and nutrients of a growing tumor is greater, than the blood supply is able to provide. At first, cells respond by induction of hypoxia-inducible factor (HIF1α), which finally modifies glucose metabolism, angiogenesis and furthermore cell survival and invasion (Staller et al. 2003). Hypoxic conditions also result in necrotic cell death in the core of the tumor. These dying cells induce the proin-flammatory mediators IL-1 and HMGB1, which additionally trigger angiogenesis and deliver further growth factors to the tumor environment (Vakkila and Lotze 2004). However, besides hypoxia, tumor neoangiogenesis is moreover driven by TAMs, which recognize hypoxia and respond with the release of angiopoetin 2 and VEGF. It is important to note, that proangiogenic genes in TAMs and other cell types are regulated by AP1, STAT3, and the NF-κB-pathway (Kujawski et al. 2008; Rius et al. 2008). So far, it cannot be finally concluded, whether hypoxia is sufficient to induce neoangiogenesis or whether the inflammatory mediators activated by hypoxia are the key drivers of angiogenesis.

2.5   T Cell Exhaustion in Chronic Viral Infection and Malignant Diseases

CD8 T cells play a pivotal role in antitumoral immune responses. Mature cytotoxic T cells are able to invade infected or tumor tissue and directly lyse target cells in an antigen-specific manner. Specificity is mediated by the $\alpha\beta$T cell receptor recognizing mutated or overexpressed self-antigens or antigens of viral origin presented on major histocompatibility complex (MHC) molecules of the putative target cell. Naïve antigen-specific CD8 T cells proliferate extensively following acute viral infection. First, they are activated by professional antigen-presenting cells (APCs), such as dendritic cells (DCs), which cross-present captured viral peptides and provide all necessary signals of costimulation. The expansion of naïve virus specific T cells after initial antigen stimulation is accompanied by the acquisition of effector functions, chemokine production and the ability to migrate towards the site of infection. The expanded population of pathogen-specific CD8 T cells normally facilitates clearance of the virus and after reaching the peak of clonal expansion, the virus-specific T cell population is subject to a rapid contraction, where most of the T cells undergo apoptosis. A small percentage of this cell population survives and establishes a pool of long-lived memory CD8 T cells that can be subdivided into effector memory (EM) and central memory (CM) T cells, depending on phenotypic markers (CD62L and CCR7). Upon rechallenge, memory cells can rapidly proliferate and exert effector functions thus providing protective immunity.

In contrast to the development of protective CD8 T cell immunity after acute infections, behaviour of CD8 T cells in chronic infections and cancer is dramatically altered. In the early course of chronic diseases, naïve pathogen- or disease-specific T cells are primed and initially gain effector functions. However, they are often incapable to differentiate into functional memory cells and effector functions deteriorate. This loss of function is also called T cell exhaustion. The process follows a hierarchical order, whereby exhausted T cells have impaired IL-2 production and proliferation, then acquiring improper cytolytic function and produce less TNF. Finally, fully exhausted T cells may even lose the ability to secrete IFN-γ (Zajac et al. 1998; Fuller and Zajac 2003; Wherry et al. 2003).

During exhaustion, CD8 T cells continually upregulate inhibitory receptors, including PD-1, Tim-3, LAG3, CD160 and 2B4 (Barber et al. 2006; Blackburn et al. 2009; Fourcade et al. 2010). Commonly, the need of T cell exhaustion is explained by the adjustment of the inflammatory response of the immune system towards viral antigens in order to limit viral replication on the one hand and on the other hand to avoid durable inflammation-mediated tissue damage in permanently infected tissue, if the host cannot fully eliminate the virus.

CD8 T cell responses in malignant diseases, directed against antigens expressed by the tumor, are frequently found in melanoma patients (Boon et al. 2006). The coexistence of spontaneously arising tumor-specific immune responses with progressive disease demonstrates a tumor-induced dysfunction in T cells as well. In cancer patients, the induction of tumor-specific T cells does not follow the mechanisms of acute viral infections. Tumor-tolerance mechanisms actively interfere with antitumoral immune responses and tumor-antigens further lack pathogen-associated molecular patterns (PAMPs) in malignant cells. This might lead to incomplete activation and licensing of cytotoxic T cells. Additionally, the constitutive antigenic challenge and the presence of an inflammatory tolerance-mediating tumor-microenvironment might further contribute to the malfunction of exhausted T cells (Fourcade et al. 2010; Baitsch et al. 2011).

The tumor-associated antigens (TAA) investigated in patients are often cancer-germline antigens (CGAs), which are expressed by tumor cells of different origins, but not by non-malignant cells, except the testis. The CGA NY-ESO-1 is often subject to spontaneous cellular and humoral responses that are detectable in patients with advanced NY-ESO-1-expressing tumors (Stockert et al. 1998; Mandic et al. 2005; Fourcade et al. 2008). Recent studies of spontaneous tumorantigen-specific CD8 T cells revealed upregulation of PD-1 and Tim-3 in mice and melanoma patients (Fourcade et al. 2010; Sakuishi et al. 2010; Baitsch et al. 2011). These two negative regulatory or immune checkpoint molecules are very important markers of T cell exhaustion in malignant diseases. PD-1 abrogates T cell receptor signaling and Tim-3 plays a role in promoting MDSCs and in regulating cytokine responses of myeloid cells (Dardalhon et al. 2010; Zhang et al. 2012).

Furthermore, PD-1 controls expansion of NY-ESO-1-specific CD8 T cells and PD-1/Tim-3 double positive cells have a more exhausted phenotype by production of less IFN-γ, IL-2 and TNF compared to single positive CD8 T cells in patients (Fourcade et al. 2010). Blockade of PD-1 and Tim-3 in animal models has been shown to partially restore T cell function and that combined targeting of Tim-3 and PD-1 pathways is more effective in tumor growth inhibition than either pathway alone (Sakuishi et al. 2010).

Interestingly, coexistence of TAA-specific CD8 cells with an effector profile were found in the circulation and exhausted cells were present in the tumor environment of metastases from melanoma patients after vaccination with CpG and Melan–A/MART–1 (Baitsch et al. 2011).

3   Inflammation and Metastases

Metastases are a central aspect in clinical oncology because over 90 % of cancer mortality is attributable to metastases. Formation of metastases is a highly complex process including tumor cell motility, intravasation into the circulation, spread through the blood or the lymphatic system, extravasation and finally establishment and outgrowth of tumor-nodules in new tissues and organs. Approximately only 0.01 % of cancer cells that enter the circulation will successfully develop micrometastases (Joyce and Pollard 2009). The initially increased motility and invasiveness of metastatic tumor cells are caused by the epithelial-mesenchymal transition (EMT). In the process of EMT, epithelial cells aquire a fibroblast-like properties that increase their motility and facilitates to invade epithelial barriers and to cross the basal membrane towards the circulation (Kalluri and Weinberg 2009). The extravasation of premetastatic cells is mediated by integrins, followed by crosstalk with immune and stromal cells, which allow them to proliferate (Polyak and Weinberg 2009). It has been shown, that leukocytes pave the way for the construction of a premetastatic niche (Erler et al. 2009; Hiratsuka et al. 2006; Kaplan et al. 2006; Kaplan et al. 2005; Padua et al. 2008) and that the inflammatory stimuli is furthermore mediated by the extracellular matrix component versican, which in turn activates macrophages and leads to the secretion of the prometastatic cytokine TNF-α (Kim et al. 2009). Another prometastatic and anti-inflammatory cytokine is TGF-β. It mediates trans-endothelial migration and metastasis by induction of angiopoietin 4 and is secreted by cancer cells, myeloid cells, and T lymphocytes. Elevated levels of TGF-β therefore often indicate a poor prognosis (Yang and Weinberg 2008).

4   Prevention of Inflammation-Associated Cancer

Cancers that are related to chronic viral infections, can be addressed at different stages during tumorigenesis (feasible interventions are summarized in Fig. 1). First of all, effective vaccinations are fundamental, not only to prevent acute infection-related illness, but also to finally prevent the development of cancer as late complication of the underlying viral infection. Patients that have suffered from an infection once in their life by viruses that are known to cause enhanced cancer risk must be routinely monitored for chronification. Chronically-infected patients need state of the art pharmacologic treatments to reduce viral loads to the most possible extend or, at its best, to finally heal up the infection. Furthermore, patients with chronic infections have to be carefully investigated for signs of precancerous lesions of the affected tissue. In this case, therapeutic interventions to inhibit or

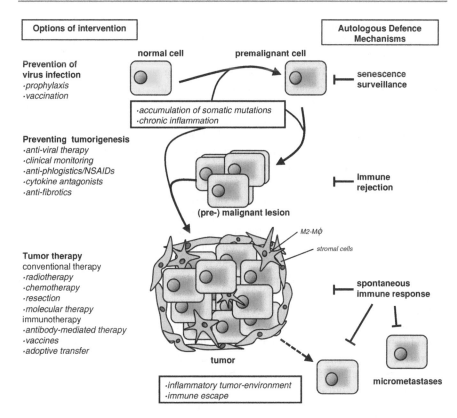

Fig. 1 The figure illustrates different stages of the development of inflammation-associated cancer, the corresponding host's immune defense mechanisms involved in tumor prevention and applicable interventions to prevent or treat cancer

even revert tissue organization should be considered in particular for those individuals that have a genetic predisposition for cancer. Since chronic inflammation promotes carcinogenesis and dissemination, long-term antiphlogistic treatment could represent a suitable cancer-preventive regimen for infected patients. Concrete means to address the above listed options for prevention of virus-mediated cancers are presented in the following section.

4.1   Antiviral Vaccinations and Therapies

HBV, HCV, or HPV are causative for the vast majority of solid cancers that are associated with chronic virus infections and infection-related inflammation.

The options to prevent HBV-related HCC is to ward off acquisition of chronic HBV infection for risk groups like intravenous drug-addicts or recipients of blood-infusions. Furthermore, effective vaccines against HBV exist since more than

20 years. Vaccination programs, including passive immunizations as part of a post-exposition-prophylaxis for children of HBV infected women, have led to a dramatic reduction of HBV carriers, and to a reduction of HCC during childhood (Chang et al. 1997). HBV vaccination is now part of the National Infant Immunization Schedule in 162 countries and represents the first example of cancer preventive vaccination. However, the actual clinical benefit of the ongoing vaccination programs on mortality due to HBV-related HCC will become apparent after further decades (Zanetti et al. 2008). Additionally, routine HBsAg-screening of blood donors will help to reduce blood transfusion-associated transmission of HBV (Schreiber et al. 1996). For the multitude of chronically HBV-infected persons, HBV vaccinations would not be effective in preventing HCC. Here, control of HBV viremia is the major therapeutic goal to reduce hepatic inflammation and disease progression to cirrhosis and HCC. If liver histologic diagnosis in chronic HBV carriers indicates the requirement of a therapeutic intervention, interferon-α and/or nucleoside analoga are applied to control the viral burden, a management that additionally has the beneficial effect to avoid HCC development. However, resistance development to nucleoside analoga remains a problem (Colombo and Donato 2005).

Unfortunately, no functional vaccines against HCV are available to prevent HCV-associated HCC. Only a few vaccine candidates have proceeded to clinical phase I/II trials but it is not yet clear whether these will reach clinical applicability (Torresi et al. 2011). Whereas effective vaccination strategies are still under investigation, significant advances have been made in the treatment of both acute and chronic HCV infection. With current medical therapy, based on pegylated IFN-α and ribavirin, approximately 50 % of patients can be cured. However, the recently developed 'directly acting viral agents' (DAAs) such as inhibitors of the NS3/4A protease or cyclophilin B inhibitors, promise further improvements to achieve sustained virologic responses in treatment of HCV infection (Patel and Heathcote 2011; Kronenberger and Zeuzem 2012; McHutchison et al. 2009).

In case of HPV, the availability of an effective vaccine represents a milestone for prevention of cervical cancer worldwide. Bivalent and quadrivalent vaccines prepared from empty virus shells (virus-like-particles) effectively prevent infection by high-risk HPV genotypes 16 and 18 that account for 70 % of cervical cancer (Garland et al. 2007; Paavonen et al. 2007). Some protection against other genotypes related to 16/18 has been reported (Joura et al. 2007).

4.2 Antifibrotic Therapy

Transient fibrogenesis and later reversal of fibrotic scar tissue is a hallmark of wound healing. Though not fully understood, the mechanisms of perpetuating fibrogenesis in case of chronic tissue damage and their fatal consequences have been intensively studied in the liver. Liver cirrhosis, the end stage of fibrotic reorganization of liver tissue, is a frequent complication of chronic HBV or HCV infections. Liver cirrhosis is associated with portal hypertension, reduced liver functions, and with an increased risk at developing HCC. A heterogenous cell population of profibrogenic myofibroblasts (MFB), originating from hepatic stellate cells (HSC) is known to orchestrate liver fibrogenesis (Lee and Friedman 2011). Upon liver injury and inflammation, a key event in the onset of fibrosis is the activation of quiescent HSCs, that can be triggered by several factors including reactive oxygen species, TLR4 ligands, uptake of apoptotic bodies and paracrine stimulation by adjacent hepatocytes, Kupffer cells and liver sinusoidal endothelial cells (LSEC). The perpetuation of the fibrogenic phenotype of MFB mainly results from a microenvironment wherein profibrogenic cytokines and growth factors such as TGF-β and PDGF are dominating. Activated cholangiocytes have been identified as an important source of these profibrogenic mediators. Phenotypic markers of activated MFBs are expression of α-smooth muscle actin (SMA), excessive collagen production, enhanced proliferation and reduced lipid content contributing to tissue stiffness. With regard to specific gene regulation, it has been shown that angiotensin 2 activates the transcription factor NF-κB thus rendering MFBs less sensitive against induction of apoptosis and promoting cell survival (Oakley et al. 2009). Recent data demonstrated that activation of the NF-κB pathway in hepatocytes induces liver fibrosis in mice (Sunami et al. 2012). Accumulation of extracellular matrix (ECM) is a further characteristic of fibrogenesis which contributes to tissue stiffening. Major regulators of ECM are matrix metalloproteinases (MMP), enzymes that are responsible for matrix degradation and removal of scar tissue. Net production of ECM results from a dysbalance of MMP expression, and secretion of their specific inhibitors (tissue inhibitors of metalloproteinases or TIMPs) by MFBs. During physiological regeneration and repair of liver damage, mechanisms exist that can mediate regression of fibrotic scar tissue once the initial cause of tissue damage has been resolved. In this process, MFBs finally disappear from the hepatic scar by phenotypic reversion, senescence, and deletion by natural killer cells. Thereby, fibrotic tissue is removed by enhanced MMP activity and reduced TIMP levels. Hepatic macrophages have been demonstrated to be involved in this regulation and scar tissue remodeling.

The ideal anti-fibrotic therapy consists of the withdrawal of the underlying disease trigger. Virus elimination in chronic HBV or HCV infection can lead to regression of fibrosis and improved liver function even in cirrhotic patients. For patients that do not sufficiently respond to antiviral treatments, specific antifibrotic therapies are urgently needed to slow down the progress of fibrosis and to reduce the risk of late stage complications and HCC development. Patients with increased risk at developing fibrosis should be determined by targeted screenings. As example, a 'seven gene signature' with specific polymorphisms in genes with relevance to fibrosis risk such as e.g. TGF-β, TNF-α, or IL-10, has a significant predictive value (cirrhosis risk score) for fibrosis progression in patients with chronic HCV (Huang et al. 2007). Consequent monitoring of those patients to determine progression of fibrosis should be performed. Non-invasive methods like Fibroscan, a sonographic evaluation of liver stiffness, could be valuable alternatives to liver biopsy in the future. Combined with serologic markers like α-fetoprotein this allows for detecting HCC at an early, potentially curable stage (Mok et al. 2005). For antifibrotic therapies, all molecular factors or mechanisms that contribute to progression or regression of fibrosis can be considered suitable targets (Fallowfield 2011). Liver damage due to inflammation and oxidative stress can be addressed by so-called hepatoprotectants. The hepatoprotective activity of some of these substances can be attributed to their ability to suppress the inflammatory activation of NFκB. A number of natural substances that can be long-term consumed in significant amounts have been investigated for their anti-oxidant and hepatoprotective effect such as resveratrol (from red wine), silymarin (from milk thistle), coffee and vitamin E.

Hepatocyte growth factor (HGF) has been identified to stimulate hepatocyte generation and demonstrated promising results in inhibition of experimental fibrosis in animal models (Xia et al. 2006). However, since HGF is a potent mitogen for hepatocytes, concerns about potential oncogenesis remain. A further potential target is the reduction of apoptotic cell turnover during liver inflammation. Apoptosis of hepatocytes activates HSC directly or via Kupffer cells. Consequently, the pan-caspase inhibitor VX-166 demonstrated antifibrotic activity in animal models (Witek et al. 2009). In contrast, pro-apoptotic strategies could also be an alternative if cell death can be selectively induced in activated HSC. To this end, the proapoptotic IkB inhibitor gliotoxin has been coupled to a single-chain antibody against synaptophysin which is selectively expressed on these cells. This approach reduced fibrosis in a rat model of CCl_4 intoxication (Douglass et al. 2008). As an alternative strategy to induce apoptosis of activated HSC, interference with cannabinoid receptor signaling has led to promising results in fibrotic animal models (Teixeira-Clerc et al. 2006), but showed psychomimetic side effects in clinical trials. Further strategies to prevent HSC activation like ligands ("glitazones") for the peroxisome proliferator-activator receptor-γ (PPAR-γ), Farnesoid-X-receptor-agonists, or HMG-CoA reductase inhibitors ("statins") have shown antifibrotic activity in experimental models that awaits confirmation in clinical trials. Angiotensin system inhibitors are on the cusp to clinical application. Both angiotensin-converting enzyme inhibitors and angiotensin-1 receptor antagonists ("sartans") can inhibit fibrosis in animals. Importantly, the AT1R antagonist losartan slowed down progression of fibrosis in patients with chronic HCV. Further future strategies could be targeting of the TGF-β pathway. Neutralizing antibodies, decoy receptors and siRNA have been studied but are not yet ready for clinical application. Pirfenidone, an inhibitor of TGF-β production has shown promising results in a non-controlled pilot study (Armendariz-Borunda et al. 2006). $\alpha v \beta 6$ integrin activates matrix-bound, latent TGF-β in the local microenvironment. The specific inhibition of this integrin could therefore be an intriguing approach to avoid side effects of systemic TGF-β inhibition. Neutralizing antibodies against $\alpha v \beta 6$ integrin inhibited cholangiocyte activation and collagen deposition in biliary and non-biliary fibrosis models (Popov et al. 2008). Agents, that stimulate the collagenolytic activity of MMP, such as the oral alkaloid Halofuginone, showed promise in animal models but still needs investigations in patients. Notably, the clinical success achieved so far does not correlate with the antifibrotic effects in experimental animal models suggesting that the available animal fibrosis models may not fully reflect the influence of ECM maturation by crosslinking and the potential existence of a "point of no return" in fibrosis (Rockey 2008). Nevertheless, functional antifibrotic therapies are urgently needed, and first progress towards clinical application becomes visible.

4.3   Anti-Inflammatory Treatment

Since chronic inflammation contributes to carcinogenesis and dissemination, antiphlogistic drugs appear to be a promising preventive regimen for infected patients upon diagnosis of virus-mediated tissue damage. Several factors that fuel cancer-associated inflammatory processes have been identified and are therefore reasonable molecular targets to prevent cancer in risk-associated patients. Several antiphlogistic drugs have been shown to reduce the incidence of cancer when used as prophylactics or to slow down tumor growth and improve survival when given as therapeutics, e.g. in colon cancer (Gupta and DuBois 2001). At first, preventive use of classical non-steroidal anti-inflammatory drugs (NSAIDs) such as aspirin could be a method of choice (Cuzick et al. 2009; Langley et al. 2011). NSAIDs are cheap, off-patent, well established and long-term clinical experience exists for low-dose usage as anti-thrombotic prophylaxis. It has been demonstrated that preventive Aspirin treatment can significantly reduce the long-term risk to develop colorectal cancer (Flossmann and Rothwell 2007; Rothwell et al. 2010). A large meta analysis showed a reduced cancer risk in several solid, mainly gastrointestinal tumors, such as gastric, oesophageal, colorectal and pancreatic cancer, but also in lung cancer (Rothwell et al. 2011). Benefit correlated with duration of aspirin uptake and with increasing age of the individual. Notably, no benefit was seen in hematological malignancies consistent with the hypothesis that the anti-inflammatory treatment is mainly directed against the chronic inflammatory microinvironment of a solid tumor. Regarding the risk and benefit balance, the Rothwell study suggests that the additional risk of aspirin-associated gastrointestinal bleedings (including fatal bleedings) is outweighed by a gain of 10 % reduction in all-cause mortality after 5–10 years daily aspirin consumption. It has to be pointed out, that the patients in these studies were not stratified whether they actually had any inflammatory burden or not. It has been shown that aspirin reduces the risk of prostate cancers, but only in individuals carrying a particular polymorphic allele of the lymphotoxin α gene which leads to increased lymphotoxin expression (Liu et al. 2006). This observation illustrates the need to early identify those patients with elevated cancer risk by genetic screenings and/or by monitoring inflammatory parameters, which would likely profit most from anti-inflammatory treatment. To reduce gastrointestinal complications in long-term treatments, selective COX-2 inhibitors could be an alternative as preventive cancer treatment (Kawamori et al. 1998; Reddy et al. 2000). In studies of several preclinical models of inherited or carcinogen-induced cancers of colon, intestine, skin and bladder it could be observed that celecoxib could significantly reduce cancer incidence (Fischer et al. 2011). Also in studies in human patients, celecoxib showed promise in prevention of spontaneous colorectal adenomas and reduced the number of polyps in familial adenomatous polyposis (Bertagnolli et al. 2006; Arber et al. 2006; Steinbach et al. 2000). However, it has to be considered that celecoxib has been associated with adverse cardiovascular events (Mukherjee et al. 2001).

Drugs that address specific molecular targets in cytokine or chemokine signaling involved in inflammatory processes also represent promising alternatives. However, most of these agents are already in use for the treatment of chronic inflammatory diseases such as rheumatoid arthritis, psoriasis, or inflammatory bowel diseases (IBD) or are in clinical development for cancer therapy. It is yet unclear whether these agents are suitable for long-term treatments. The structural analogue of thalidomide, lenalidomide, is known to suppress the production of several inflammation-associated cytokines and has shown to be active against melanoma in combination with dexamethasone (Weber et al. 2007).

As already mentioned, NFκB and STAT3 pathways play a critical role in both carcinogenesis and resistance to therapy suggesting that targeted inhibitors could have significant preventive or therapeutic potential. However, pharmacologic long-term inhibition of NFκB can lead to severe immunodeficiency, neutrophilia and enhanced acute inflammation due to increased IL-1β levels (Greten et al. 2007). Alternatively, it appears to be more promising to address defined upstream targets of NFκB since multiple extrinsic and intrinsic pathways converge to activate this central transcription factor. Inhibitors for STAT3 and JAK2, which is upstream of STAT3, have been developed and showed oncostatic activity in solid tumors in mouse xenograft models (Hedvat et al. 2009). Furthermore, receptor antagonists and blocking antibodies addressing IL-6, IL-6 receptor, CCR2, CCR4, and CXCR4 are in clinical development for the treatment of several tumor entities. TNFα-antibodies such as infliximab has been a breakthrough for the treatment of IBD and, importantly, long term application appears to be safe (Fidder et al. 2009). First clinical studies of TNFα antagonists in patients with advanced cancer have resulted in stable disease or partial responses (Brown et al. 2008; Harrison et al. 2007). It is further known that infliximab also reduced colitis-associated cancer in mice (Kim et al. 2010) suggesting a cancer preventive benefit in patients. On the other hand, TNFα has been implicated in the clearance of virus infections (Trevejo et al. 2001) and has been shown to be critical for induction of antitumoral T cell responses in animals (Calzascia et al. 2007). Due to this complex and sometimes contradictory biology of TNFα it is currently difficult to estimate whether TNFα antagonists could be promising as preventive mean for cancer that is caused by virus-mediated inflammation. From the clinical point of view, metastases are the much more challenging manifestation of a malignant disease compared to the primary nodule. Since inflammation is a central driving force of dissemination, application of anti-inflammatory drugs could be an effective intervention to prevent the development of metastases. Recently, the antagonistic RANKL antibody denosumab, initially developed for the treatment of osteoporosis, delayed the development of bone metastases in advanced clinical studies in prostate and breast cancer (Stopeck et al. 2010; Smith et al. 2012). For prevention of metastases, inhibition of TNFα could also be promising, since mouse experiments have shown that blocking TNFα can convert inflammation-promoted metastatic tumor growth to TRAIL-mediated tumor regression (Luo et al. 2004). Altogether, these examples show that molecular mechanisms and signaling pathways involved in tumor- and metastases-associated inflammation are promising targets for preventive and therapeutic interventions.

5   Outlook: Perspectives in Cancer Immunotherapy

Not only for prevention but also in therapy of cancer, antitumoral immune responses are coming more and more into focus. Antitumoral immune responses can even be triggered by conventional chemotherapeutic treatments and may significantly contribute to the therapeutic outcome (Casares et al. 2005; Apetoh et al. 2008). Our current knowledge about tumor-immune responses basically suggests two different strategies for tumor immunotherapy. One strategy is to support or reactivate innate or preexisting adaptive immune responses, the other is *de novo* induction of adaptive responses by therapeutic interventions. Both are about equally challenging, since tumors have developed numerous ways to escape immune attacks. Even if we are able to elucidate the underlying mechanisms and have many therapies at hand, we still cannot be sure, whether a certain therapy could be undermined by the tumor's mechanisms of immune suppression in the individual case. Therefore, the ideal immunotherapeutic strategy aims at the achilles heel of the tumor and would be additionally applicable for all solid tumor entities.

Recent findings on the importance of the chronic inflammatory tumor-environment for tumor-maintainance and –progression have lead to the idea to use anti-inflammatory drugs for cancer therapy. The fundamental advantage of this approach is, that immune cells are not prone to develop drug resistance. However, it is likely, that anti-inflammatory therapy does not exhibit sufficient cytotoxic effects on cancer cells and thus must be combined with additional therapies to manifest effective therapeutic effects. As already mentioned above, some phase I/II clinical trials investigate efficacy of anti-IL6 and anti-TNF-α drugs in various cancers (Balkwill 2009).

Among adaptive immune responses, cytotoxic T cells, which are able to mediate direct lysis of transformed cells, are commonly regarded as the most promising cell type for cancer immunotherapy. However, effective tumor-specific CD8 T cell responses are difficult to induce in tumor-bearing hosts. Paracrine mediators, like adenosine, prostaglandin E2, VEGF-A, and TGF-ß mediate direct and indirect immunosuppressive activities and act on different levels of the immune system. It has been shown that immunosuppressive activities lead to blunted spontaneous adaptive responses by T cell exhaustion (Fourcade et al. 2010; Sakuishi et al. 2010; Baitsch et al. 2011) or take effect on dendritic cells, inducing defective maturation by inhibiting costimulatory signals, which in turn fails to prime a tumor-directed CD8

T cell response (Sharma et al. 2010). Therefore, a careful design of a vaccine to elicit an effective tumor-directed CD8 T cell response must encompass a strategy to block the tumors counteractions as well. In a melanoma mouse model inhibition of IDO (indoleamine 2,3-dioxygenase)-mediated immune suppression by use of 1-methyl-tryptophan allowed for induction of potent cytotoxic responses with a CD8 T cell-vaccine by protecting DCs in TDLN from malfunction (Sharma et al. 2010). Another approach to circumvent tumor-mediated immunosuppression is an infection mediated by a lytic virus in a tumor nodule. The infection disrupts tumor architecture and leads to an inflammation, which is accompanied by leukocytic infiltration and abundant tumor-cell death. When a tumor-directed DC-vaccine is applied at this time point of virus-mediated inflammation, it triggers a strong cytotoxic CD8 T cell response, whereas other timepoints of vaccination in combination with virotherapy does not exhibit a significant therapeutic effect. Interestingly, compared to a true virus infection, inflammations mediated by toll-like receptor ligands failed to support effective induction of CD8 T cell response by DC-vaccination, most likely due to lacking cross-presentation of tumor-associated antigens by dendritic cells within the infected tumor (Woller et al. 2011).

The clinical translation of immunotherapeutic strategies in the recent years raises hope to establish a new field in the treatment of cancer, a field, which might be also combined with conventional therapies. A GM-CSF-armed, tk-deficient poxvirus showed promising results in clinical studies, where significant responses were observed in hepatocellular carcinoma (Breitbach et al. 2011; Park et al. 2008). In a clinical study phase I/II patients with advanced solid tumors of different origin received anti-PD-1 antibodies to reverse exhausted T cells. Some observable responses were durable for more then one year. However, treatment with a blocking PD-1 antibody only showed objective responses in patients with a positive staining of PD-L1 in tumor-tissue (Topalian et al. 2012).

These studies, among many others, elucidate novel interventions and possibilities for immunotherapy of cancer. They are based on accumulating knowledge about the immune system, cancer biology, and interactions between both. Induction of specific immune reactions redirected to the tumor appears to open up an exiting new field in the treatment of malignant diseases. On the other hand, many lessons have to be learned and it is still a long way to go from the more or less experimental approach of immunotherapy towards well-established therapies.

References

Aggarwal BB, Vijayalekshmi RV, Sung B (2009) Targeting inflammatory pathways for prevention and therapy of cancer: short-term friend, long-term foe. Clin Cancer Res 15:425–430

Andreu P, Johansson M, Affara NI, Pucci F, Tan T, Junankar S, Korets L, Lam J, Tawfik D, DeNardo DG, Naldini L, de Visser KE, De PM, Coussens LM (2010) FcRgamma activation regulates inflammation-associated squamous carcinogenesis. Cancer Cell 17:121–134

Apetoh L, Tesniere A, Ghiringhelli F, Kroemer G, Zitvogel L (2008) Molecular interactions between dying tumor cells and the innate immune system determine the efficacy of conventional anticancer therapies. Cancer Res 68:4026–4030

Arber N, Eagle CJ, Spicak J, Racz I, Dite P, Hajer J, Zavoral M, Lechuga MJ, Gerletti P, Tang J, Rosenstein RB, Macdonald K, Bhadra P, Fowler R, Wittes J, Zauber AG, Solomon SD, Levin B (2006) Celecoxib for the prevention of colorectal adenomatous polyps. N Engl J Med 355:885–895

Armendariz-Borunda J, Islas-Carbajal MC, Meza-Garcia E, Rincon AR, Lucano S, Sandoval AS, Salazar A, Berumen J, Alvarez A, Covarrubias A, Arechiga G, Garcia L (2006) A pilot study in patients with established advanced liver fibrosis using pirfenidone. Gut 55:1663–1665

Baitsch L, Baumgaertner P, Devevre E, Raghav SK, Legat A, Barba L, Wieckowski S, Bouzourene H, Deplancke B, Romero P, Rufer N, Speiser DE (2011) Exhaustion of tumor-specific CD8(+) T cells in metastases from melanoma patients. J Clin Invest 121:2350–2360

Balkwill F (2009) Tumour necrosis factor and cancer. Nat Rev Cancer 9:361–371

Balkwill F, Mantovani A (2001) Inflammation and cancer: back to Virchow? Lancet 357:539–545

Barber DL, Wherry EJ, Masopust D, Zhu B, Allison JP, Sharpe AH, Freeman GJ, Ahmed R (2006) Restoring function in exhausted CD8 T cells during chronic viral infection. Nature 439:682–687

Barth H, Robinet E, Liang TJ, Baumert TF (2008) Mouse models for the study of HCV infection and virus-host interactions. J Hepatol 49:134–142

Bertagnolli MM, Eagle CJ, Zauber AG, Redston M, Solomon SD, Kim K, Tang J, Rosenstein RB, Wittes J, Corle D, Hess TM, Woloj GM, Boisserie F, Anderson WF, Viner JL, Bagheri D, Burn J, Chung DC, Dewar T, Foley TR, Hoffman N, Macrae F, Pruitt RE, Saltzman JR, Salzberg B, Sylwestrowicz T, Gordon GB, Hawk ET (2006) Celecoxib for the prevention of sporadic colorectal adenomas. N Engl J Med 355:873–884

Bielas JH, Loeb KR, Rubin BP, True LD, Loeb LA (2006) Human cancers express a mutator phenotype. Proc Natl Acad Sci U S A 103:18238–18242

Biswas SK, Mantovani A (2010) Macrophage plasticity and interaction with lymphocyte subsets: cancer as a paradigm. Nat Immunol 11:889–896

Blackburn SD, Shin H, Haining WN, Zou T, Workman CJ, Polley A, Betts MR, Freeman GJ, Vignali DA, Wherry EJ (2009) Coregulation of CD8+T cell exhaustion by multiple inhibitory receptors during chronic viral infection. Nat Immunol 10:29–37

Boon T, Coulie PG, Van den Eynde BJ, van der Bruggen P (2006) Human T cell responses against melanoma. Annu Rev Immunol 24:175–208

Breitbach CJ, Burke J, Jonker D, Stephenson J, Haas AR, Chow LQ, Nieva J, Hwang TH, Moon A, Patt R, Pelusio A, Le Boeuf F, Burns J, Evgin L, De Silva N, Cvancic S, Robertson T, Je JE, Lee YS, Parato K, Diallo JS, Fenster A, Daneshmand M, Bell JC, Kirn DH (2011) Intravenous delivery of a multi-mechanistic cancer-targeted oncolytic poxvirus in humans. Nature 477:99–102

Brown ER, Charles KA, Hoare SA, Rye RL, Jodrell DI, Aird RE, Vora R, Prabhakar U, Nakada M, Corringham RE, DeWitte M, Sturgeon C, Propper D, Balkwill FR, Smyth JF (2008) A clinical study assessing the tolerability and biological effects of infliximab, a TNF-alpha inhibitor, in patients with advanced cancer. Ann Oncol 19:1340–1346

Calzascia T, Pellegrini M, Hall H, Sabbagh L, Ono N, Elford AR, Mak TW, Ohashi PS (2007) TNF-alpha is critical for antitumor but not antiviral T cell immunity in mice. J Clin Invest 117:3833–3845

Casares N, Pequignot MO, Tesniere A, Ghiringhelli F, Roux S, Chaput N, Schmitt E, Hamai A, Hervas-Stubbs S, Obeid M, Coutant F, Metivier D, Pichard E, Aucouturier P, Pierron G, Garrido C, Zitvogel L, Kroemer G (2005) Caspase-dependent immunogenicity of doxoru-bicin-induced tumor cell death. J Exp Med 202:1691–1701

Cassatella MA, Locati M, Mantovani A (2009) Never underestimate the power of a neutrophil. Immunity 31:698–700

Chang MH, Chen CJ, Lai MS, Hsu HM, Wu TC, Kong MS, Liang DC, Shau WY, Chen DS (1997) Universal hepatitis B vaccination in Taiwan and the incidence of hepatocellular carcinoma in children. Taiwan Childhood Hepatoma Study Group [see comments]. N Engl J Med 336:1855–1859

Chen CJ, Yang HI, Su J, Jen CL, You SL, Lu SN, Huang GT, Iloeje UH (2006) Risk of hepatocellular carcinoma across a biological gradient of serum hepatitis B virus DNA level. JAMA 295:65–73

Colombo M, Donato MF (2005) Prevention of hepatocellular carcinoma. Semin Liver Dis 25:155–161

Colotta F, Allavena P, Sica A, Garlanda C, Mantovani A (2009) Cancer-related inflammation, the seventh hallmark of cancer: links to genetic instability. Carcinogenesis 30:1073–1081

Condeelis J, Pollard JW (2006) Macrophages: obligate partners for tumor cell migration, invasion, and metastasis. Cell 124:263–266

Custer B, Sullivan SD, Hazlet TK, Iloeje U, Veenstra DL, Kowdley KV (2004) Global epidemiology of hepatitis B virus. J Clin Gastroenterol 38:S158–S168

Cuzick J, Otto F, Baron JA, Brown PH, Burn J, Greenwald P, Jankowski J, La VC, Meyskens F, Senn HJ, Thun M (2009) Aspirin and non-steroidal anti-inflammatory drugs for cancer prevention: an international consensus statement. Lancet Oncol 10:501–507

Dardalhon V, Anderson AC, Karman J, Apetoh L, Chandwaskar R, Lee DH, Cornejo M, Nishi N, Yamauchi A, Quintana FJ, Sobel RA, Hirashima M, Kuchroo VK (2010) Tim-3/galectin-9 pathway: regulation of Th1 immunity through promotion of CD11b + Ly-6G + myeloid cells. J Immunol 185:1383–1392

de Martel C, Franceschi S (2009) Infections and cancer: established associations and new hypotheses. Crit Rev Oncol Hematol 70:183–194

DeNardo DG, Barreto JB, Andreu P, Vasquez L, Tawfik D, Kolhatkar N, Coussens LM (2009) CD4(+) T cells regulate pulmonary metastasis of mammary carcinomas by enhancing protumor properties of macrophages. Cancer Cell 16:91–102

Donato F, Boffetta P, Puoti M (1998) A meta-analysis of epidemiological studies on the combined effect of hepatitis B and C virus infections in causing hepatocellular carcinoma. Int J Cancer 75:347–354

Douglass A, Wallace K, Parr R, Park J, Durward E, Broadbent I, Barelle C, Porter AJ, Wright MC (2008) Antibody-targeted myofibroblast apoptosis reduces fibrosis during sustained liver injury. J Hepatol 49:88–98

Dvorak HF (1986) Tumors: wounds that do not heal. Similarities between tumor stroma generation and wound healing. N Engl J Med 315:1650–1659

Erdman SE, Sohn JJ, Rao VP, Nambiar PR, Ge Z, Fox JG, Schauer DB (2005) CD4 + CD25 + regulatory lymphocytes induce regression of intestinal tumors in ApcMin/ + mice. Cancer Res 65:3998–4004

Erez N, Truitt M, Olson P, Arron ST, Hanahan D (2010) Cancer-Associated Fibroblasts Are Activated in Incipient Neoplasia to Orchestrate Tumor-Promoting Inflammation in an NF-kappaB-Dependent Manner. Cancer Cell 17:135–147

Erler JT, Bennewith KL, Cox TR, Lang G, Bird D, Koong A, Le QT, Giaccia AJ (2009) Hypoxia-induced lysyl oxidase is a critical mediator of bone marrow cell recruitment to form the premetastatic niche. Cancer Cell 15:35–44

Fallowfield JA (2011) Therapeutic targets in liver fibrosis. Am J Physiol Gastrointest Liver Physiol 300:G709–G715

Farazi PA, DePinho RA (2006) Hepatocellular carcinoma pathogenesis: from genes to environment. Nat Rev Cancer 6:674–687

Fidder H, Schnitzler F, Ferrante M, Noman M, Katsanos K, Segaert S, Henckaerts L, Van AG, Vermeire S, Rutgeerts P (2009) Long-term safety of infliximab for the treatment of inflammatory bowel disease: a single-centre cohort study. Gut 58:501–508

Fischer SM, Hawk ET, Lubet RA (2011) Coxibs and other nonsteroidal anti-inflammatory drugs in animal models of cancer chemoprevention. Cancer Prev Res (Phila) 4:1728–1735

Flossmann E, Rothwell PM (2007) Effect of aspirin on long-term risk of colorectal cancer: consistent evidence from randomised and observational studies. Lancet 369:1603–1613

Fourcade J, Kudela P, Andrade Filho PA, Janjic B, Land SR, Sander C, Krieg A, Donnenberg A, Shen H, Kirkwood JM, Zarour HM (2008) Immunization with analog peptide in combination with CpG and montanide expands tumor antigen-specific CD8 + T cells in melanoma patients. J Immunother 31:781–791

Fourcade J, Sun Z, Benallaoua M, Guillaume P, Luescher IF, Sander C, Kirkwood JM, Kuchroo V, Zarour HM (2010) Upregulation of Tim-3 and PD-1 expression is associated with tumor antigen-specific CD8 + T cell dysfunction in melanoma patients. J Exp Med 207:2175–2186

Fridlender ZG, Sun J, Kim S, Kapoor V, Cheng G, Ling L, Worthen GS, Albelda SM (2009) Polarization of tumor-associated neutrophil phenotype by TGF-beta: "N1" versus "N2" TAN. Cancer Cell 16:183–194

Fukuda A, Wang SC, Morris JP, Folias AE, Liou A, Kim GE, Akira S, Boucher KM, Firpo MA, Mulvihill SJ, Hebrok M (2011) Stat3 and MMP7 contribute to pancreatic ductal adenocarcinoma initiation and progression. Cancer Cell 19:441–455

Fuller MJ, Zajac AJ (2003) Ablation of CD8 and CD4 T cell responses by high viral loads. J Immunol 170:477–486

Gabrilovich DI, Nagaraj S (2009) Myeloid-derived suppressor cells as regulators of the immune system. Nat Rev Immunol 9:162–174

Gallimore AM, Simon AK (2008) Positive and negative influences of regulatory T cells on tumour immunity. Oncogene 27:5886–5893

Galon J, Costes A, Sanchez-Cabo F, Kirilovsky A, Mlecnik B, Lagorce-Pages C, Tosolini M, Camus M, Berger A, Wind P, Zinzindohoue F, Bruneval P, Cugnenc PH, Trajanoski Z, Fridman WH, Pages F (2006) Type, density, and location of immune cells within human colorectal tumors predict clinical outcome. Science 313:1960–1964

Garland SM, Hernandez-Avila M, Wheeler CM, Perez G, Harper DM, Leodolter S, Tang GW, Ferris DG, Steben M, Bryan J, Taddeo FJ, Railkar R, Esser MT, Sings HL, Nelson M, Boslego J, Sattler C, Barr E, Koutsky LA (2007) Quadrivalent vaccine against human papillomavirus to prevent anogenital diseases. N Engl J Med 356:1928–1943

Gordon S, Martinez FO (2010) Alternative activation of macrophages: mechanism and functions. Immunity 32:593–604

Greten FR, Arkan MC, Bollrath J, Hsu LC, Goode J, Miething C, Goktuna SI, Neuenhahn M, Fierer J, Paxian S, Van RN, Xu Y, O'Cain T, Jaffee BB, Busch DH, Duyster J, Schmid RM, Eckmann L, Karin M (2007) NF-kappaB is a negative regulator of IL-1beta secretion as revealed by genetic and pharmacological inhibition of IKKbeta. Cell 130:918–931

Grivennikov S, Karin E, Terzic J, Mucida D, Yu GY, Vallabhapurapu S, Scheller J, Rose-John S, Cheroutre H, Eckmann L, Karin M (2009) IL-6 and Stat3 are required for survival of intestinal epithelial cells and development of colitis-associated cancer. Cancer Cell 15:103–113

Guerra C, Schuhmacher AJ, Canamero M, Grippo PJ, Verdaguer L, Perez-Gallego L, Dubus P, Sandgren EP, Barbacid M (2007) Chronic pancreatitis is essential for induction of pancreatic ductal adenocarcinoma by K-Ras oncogenes in adult mice. Cancer Cell 11:291–302

Gupta RA, DuBois RN (2001) Colorectal cancer prevention and treatment by inhibition of cyclooxygenase-2. Nat Rev Cancer 1:11–21

Haddow A (1972) Molecular repair, wound healing, and carcinogenesis: tumor production a possible overhealing? Adv Cancer Res 16:181–234

Hanahan D, Weinberg RA (2011) Hallmarks of cancer: the next generation. Cell 144:646–674

Harrison ML, Obermueller E, Maisey NR, Hoare S, Edmonds K, Li NF, Chao D, Hall K, Lee C, Timotheadou E, Charles K, Ahern R, King DM, Eisen T, Corringham R, DeWitte M, Balkwill F, Gore M (2007) Tumor necrosis factor alpha as a new target for renal cell carcinoma: two sequential phase II trials of infliximab at standard and high dose. J Clin Oncol 25:4542–4549

Hedvat M, Huszar D, Herrmann A, Gozgit JM, Schroeder A, Sheehy A, Buettner R, Proia D, Kowolik CM, Xin H, Armstrong B, Bebernitz G, Weng S, Wang L, Ye M, McEachern K, Chen H, Morosini D, Bell K, Alimzhanov M, Ioannidis S, McCoon P, Cao ZA, Yu H, Jove R, Zinda M (2009) The JAK2 inhibitor AZD1480 potently blocks Stat3 signaling and oncogenesis in solid tumors. Cancer Cell 16:487–497

Hiratsuka S, Watanabe A, Aburatani H, Maru Y (2006) Tumour-mediated upregulation of chemoattractants and recruitment of myeloid cells predetermines lung metastasis. Nat Cell Biol 8:1369–1375

Huang H, Shiffman ML, Friedman S, Venkatesh R, Bzowej N, Abar OT, Rowland CM, Catanese JJ, Leong DU, Sninsky JJ, Layden TJ, Wright TL, White T, Cheung RC (2007) A 7 gene signature identifies the risk of developing cirrhosis in patients with chronic hepatitis C. Hepatology 46:297–306

Hussain SP, Harris CC (2007) Inflammation and cancer: an ancient link with novel potentials. Int J Cancer 121:2373–2380

Joura EA, Leodolter S, Hernandez-Avila M, Wheeler CM, Perez G, Koutsky LA, Garland SM, Harper DM, Tang GW, Ferris DG, Steben M, Jones RW, Bryan J, Taddeo FJ, Bautista OM, Esser MT, Sings HL, Nelson M, Boslego JW, Sattler C, Barr E, Paavonen J (2007) Efficacy of a quadrivalent prophylactic human papillomavirus (types 6, 11, 16, and 18) L1 virus-like-particle vaccine against high-grade vulval and vaginal lesions: a combined analysis of three randomised clinical trials. Lancet 369:1693–1702

Joyce JA, Pollard JW (2009) Microenvironmental regulation of metastasis. Nat Rev Cancer 9:239–252

Kalluri R, Weinberg RA (2009) The basics of epithelial-mesenchymal transition. J Clin Invest 119:1420–1428

Kang TW, Yevsa T, Woller N, Hoenicke L, Wuestefeld T, Dauch D, Hohmeyer A, Gereke M, Rudalska R, Potapova A, Iken M, Vucur M, Weiss S, Heikenwalder M, Khan S, Gil J, Bruder D, Manns M, Schirmacher P, Tacke F, Ott M, Luedde T, Longerich T, Kubicka S, Zender L (2011) Senescence surveillance of pre-malignant hepatocytes limits liver cancer development. Nature 479:547–551

Kaplan RN, Riba RD, Zacharoulis S, Bramley AH, Vincent L, Costa C, MacDonald DD, Jin DK, Shido K, Kerns SA, Zhu Z, Hicklin D, Wu Y, Port JL, Altorki N, Port ER, Ruggero D, Shmelkov SV, Jensen KK, Rafii S, Lyden D (2005) VEGFR1-positive haematopoietic bone marrow progenitors initiate the pre-metastatic niche. Nature 438:820–827

Kaplan RN, Rafii S, Lyden D (2006) Preparing the "soil": the premetastatic niche. Cancer Res 66:11089–11093

Karin M (2006) Nuclear factor-kappaB in cancer development and progression. Nature 441:431–436

Kawamori T, Rao CV, Seibert K, Reddy BS (1998) Chemopreventive activity of celecoxib, a specific cyclooxygenase-2 inhibitor, against colon carcinogenesis. Cancer Res 58:409–412

Kim S, Takahashi H, Lin WW, Descargues P, Grivennikov S, Kim Y, Luo JL, Karin M (2009) Carcinoma-produced factors activate myeloid cells through TLR2 to stimulate metastasis. Nature 457:102–106

Kim YJ, Hong KS, Chung JW, Kim JH, Hahm KB (2010) Prevention of colitis-associated carcinogenesis with infliximab. Cancer Prev Res (Phila) 3:1314–1333

Koebel CM, Vermi W, Swann JB, Zerafa N, Rodig SJ, Old LJ, Smyth MJ, Schreiber RD (2007) Adaptive immunity maintains occult cancer in an equilibrium state. Nature 450:903–907

Kohrt HE, Nouri N, Nowels K, Johnson D, Holmes S, Lee PP (2005) Profile of immune cells in axillary lymph nodes predicts disease-free survival in breast cancer. PLoS Med 2:e284

Kronenberger B, Zeuzem S (2012) New developments in HCV therapy. J Viral Hepat 19(Suppl 1):48–51

Kujawski M, Kortylewski M, Lee H, Herrmann A, Kay H, Yu H (2008) Stat3 mediates myeloid cell-dependent tumor angiogenesis in mice. J Clin Invest 118:3367–3377

Laghi L, Bianchi P, Miranda E, Balladore E, Pacetti V, Grizzi F, Allavena P, Torri V, Repici A, Santoro A, Mantovani A, Roncalli M, Malesci A (2009) CD3+cells at the invasive margin of deeply invading (pT3-T4) colorectal cancer and risk of post-surgical metastasis: a longitudinal study. Lancet Oncol 10:877–884

Langley RE, Burdett S, Tierney JF, Cafferty F, Parmar MK, Venning G (2011) Aspirin and cancer: has aspirin been overlooked as an adjuvant therapy? Br J Cancer 105:1107–1113

Langowski JL, Zhang X, Wu L, Mattson JD, Chen T, Smith K, Basham B, McClanahan T, Kastelein RA, Oft M (2006) IL-23 promotes tumour incidence and growth. Nature 442:461–465

Lee UE, Friedman SL (2011) Mechanisms of hepatic fibrogenesis. Best Pract Res Clin Gastroenterol 25:195–206

Lesina M, Kurkowski MU, Ludes K, Rose-John S, Treiber M, Kloppel G, Yoshimura A, Reindl W, Sipos B, Akira S, Schmid RM, Algul H (2011) Stat3/Socs3 activation by IL-6 transsignaling promotes progression of pancreatic intraepithelial neoplasia and development of pancreatic cancer. Cancer Cell 19:456–469

Liu X, Plummer SJ, Nock NL, Casey G, Witte JS (2006) Nonsteroidal antiinflammatory drugs and decreased risk of advanced prostate cancer: modification by lymphotoxin alpha. Am J Epidemiol 164:984–989

Luo JL, Maeda S, Hsu LC, Yagita H, Karin M (2004) Inhibition of NF-kappaB in cancer cells converts inflammation- induced tumor growth mediated by TNFalpha to TRAIL-mediated tumor regression. Cancer Cell 6:297–305

Mandic M, Castelli F, Janjic B, Almunia C, Andrade P, Gillet D, Brusic V, Kirkwood JM, Maillere B, Zarour HM (2005) One NY-ESO-1-derived epitope that promiscuously binds to multiple HLA-DR and HLA-DP4 molecules and stimulates autologous CD4 + T cells from patients with NY-ESO-1-expressing melanoma. J Immunol 174:1751–1759

Mantovani A (2009) The yin-yang of tumor-associated neutrophils. Cancer Cell 16:173–174

Mantovani A, Sozzani S, Locati M, Allavena P, Sica A (2002) Macrophage polarization: tumor-associated macrophages as a paradigm for polarized M2 mononuclear phagocytes. Trends Immunol 23:549–555

Mantovani A, Allavena P, Sica A, Balkwill F (2008) Cancer-related inflammation. Nature 454:436–444

McHutchison JG, Everson GT, Gordon SC, Jacobson IM, Sulkowski M, Kauffman R, McNair L, Alam J, Muir AJ (2009) Telaprevir with peginterferon and ribavirin for chronic HCV genotype 1 infection. N Engl J Med 360:1827–1838

Mok TS, Yeo W, Yu S, Lai P, Chan HL, Chan AT, Lau JW, Wong H, Leung N, Hui EP, Sung J, Koh J, Mo F, Zee B, Johnson PJ (2005) An intensive surveillance program detected a high incidence of hepatocellular carcinoma among hepatitis B virus carriers with abnormal alpha-fetoprotein levels or abdominal ultrasonography results. J Clin Oncol 23:8041–8047

Mukherjee D, Nissen SE, Topol EJ (2001) Risk of cardiovascular events associated with selective COX-2 inhibitors. JAMA 286:954–959

Murdoch C, Muthana M, Coffelt SB, Lewis CE (2008) The role of myeloid cells in the promotion of tumour angiogenesis. Nat Rev Cancer 8:618–631

Oakley F, Teoh V, Ching AS, Bataller R, Colmenero J, Jonsson JR, Eliopoulos AG, Watson MR, Manas D, Mann DA (2009) Angiotensin II activates I kappaB kinase phosphorylation of RelA at Ser 536 to promote myofibroblast survival and liver fibrosis. Gastroenterology 136:2334–2344

Paavonen J, Jenkins D, Bosch FX, Naud P, Salmeron J, Wheeler CM, Chow SN, Apter DL, Kitchener HC, Castellsague X, de Carvalho NS, Skinner SR, Harper DM, Hedrick JA, Jaisamrarn U, Limson GA, Dionne M, Quint W, Spiessens B, Peeters P, Struyf F, Wieting SL, Lehtinen MO, Dubin G (2007) Efficacy of a prophylactic adjuvanted bivalent L1 virus-like-particle vaccine against infection with human papillomavirus types 16 and 18 in young women: an interim analysis of a phase III double-blind, randomised controlled trial. Lancet 369:2161–2170

Padua D, Zhang XH, Wang Q, Nadal C, Gerald WL, Gomis RR, Massague J (2008) TGFbeta primes breast tumors for lung metastasis seeding through angiopoietin-like 4. Cell 133:66–77

Park BH, Hwang T, Liu TC, Sze DY, Kim JS, Kwon HC, Oh SY, Han SY, Yoon JH, Hong SH, Moon A, Speth K, Park C, Ahn YJ, Daneshmand M, Rhee BG, Pinedo HM, Bell JC, Kirn DH (2008) Use of a targeted oncolytic poxvirus, JX-594, in patients with refractory primary or metastatic liver cancer: a phase I trial. Lancet Oncol 9:533–542

Parkin DM (2006) The global health burden of infection-associated cancers in the year 2002. Int J Cancer 118:3030–3044

Parkin DM, Bray F (2006) Chapter 2: The burden of HPV-related cancers. Vaccine 24 (Suppl 3):S3-11–S3/25

Patel H, Heathcote EJ (2011) Sustained virological response with 29 days of Debio 025 monotherapy in hepatitis C virus genotype 3. Gut 60:879

Pedroza-Gonzalez A, Xu K, Wu TC, Aspord C, Tindle S, Marches F, Gallegos M, Burton EC, Savino D, Hori T, Tanaka Y, Zurawski S, Zurawski G, Bover L, Liu YJ, Banchereau J, Palucka AK (2011) Thymic stromal lymphopoietin fosters human breast tumor growth by promoting type 2 inflammation. J Exp Med 208:479–490

Plummer M, Schiffman M, Castle PE, Maucort-Boulch D, Wheeler CM (2007) A 2-year prospective study of human papillomavirus persistence among women with a cytological diagnosis of atypical squamous cells of undetermined significance or low-grade squamous intraepithelial lesion. J Infect Dis 195:1582–1589

Polyak K, Weinberg RA (2009) Transitions between epithelial and mesenchymal states: acquisition of malignant and stem cell traits. Nat Rev Cancer 9:265–273

Popov Y, Patsenker E, Stickel F, Zaks J, Bhaskar KR, Niedobitek G, Kolb A, Friess H, Schuppan D (2008) Integrin alphavbeta6 is a marker of the progression of biliary and portal liver fibrosis and a novel target for antifibrotic therapies. J Hepatol 48:453–464

Rakoff-Nahoum S, Medzhitov R (2009) Toll-like receptors and cancer. Nat Rev Cancer 9:57–63

Reddy BS, Hirose Y, Lubet R, Steele V, Kelloff G, Paulson S, Seibert K, Rao CV (2000) Chemoprevention of colon cancer by specific cyclooxygenase-2 inhibitor, celecoxib, administered during different stages of carcinogenesis. Cancer Res 60:293–297

Rius J, Guma M, Schachtrup C, Akassoglou K, Zinkernagel AS, Nizet V, Johnson RS, Haddad GG, Karin M (2008) NF-kappaB links innate immunity to the hypoxia response through transcriptional regulation of HIF-1alpha. Nature 453:807–811

Roberts SJ, Ng BY, Filler RB, Lewis J, Glusac EJ, Hayday AC, Tigelaar RE, Girardi M (2007) Characterizing tumor-promoting T cells in chemically induced cutaneous carcinogenesis. Proc Natl Acad Sci U S A 104:6770–6775

Roca H, Varsos ZS, Sud S, Craig MJ, Ying C, Pienta KJ (2009) CCL2 and interleukin-6 promote survival of human CD11b+peripheral blood mononuclear cells and induce M2-type macrophage polarization. J Biol Chem 284:34342–34354

Rockey DC (2008) Current and future anti-fibrotic therapies for chronic liver disease. Clin Liver Dis 12:939–962, xi

Rodier F, Coppe JP, Patil CK, Hoeijmakers WA, Munoz DP, Raza SR, Freund A, Campeau E, Davalos AR, Campisi J (2009) Persistent DNA damage signalling triggers senescence-associated inflammatory cytokine secretion. Nat Cell Biol 11:973–979

Rothwell PM, Wilson M, Elwin CE, Norrving B, Algra A, Warlow CP, Meade TW (2010) Long-term effect of aspirin on colorectal cancer incidence and mortality: 20-year follow-up of five randomised trials. Lancet 376:1741–1750

Rothwell PM, Fowkes FG, Belch JF, Ogawa H, Warlow CP, Meade TW (2011) Effect of daily aspirin on long-term risk of death due to cancer: analysis of individual patient data from randomised trials. Lancet 377:31–41

Ryan KM, Ernst MK, Rice NR, Vousden KH (2000) Role of NF-kappaB in p53-mediated programmed cell death. Nature 404:892–897

Sakuishi K, Apetoh L, Sullivan JM, Blazar BR, Kuchroo VK, Anderson AC (2010) Targeting Tim-3 and PD-1 pathways to reverse T cell exhaustion and restore anti-tumor immunity. J Exp Med 207:2187–2194

Sato Y, Takahashi S, Kinouchi Y, Shiraki M, Endo K, Matsumura Y, Kakuta Y, Tosa M, Motida A, Abe H, Imai G, Yokoyama H, Nomura E, Negoro K, Takagi S, Aihara H, Masumura K, Nohmi T, Shimosegawa T (2006) IL-10 deficiency leads to somatic mutations in a model of IBD. Carcinogenesis 27:1068–1073

Schioppa T, Moore R, Thompson RG, Rosser EC, Kulbe H, Nedospasov S, Mauri C, Coussens LM, Balkwill FR (2011) B regulatory cells and the tumor-promoting actions of TNF-alpha during squamous carcinogenesis. Proc Natl Acad Sci U S A 108:10662–10667

Schreiber GB, Busch MP, Kleinman SH, Korelitz JJ (1996) The risk of transfusion-transmitted viral infections. The Retrovirus Epidemiology Donor Study. N Engl J Med 334:1685–1690

Sharma MD, Hou DY, Baban B, Koni PA, He Y, Chandler PR, Blazar BR, Mellor AL, Munn DH (2010) Reprogrammed foxp3(+) regulatory T cells provide essential help to support cross-presentation and CD8(+) T cell priming in naive mice. Immunity 33:942–954

Shepard CW, Finelli L, Alter MJ (2005) Global epidemiology of hepatitis C virus infection. Lancet Infect Dis 5:558–567

Shepard CW, Simard EP, Finelli L, Fiore AE, Bell BP (2006) Hepatitis B virus infection: epidemiology and vaccination. Epidemiol Rev 28:112–125

Sinha P, Clements VK, Bunt SK, Albelda SM, Ostrand-Rosenberg S (2007) Cross-talk between myeloid-derived suppressor cells and macrophages subverts tumor immunity toward a type 2 response. J Immunol 179:977–983

Smith JS, Lindsay L, Hoots B, Keys J, Franceschi S, Winer R, Clifford GM (2007) Human papillomavirus type distribution in invasive cervical cancer and high-grade cervical lesions: a meta-analysis update. Int J Cancer 121:621–632

Smith MR, Saad F, Coleman R, Shore N, Fizazi K, Tombal B, Miller K, Sieber P, Karsh L, Damiao R, Tammela TL, Egerdie B, Van PH, Chin J, Morote J, Gomez-Veiga F, Borkowski T, Ye Z, Kupic A, Dansey R, Goessl C (2012) Denosumab and bone-metastasis-free survival in men with castration-resistant prostate cancer: results of a phase 3, randomised, placebo-controlled trial. Lancet 379:39–46

Staller P, Sulitkova J, Lisztwan J, Moch H, Oakeley EJ, Krek W (2003) Chemokine receptor CXCR4 downregulated by von Hippel-Lindau tumour suppressor pVHL. Nature 425:307–311

Steinbach G, Lynch PM, Phillips RK, Wallace MH, Hawk E, Gordon GB, Wakabayashi N, Saunders B, Shen Y, Fujimura T, Su LK, Levin B (2000) The effect of celecoxib, a cyclooxygenase-2 inhibitor, in familial adenomatous polyposis. N Engl J Med 342:1946–1952

Stockert E, Jager E, Chen YT, Scanlan MJ, Gout I, Karbach J, Arand M, Knuth A, Old LJ (1998) A survey of the humoral immune response of cancer patients to a panel of human tumor antigens. J Exp Med 187:1349–1354

Stopeck AT, Lipton A, Body JJ, Steger GG, Tonkin K, de Boer RH, Lichinitser M, Fujiwara Y, Yardley DA, Viniegra M, Fan M, Jiang Q, Dansey R, Jun S, Braun A (2010) Denosumab compared with zoledronic acid for the treatment of bone metastases in patients with advanced breast cancer: a randomized, double-blind study. J Clin Oncol 28:5132–5139

Sunami Y, Leithauser F, Gul S, Fiedler K, Guldiken N, Espenlaub S, Holzmann KH, Hipp N, Sindrilaru A, Luedde T, Baumann B, Wissel S, Kreppel F, Schneider M, Scharffetter-Kochanek K, Kochanek S, Strnad P, Wirth T (2012) Hepatic activation of IKK/NF-kappaB signaling induces liver fibrosis via macrophage-mediated chronic inflammation. Hepatology 56:1117–1128

Swann JB, Smyth MJ (2007) Immune surveillance of tumors. J Clin Invest 117:1137–1146

Teixeira-Clerc F, Julien B, Grenard P, Van Tran NJ, Deveaux V, Li L, Serriere-Lanneau V, Ledent C, Mallat A, Lotersztajn S (2006) CB1 cannabinoid receptor antagonism: a new strategy for the treatment of liver fibrosis. Nat Med 12:671–676

Topalian SL, Hodi FS, Brahmer JR, Gettinger SN, Smith DC, McDermott DF, Powderly JD, Carvajal RD, Sosman JA, Atkins MB, Leming PD, Spigel DR, Antonia SJ, Horn L, Drake CG, Pardoll DM, Chen L, Sharfman WH, Anders RA, Taube JM, McMiller TL, Xu H, Korman AJ, Jure-Kunkel M, Agrawal S, McDonald D, Kollia GD, Gupta A, Wigginton JM, Sznol M (2012) Safety, Activity, and Immune Correlates of Anti-PD-1 Antibody in Cancer. N Engl J Med 366:2443–2454

Torresi J, Johnson D, Wedemeyer H (2011) Progress in the development of preventive and therapeutic vaccines for hepatitis C virus. J Hepatol 54:1273–1285

Trevejo JM, Marino MW, Philpott N, Josien R, Richards EC, Elkon KB, Falck-Pedersen E (2001) TNF-alpha -dependent maturation of local dendritic cells is critical for activating the adaptive immune response to virus infection. Proc Natl Acad Sci U S A 98:12162–12167

Vakkila J, Lotze MT (2004) Inflammation and necrosis promote tumour growth. Nat Rev Immunol 4:641–648

Voronov E, Shouval DS, Krelin Y, Cagnano E, Benharroch D, Iwakura Y, Dinarello CA, Apte RN (2003) IL-1 is required for tumor invasiveness and angiogenesis. Proc Natl Acad Sci U S A 100:2645–2650

Weber DM, Chen C, Niesvizky R, Wang M, Belch A, Stadtmauer EA, Siegel D, Borrello I, Rajkumar SV, Chanan-Khan AA, Lonial S, Yu Z, Patin J, Olesnyckyj M, Zeldis JB, Knight RD (2007) Lenalidomide plus dexamethasone for relapsed multiple myeloma in North America. N Engl J Med 357:2133–2142

Wherry EJ, Teichgraber V, Becker TC, Masopust D, Kaech SM, Antia R, von Andrian UH, Ahmed R (2003) Lineage relationship and protective immunity of memory CD8 T cell subsets. Nat Immunol 4:225–234

Witek RP, Stone WC, Karaca FG, Syn WK, Pereira TA, Agboola KM, Omenetti A, Jung Y, Teaberry V, Choi SS, Guy CD, Pollard J, Charlton P, Diehl AM (2009) Pan-caspase inhibitor VX-166 reduces fibrosis in an animal model of nonalcoholic steatohepatitis. Hepatology 50:1421–1430

Woller N, Knocke S, Mundt B, Gurlevik E, Struver N, Kloos A, Boozari B, Schache P, Manns MP, Malek NP, Sparwasser T, Zender L, Wirth TC, Kubicka S, Kuhnel F (2011) Virus-induced tumor inflammation facilitates effective DC cancer immunotherapy in a Treg-dependent manner in mice. J Clin Invest 121:2570–2582

Xia JL, Dai C, Michalopoulos GK, Liu Y (2006) Hepatocyte growth factor attenuates liver fibrosis induced by bile duct ligation. Am J Pathol 168:1500–1512

Yang J, Weinberg RA (2008) Epithelial-mesenchymal transition: at the crossroads of development and tumor metastasis. Dev Cell 14:818–829

Zajac AJ, Blattman JN, Murali-Krishna K, Sourdive DJ, Suresh M, Altman JD, Ahmed R (1998) Viral immune evasion due to persistence of activated T cells without effector function. J Exp Med 188:2205–2213

Zanetti AR, Van DP, Shouval D (2008) The global impact of vaccination against hepatitis B: a historical overview. Vaccine 26:6266–6273

Zhang Y, Ma CJ, Wang JM, Ji XJ, Wu XY, Moorman JP, Yao ZQ (2012) Tim-3 regulates pro- and anti-inflammatory cytokine expression in human CD14 + monocytes. J Leukoc Biol 91:189–196

The Oncogenic Role of Hepatitis B Virus

Lise Rivière, Aurélie Ducroux and Marie Annick Buendia

Abstract

The hepatitis B virus (HBV) is a small enveloped DNA virus that causes acute and chronic hepatitis. HBV infection is a world health problem, with 350 million chronically infected people at increased risk of developing liver disease and hepatocellular carcinoma (HCC). HBV has been classified among human tumor viruses by virtue of a robust epidemiologic association between chronic HBV carriage and HCC occurrence. In the absence of cytopathic effect in infected hepatocytes, the oncogenic role of HBV might involve a combination of direct and indirect effects of the virus during the multistep process of liver carcinogenesis. Liver inflammation and hepatocyte proliferation driven by host immune responses are recognized driving forces of liver cell transformation. Genetic and epigenetic alterations can also result from viral DNA integration into host chromosomes and from prolonged expression of viral gene products. Notably, the transcriptional regulatory protein HBx encoded by the X gene is endowed with tumor promoter activity. HBx has pleiotropic activities and plays a major role in HBV pathogenesis and in liver carcinogenesis. Because hepatic tumors carry a dismal prognosis, there is urgent need to develop early

M. A. Buendia (✉)
Inserm U785, University Paris-Sud, 12 Avenue Paul Vaillant Couturier, 94800, Villejuif, France
e-mail: marie-annick.buendia@inserm.fr; marie-annick.buendia@pasteur.fr

L. Rivière · A. Ducroux
Institut Pasteur, Hepacivirus and Innate Immunity Unit, 28 rue du Dr Roux, 75015, Paris, France
e-mail: lise.riviere@pasteur.fr

A. Ducroux
e-mail: aurelie.ducroux@pasteur.fr

M. H. Chang and K.-T. Jeang (eds.), *Viruses and Human Cancer*,
Recent Results in Cancer Research 193, DOI: 10.1007/978-3-642-38965-8_4,
© Springer-Verlag Berlin Heidelberg 2014

diagnostic markers of HCC and effective therapies against chronic hepatitis B. Deciphering the oncogenic mechanisms that underlie HBV-related tumorigenesis might help developing adapted therapeutic strategies.

Keywords

Inflammation · Liver disease · Hepatocellular carcinoma · HBx · Transcription · Genetic · Epigenetic

Contents

1   Introduction

The hepatitis B virus (HBV) is a small, enveloped DNA virus with preferential tropism for the hepatocyte. HBV is highly infectious, and its spread in all countries over the world can be easily followed using the surface antigen HBsAg as serum marker (El-Serag 2012). Chronic HBV carriage has been linked to the development of hepatocellular carcinoma (HCC) by robust epidemiologic evidence: liver cancer is more frequent in HBV endemic areas, the rate of HBV infection is higher among liver cancer patients than in the general population, and the risk of developing hepatic tumors is markedly increased for HBV carriers (Beasley et al. 1981; Pagano et al. 2004). Moreover, mammalian hepadnaviruses closely related to HBV, including the woodchuck hepatitis virus (WHV) and the ground squirrel hepatitis virus (GSHV) induce liver tumors in their hosts (reviewed in Benhenda et al. 2009b). While the risk of developing HCC is strongly increased for people chronically infected with HBV, significant differences in disease outcome have been associated with host and viral parameters, including male gender, infection at early age, cooperative effects of different risk factors such as Aflatoxin B1, alcohol, diabetes and obesity, as well as high viral load, HBeAg positivity, and some HBV genotypes (for review, see Fallot et al. 2012). Importantly, in a meta-analysis including a total of 11,582 HBV-infected individuals, it has been shown that HBV mutations accumulate at specific positions during chronic hepatitis from healthy carrier state to HCC. Notably, high probability of HCC development has been correlated with mutations in the PreS domain combined with mutations C1653T, T1753 V, and A1762T/G1764A in the basal core promoter that overlaps with the $3'$ end of the X gene (Liu et al. 2009).

While chronic HBV infection is currently associated with more than one half of HCC cases worldwide, only 15–20 % of the tumors occur in HBsAg-positive patients in Western countries and Japan (Parkin 2006). Epidemiological studies have shown a high prevalence of anti-HBs and anti-HBc antibodies in HBsAg-negative patients with HCC. Furthermore, HBV DNA has been detected in HCCs developing on non-cirrhotic livers from HBsAg-negative patients (Matsuzaki et al. 1997). Thus, past and resolved HBV infection might retain pro-oncogenic properties. Accordingly, increased risk of tumor development has been noted in patients that display characteristics of "occult" HBV infection, including absence of HBV serological markers but low levels of HBV DNA detected by polymerase chain reaction (PCR) in the serum and/or liver (Brechot 2004; Raimondo et al. 2008).

Therapeutic options for patients with chronic HBV infection consist mainly in treatment with pegylated interferon-alpha (IFNα), which has antiviral, immuno-modulatory, and probably antitumoral activities, and treatment with oral nucleoside or nucleotide analogs, including lamivudine and adefovir (Scaglione and Lok 2012). Although these protocols rarely lead to complete elimination of the virus, they might prevent further liver damage and even reverse previous damage. Interestingly, a significantly lower incidence of HCC has been noted for chronic hepatitis B patients receiving therapy for two years compared to untreated patients (Papatheodoridis et al. 2010).

The hepatitis B vaccine, which prevents chronic HBV infection and related liver diseases, has been introduced into national immunization programs for children in most countries. The efficacy of this vaccine in lowering the rate of chronic infections, and the incidence of liver cancer, particularly in children, has been fully demonstrated in endemic countries (Ott et al. 2012). However, the total number of chronic HBV carriers worldwide is still increasing, representing 6 % of the total world population.

2   The HBV Genome and Replication

The HBV genome, 3.2 kb in size, is the smallest among human DNA viruses. Eight HBV genotypes (genotypes A–H) with greater than 8 % nucleotide sequence divergence have been identified, and further categorized into subgenotypes (Schaefer 2005). The viral genome presents a highly compact organization with 4 overlapping open reading frames (ORFs) that cover the entire genome and encode the capsid and surface proteins, the polymerase and the HBx transactivator (Fig. 1). In addition, a protein generated from a spliced mRNA through a frame shift within the polymerase gene has been termed HBSP (Soussan et al. 2000). Consequently, the regulatory regions (promoters, enhancers, transcription start sites, and polyA signal) are embedded within coding sequences.

Upon HBV infection, the partially double-stranded DNA genome is delivered to the nucleus where it is converted into a covalently closed circular DNA (cccDNA). The cccDNA serves as the template for transcription of all viral RNAs

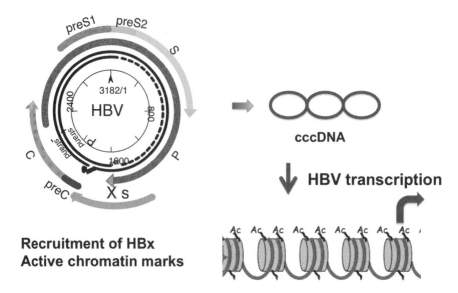

Fig. 1 The hepatitis B virus genome and the covalently closed circular DNA (cccDNA). The viral genome (*left panel*) is made of two linear DNA strands maintained in circular configuration by their cohesive ends. The minus strand is around 3.182 kb in length, while the plus strand has variable length with fixed 5' end position and variable 3' end. The broad arrows surrounding the viral genome represent four open reading frames that encode the viral proteins. After viral entry, the HBV genome is delivered into the nucleus where it is converted into cccDNA by the cellular machinery. The cccDNA is organized into a minichromosome with bound histone and non-histone proteins including HBx and the core protein. By recruiting different chromatin modifying factors such as CBP/p300, HBx plays a pivotal role in the modulation of epigenetic marks such as histone tail acetylation and methylation that results into active HBV transcription

by the host RNA polymerase II (Fig. 1). The production of the major HBV transcripts of 3.5-, 2.4-, 2.1-, and 0.7-kb is controlled by 4 distinct viral promoters and by 2 enhancers, Enh I and Enh II (Doitsh and Shaul 2004; Moolla et al. 2002). Several binding sites for ubiquitous and liver-specific transcription factors that likely regulate HBV transcription in vivo have been identified in the cis-acting sequences of HBV promoters and enhancers (Fallot et al. 2012; Moolla et al. 2002). Moreover, it has been shown that HBV cccDNA transcription is regulated by epigenetic mechanisms such as DNA methylation and histone acetylation (Pollicino et al. 2006; Vivekanandan et al. 2009). Because persistence of the cccDNA might be responsible for the failure of most antiviral treatments to completely eradicate the virus, it represents an interesting target for effective therapy of chronic hepatitis B.

Following transport of HBV RNAs to the cytoplasm and their translation into viral proteins, the pregenomic RNA is selectively packaged with the polymerase into progeny capsids and reverse-transcribed into RC-DNA by the viral polymerase. The HBV polymerase harbors RNA-dependent DNA polymerase activity and sequence homology with the RTs of retroviruses and cauliflower mosaic virus

(CaMV) (Toh et al. 1983). HBV replication is initiated by the binding of the viral RT to the encapsidation signal epsilon located at the $5'$ end of pregenomic RNA (Bartenschlager and Schaller 1992). After asymmetrical replication of the two HBV DNA strands, capsids containing mature RC-DNA are then assembled with the viral envelope in the endoplasmic reticulum, and subsequent budding of HBV is thought to occur at intracellular membranes (Bruss 2007). Additionally, empty envelope particles are produced in large excess, behaving probably as decoys for the host immune system. A minor portion of mature capsids can re-cycle RC-DNA to the nucleus and contribute to the amplification of the nuclear cccDNA pool. The availability of envelope proteins in adequate amounts and their efficiency to package nucleocapsids are critical for coordinate regulation of virion morpho-genesis and accumulation of cccDNA (Summers et al. 1991). While huge amounts of viral particles can be produced in healthy liver from asymptomatic HBV car-riers, liver damage can result from disruption of the normal replication process leading to intrahepatic accumulation of the large envelope protein or excessive amounts of cccDNA (Wang et al. 2006).

3   HBx: A Promiscuous Transactivator

The HBx (or pX) protein encoded by the X gene is a small polypeptide of 154 amino acids (16.5 kDa) that is expressed at low levels in acute and chronic infections and induces humoral and cellular immune responses (Su et al. 1998). The X gene is conserved in mammalian hepadnaviruses, but curiously, it is absent from avian hepadnavirus genomes. Different domains associated with specific functions have been identified in this small protein (Fig. 2). However, little is known about the three-dimensional structure of HBx, as it has not been possible so far to produce sufficient amounts of soluble protein. HBx appears as an unstruc-tured protein that might be folded and gain secondary structure through its interaction with partner proteins (Rui et al. 2005). Such flexibility could account for the large array of HBx activities.

An essential role of HBx is virus replication in vivo, but this role is far from being completely understood, and it is likely multifactorial. Indeed, HBx has been endowed with multiple activities. The HBx protein is described as a promiscuous transactivator of viral and cellular genes, acting both from cytoplasmic and nuclear locations. HBx activates the transcription from cellular and viral promoters including HBV promoters, and it subverts different cellular functions and signal transduction pathways through modulation of cytoplasmic calcium, cell prolifer-ation, and apoptosis. While these cellular activities modulated by HBx likely contribute to increase virus replication, evidence for a more direct role of HBx in stimulating HBV transcription has been provided by different studies in cell lines and murine models. HBx strongly activates HBV replication in HBV transgenic mice and in mice hydrodynamically injected with a plasmid encoding the HBV genome, and this effect is less pronounced in hepatoma cell lines (Keasler et al.

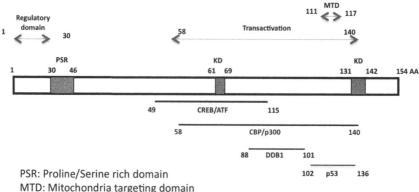

PSR: Proline/Serine rich domain
MTD: Mitochondria targeting domain
KD: Kunitz domain-like sequences

Fig. 2 Schematic representation of the small HBx protein with important functional domains. The N-terminal domain encodes a regulatory region (Murakami 1999) and a proline/serine-rich region. The domain carrying most HBx transcriptional transactivation activity has been mapped by Kumar and colleagues (Kumar et al. 1996). It includes the mitochondrial targeting region (Li et al. 2008). Binding domains with cellular partners are shown at the *bottom* part, including CREB/ATF (Maguire et al. 1991), CBP/p300 (Cougot et al. 2007), DDB1 (Becker et al. 1998; Lin-Marq et al. 2001), and p53 (Lin et al. 1997)

2007; Leupin et al. 2005; Tang et al. 2005). HBx is also required to initiate and maintain HBV replication as a key regulator during natural HBV infection (Lucifora et al. 2011). It has been shown recently that HBx interferes with epigenetic regulation of the HBV minichromosome by increasing the active chromatin marks and antagonizing the inhibitory factors (Belloni et al. 2009; Cougot et al. 2012).

HBx exerts its activities by interacting with a large number of cellular partners that are located either in the cytoplasm or in the nucleus, in agreement with the dual location of HBx. Consistent with its role in transcription, HBx has been shown to interact with components of the basal transcriptional machinery (RPB5, TFIIB, TBP, TFIIH), with transcription factors (ATF/CREB, c/EBP, NF-IL6, RXR receptor) as well as coactivators (CBP/P300, ASC-2) or repressor (DNMT) (reviewed in Benhenda et al. 2009a). We have shown that HBx is recruited to the promoters of CREB target genes and increases the recruitment of CBP/P300 on CREB, thereby activating transcription (Cougot et al. 2007). Thus, HBx could be directly involved in assembly of enhancer–transcription factor complexes. HBx clearly plays a direct role in the epigenetic regulation of HBV transcription, by modulating a network of transcription regulatory factors within large molecular weight protein complexes (Belloni et al. 2009; Cougot et al. 2012).

4 HBx and E3 Ubiquitin Ligases: A Link Beyond Protein Stability?

Manipulation of protein degradation machineries is a common strategy used by viruses to provide a favorable environment for their replication and escape protective mechanisms raised by the host cell (Barry and Fruh 2006). Interaction between HBx and damage-specific DNA binding protein 1 (DDB1), an adaptor subunit of the Cul4A-based ubiquitin E3 ligase complex, has been extensively documented (Lee et al. 1995; Leupin et al. 2003; Sitterlin et al. 1997). DDB1 is a highly conserved protein that links multiple WD40 proteins to the CUL4– ROC1– E2 catalytic core and that is implicated in DNA repair via its association with DDB2 (Jackson and Xiong 2009). The minimal DDB1 binding domain on HBx (amino acids 88–100) contains a sequence evoking the DDB1-binding WD40 (DWD) motifs shared by cellular DCAFs (He et al. 2006). Several point mutations in this region impede HBx binding to DDB1 (Bergametti et al. 2002; Hodgson et al. 2012) (Table 1). High-resolution crystal structure demonstrated that a

Table 1 Survey of recurrent HBx mutants detected in hepatocellular carcinoma and synthetic mutants used for demonstrating specific HBx functions. For each HBx mutant, reported activities such as transcriptional transactivation, deregulation of cell cycle and viability, and transformation are shown. Increased activity of mutant as compared to wild-type HBx is shown by +, equal activity by =, and reduced activity by −. Divergent data from 2 different papers are represented as double symbols (=/+ or =/−). Corresponding references can be found in the reference section

	Epidemiology	Transactivation	Growth inhibition / apoptosis / G1 arrest	Cell transformation vs HBx WT	Other	References
HBx wt 1–154		Yes	Yes	Context-dependent		
Frequent mutations in HCC						
K130M + V131I	Increased risk of HCC Correlation with fulminant hepatitis and cirrhosis		-	+	Low efficiency of T cells activation	Takahashi et al, 1998 Malmassari et al, 2007 Liu et al, 2009
L30F / S144A	detected in 32% of tumors					Wang et al, 2012
V88I	detected in 27% of					Wang et al, 2012
S31A	prevalent in HCC and cirrhosis patients	=/-	=/-			Yeh et al, 2000
C-terminal deletions	Frequently detected in HCC (integrated HBV sequences)					
1–101			-			
1–114			-	+		Poussin et al, 1999
1–120			-	+		Sirma et al, 1999
1–124			=	-		Tu et al, 2001
1–128		-	=			Ma et al, 2008
1–131		-			Less stable	Lizzano et al, 2011
1–134			-	+		Luo et al, 2012
1–136		-	-	=/+		Fujimoto et al, 2012
1–138				+		
1–140		=/-	-	+	Less stable	Sung et al, 2012
1–143		=	=	=		
1–144		=	=	=		
1–146		=	=	=		
Other mutations/deletions						
1–50		-		=	No interaction with RPB5	Cheong et al, 1995 Luo et al, 2012
43–154			=			Gottlob et al, 1998
51–154		=/-		-		Kumar et al, 1996
58–140		=				Gottlob et al, 1998
106–154		-	-	-		
R96E		-				Leupin et al, 2003
C61L					Decreased interaction with DDB1	Becker et al, 1998
C69L						Lin-Marq et al, 2001
P90V/K91L		-				
C115A					Disruption of Mito targeting	Li et al, 2008

peptide consisting of the HBx amino acids 88–100 could directly bind to DDB1 (Li et al. 2010). Moreover, the specific region on DDB1 that is bound by HBx is the common binding site of most DCAFs. These data suggest that HBx shares similar properties with cellular DCAFs and might displace other DCAFs, thereby modifying the profile of Cul4A complex substrates. So far, however, a role of HBx in regulating the stability or localization of such substrates has not been identified.

It has been shown that the HBx/DDB1 interaction is conserved among mammalian hepadnaviruses and plays essential role in virus replication and in the stability and functions of HBx (Bontron et al. 2002; Sitterlin et al. 2000a). In particular, transcriptional and pro-apoptotic properties of HBx are dependent upon its interaction with DDB1 (Lin-Marq et al. 2001; Sitterlin et al. 2000b). Mutant HBx that has lost DDB1 binding capacity also loses concomitant cytotoxicity, but regains activity when directly fused to DDB1. These studies indicate that HBx does not inhibit the normal DDB1 functions, but confers apoptotic potential to DDB1 (Bontron et al. 2002). However, in other studies using HBx mutants that no longer bind to DDB1, the transcriptional activity of HBx has been dissociated from DDB1 binding (Wentz et al. 2000) (see Table 1). Moreover, recent studies of HBx-mediated activation of HBV transcription and replication have confirmed the importance of HBx-DDB1 interaction, but also pointed to the requirement for the HBx carboxy-terminal region for maximal stimulation of HBV transcription (Hodgson et al. 2012; Luo et al. 2012).

DDB1 also belongs to the UV-DDB complex implicated in repair of UV-damaged DNA, and HBx has been shown to inhibit nucleotide excision repair (NER) (Mathonnet et al. 2004). Further investigations of the effects of HBx-DDB1 interaction on DNA repair have shown that HBx induces mitotic abnormalities indirectly, by interfering with S-phase, and that an altered S-phase can result in abnormal centrosome duplication. HBx induces lagging chromosomes in cells dividing with a normal bipolar spindle, followed by the formation of multinucleated cells with abnormal centrosome numbers and multipolar spindles, and these effects are strictly dependent upon correct binding of HBx to DDB1 (Martin-Lluesma et al. 2008). This interaction is therefore of crucial importance for the pathogenesis of chronic hepatitis B.

5   HBx and Malignant Transformation of Liver Cells

The regulatory hepatitis B virus X protein is thought to be involved in oncogenesis. Although HBx does not behave as a strong oncogene per se, it can transform SV40-immortilized murine hepatocytes and induce liver tumors or accelerate carcinogenesis induced by carcinogens or by c-Myc in HBx transgenic mice. (Höhne et al. 1990; Kim et al. 1991; Madden et al. 2001; Terradillos et al. 1997). Transactivation of cellular oncogenes and growth factors, and deregulation of cell cycle progression are two mechanisms that might account for the weak oncogenicity of this viral protein (reviewed in (Bouchard and Schneider 2004). Indeed, in other human oncogenic viruses such as HTLV-1, EBV, and HPV-16 and -18, transforming capacity is associated with the transcriptional transactivation activity of viral gene products. Comparative analyses of the different viral transactivators may help guide future work in this field. Additionally, the ability of HBx to induce aberrant activation of several signaling pathways, including the Src, Ras, MAP kinase, and CREB pathways, might lead to accumulation of cellular dysfunctions that favor malignant transformation (reviewed in Andrisani and Barnabas 1999).

Another relevant pro-carcinogenic activity of HBx has been shown to target the mitotic cell cycle. Interaction of HBx with different cellular partners implicated in centrosome formation might result in defects in centrosome dynamics and mitotic spindle formation. HBx has been shown to bind and partially inactivate BubR1, a mitotic kinase effector that specifies microtubule attachments and checkpoint functions (Kim et al. 2008). HBx also binds HBXIP, a major regulator of centrosome duplication, required for bipolar spindle formation and cytokinesis (Wen et al. 2008). Interestingly, DNA re-replication induced by HBx has been causally related to partial polyploidy, a condition known to be associated with cancer (Rakotomalala et al. 2008). Accordingly, HBx has been implicated in induction of chromosomal instability (CIN), a cancer hallmark. CIN cells with extra centrosomes display an increased frequency of lagging chromosomes during anaphase, leading to chromosome missegregation. In cells expressing HBx, prolongation of S-phase has been associated with aberrant centrosome duplication, multipolar spindle formation, chromosome segregation defects, and appearance of multinucleated cells (Studach et al. 2009, 2010). HBx-expressing polyploid cells are prone to oncogenic transformation mediated by the mitotic entry factor polo-like kinase 1 (Plk1), and in turn, Plk1 downregulates ZNF198 and SUZ12 that modulate HBV transcription and replication through epigenetic mechanisms (Wang et al. 2011). Others have implicated the interaction between HBx and DDB1 in increasing genetic instability (Li et al. 2010; Martin-Lluesma et al. 2008). Altogether, these data provide a strong link between HBx activities and chromosomal instability that could account for the high rate of chromosomal defects observed in tumors related to HBV infection compared with other risk factors (Boyault et al. 2007).

The question of whether HBx is involved in cancer progression has been raised by the observation that HBx expression is usually retained in human HCCs, correlating with HBV methylome studies that showed global hypermethylation of the HBV genome in HCC, except for the X gene (Fernandez et al. 2009).

6   Integration of HBV DNA and Mutated X Genes in HCC

Integrated HBV DNA sequences have been detected in a majority of HBV-related HCCs (Bréchot et al. 1980; Bonilla Guerrero and Roberts 2005). Recent analysis of viral insertion sites using various PCR-based methods has demonstrated that integration occurs not only in repetitive elements and fragile sites, but also frequently in actively transcribed regions, nearby or within cellular genes (Murakami et al. 2005). Evidence for oncogenic potential of the rearranged viral-host sequences has been provided conclusively in few cases in which production of chimeric HBV/cellular transcripts resulted in hybrid viral/cellular proteins. Besides "historical" targets such as the retinoic acid receptor a (RARα) and cyclin A2, a number of HBV insertions have been shown to target key regulators of cell proliferation and cell death. HBx fusion proteins have been described in many cases, including fusion with the SERCA1 gene and the MLL2, MLL3, and MLL4 genes (Chami et al. 2000; Saigo et al. 2008; Tamori et al. 2005).

On the other side, integrated HBx sequences are usually mutated, and they have been endowed with peculiar properties, leading to the notion of "transacting" oncogenic mechanisms. Because the 3' end of the X gene coincides with the recombination-prone region of the HBV genome located between the direct repeats DR1 and DR2, frequent C-terminal truncations of the X gene have been detected in HCCs (Luo et al. 2012) (Table 1). Functional studies have outlined diverse properties of tumor-derived HBx proteins compared to wild type. In some cases, these sequences retained transactivating capacity and the ability to bind p53 and block p53-mediated apoptosis, which may provide a growth advantage for neoplastic cells (Huo et al. 2001). Moreover, in vitro and in vivo studies showed that the truncated rather than the full-length HBx could effectively transform immortalized liver cells and modulate expression of key genes implicated in the control of cell cycle and apoptosis (Ma et al. 2008). Recent studies have demonstrated that the C-terminus of HBx impacts the stability of the protein, its transactivation activity and its ability to stimulate HBV replication (Lizzano et al. 2011). By contrast, in other cases, the truncated X sequences harbored a number of missense mutations and were less effective in transcriptional transactivation, apoptosis induction, and in supporting HBV replication (Sirma et al. 1999). Collectively, recent studies of the HBx protein functions have contributed to better delineate the importance of the X gene product at different steps of the malignant process induced by persistent HBV infection.

7   Conclusions and Future Prospects

HCC is among the most frequent cancers in many countries, and its incidence has been rising during the last years (Ferlay et al. 2010). Like other human cancers, the tumorigenic process in HBV carriers evolves through multiple steps resulting in the accumulation of genetic and epigenetic lesions (Fig. 3). In the absence of direct transforming capacity of the virus in cell lines and transgenic mice, and because HCC develops after a long latency period following primary HBV infection, the question may be raised of whether the hepatitis B virus is oncogenic. In a simple view, immune responses against infected hepatocytes might trigger a pro-carcinogenic inflammatory process associated with continuous necrosis and cell regeneration, fostering the accumulation of genetic and epigenetic defects. In

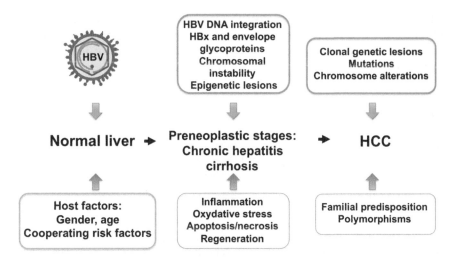

Fig. 3 Multistep process of HBV-induced liver tumorigenesis. Chronic HBV infection, alone or in combination with cofactors, triggers a cascade of events ultimately leading to transformation. At preneoplastic stages, the immune response induces a necroinflammatory disease that creates a suitable environment for the occurrence of genetic and epigenetic lesions. Integration of HBV DNA sequences into the host genome as well as long-term production of the viral HBx and surface proteins may contribute to the tumoral process via multiple mechanisms

this model, the long-term toxic effects of viral gene products during productive HBV infection might potentiate the action of exogenous carcinogenic factors, such as Aflatoxin B1 and alcohol. However, it is worth noticing that chronic HBV infection has been assigned a strong and independent causative role by extensive epidemiological studies worldwide and that evidence is increasing for oncogenic potential of the regulatory protein HBx and surface glycoproteins in native or mutant state. Thus, HBV shares a number of basic strategies with other human oncogenic viruses. It might also be speculated that HCC latency period depends on the occurrence of a decisive HBV integration event that would promote genetic instability or lead to cis- or transactivation of relevant genes. Identifying the cellular effectors connecting HBV infection and liver cell transformation is of upmost importance.

The question of whether HBV could trigger distinct tumorigenic pathways compared to other risk factors can now be addressed via novel technologies for genome-wide scans of gene expression, genetic and epigenetic alterations, and proteomics. Thus, despite substantial heterogeneity, HBV-related tumors have been recently classified into subgroups that display high chromosomal instability, stemness features, increased frequency of p53 mutation, and low rate of ß-catenin mutation (Boyault et al. 2007; Guichard et al. 2012). It will now be important to apply these technologies to the analysis of chronic hepatitis B at different stages, in order to identify suitable prognostic markers and therapeutic targets. In addition, there is need to better define the role of HBV in the occurrence of epigenetic alterations such as methylation-associated gene silencing and abnormal expression of microRNAs that play crucial part in malignant transformation. Ongoing advances in this field will undoubtedly provide adapted tools for improving the management of chronic hepatitis B and liver cancer.

References

Andrisani OM, Barnabas S (1999) The transcriptional function of the hepatitis B virus X protein and its role in hepatocarcinogenesis. Int J Oncol 15:373–379

Barry M, Fruh K (2006) Viral modulators of cullin RING ubiquitin ligases: culling the host defense. Sci STKE 335:21

Bartenschlager R, Schaller H (1992) Hepadnaviral assembly is initiated by polymerase binding to the encapsidation signal in the viral genome. EMBO J 11:3413–3420

Beasley RP, Lin CC, Hwang LY et al (1981) Hepatocellular carcinoma and hepatitis B virus: a prospective study of 22,707 men in Taiwan. Lancet 2:1129–1133

Becker SA, Lee TH, Butel JS et al (1998) Hepatitis B virus X protein interferes with cellular DNA repair. J Virol 72:266–272

Belloni L, Pollicino T, De Nicola F et al (2009) Nuclear HBx binds the HBV minichromosome and modifies the epigenetic regulation of cccDNA function. Proc Natl Acad Sci U S A 106:19975–19979

Benhenda S, Cougot D, Buendia MA et al (2009a) Hepatitis B virus X protein molecular functions and its role in virus life cycle and pathogenesis. Adv Cancer Res 103:75–109

Benhenda S, Cougot D, Neuveut C et al (2009b) Liver cell transformation in chronic HBV infection. Viruses 1:630–646

Bergametti F, Sitterlin D, Transy C (2002) Turnover of hepatitis B virus X protein is regulated by damaged DNA-binding complex. J Virol 76:6495–6501

Bonilla Guerrero R, Roberts LR (2005) The role of hepatitis B virus integrations in the pathogenesis of human hepatocellular carcinoma. J Hepatol 42:760–777

Bontron S, Lin-Marq N, Strubin M (2002) Hepatitis B virus X protein associated with UV-DDB1 induces cell death in the nucleus and is functionally antagonized by UV-DDB2. J Biol Chem 277:38847–38854

Bouchard MJ, Schneider RJ (2004) The enigmatic X gene of hepatitis B virus. J Virol 78:12725–12734

Boyault S, Rickman DS, de Reynies A et al (2007) Transcriptome classification of HCC is related to gene alterations and to new therapeutic targets. Hepatology 45:42–52

Brechot C (2004) Pathogenesis of hepatitis B virus-related hepatocellular carcinoma: old and new paradigms. Gastroenterology 127:S56–S61

Bréchot C, Pourcel C, Louise A et al (1980) Presence of integrated hepatitis B virus DNA sequences in cellular DNA of human hepatocellular carcinoma. Nature 286:533–535

Bruss V (2007) Hepatitis B virus morphogenesis. World J Gastroenterol 13:65–73

Chami M, Gozuacik D, Saigo K et al (2000) Hepatitis B virus-related insertional mutagenesis implicates SERCA1 gene in the control of apoptosis. Oncogene 19:2877–2886

Cheong J, Yi M, Lin Y et al (1995) Human RPB5, a subunit shared by eukaryotic nuclear RNA polymerases, binds human hepatitis B virus X protein and may play a role in X transactivation. EMBO J 14:143–150

Cougot D, Allemand E, Riviere L et al (2012) Inhibition of PP1 phosphatase activity by HBx: a mechanism for the activation of hepatitis B virus transcription. Sci Signal 5(250):ra1

Cougot D, Wu Y, Cairo S et al (2007) The hepatitis B virus X protein functionally interacts with CREB-binding protein/p300 in the regulation of CREB-mediated transcription. J Biol Chem 282:4277–4287

Doitsh G, Shaul Y (2004) Enhancer I predominance in hepatitis B virus gene expression. Mol Cell Biol 24:1799–1808

El-Serag HB (2012) Epidemiology of viral hepatitis and hepatocellular carcinoma. Gastroenterology 142(1264–1273):e1261

Fallot G, Neuveut C, Buendia MA (2012) Diverse roles of hepatitis B virus in liver cancer. Curr Opin Virol available online (in press)

Ferlay J, Shin HR, Bray F et al (2010) Estimates of worldwide burden of cancer in 2008: GLOBOCAN 2008. Int J Cancer 127:2893–2917

Fernandez AF, Rosales C, Lopez-Nieva P et al (2009) The dynamic DNA methylomes of double-stranded DNA viruses associated with human cancer. Genome Res 19:438–451

Fujimoto A, Totoki Y, Abe T et al (2012) Whole-genome sequencing of liver cancers identifies etiological influences on mutation patterns and recurrent mutations in chromatin regulators. Nat Genet 44:760–764

Gottlob K, Pagano S, Levrero M et al (1998) Hepatitis B virus X protein transcription activation domains are neither required nor sufficient for cell transformation. Cancer Res 58:3566–3570

Guichard C, Amaddeo G, Imbeaud S et al (2012) Integrated analysis of somatic mutations and focal copy-number changes identifies key genes and pathways in hepatocellular carcinoma. Nat Genet 44:694–698

He YJ, McCall CM, Hu J et al (2006) DDB1 functions as a linker to recruit receptor WD40 proteins to CUL4-ROC1 ubiquitin ligases. Genes Dev 20:2949–2954

Hodgson AJ, Hyser JM, Keasler VV et al (2012) Hepatitis B virus regulatory HBx protein binding to DDB1 is required but is not sufficient for maximal HBV replication. Virology 426:73–82

Höhne M, Schaefer S, Seifer M et al (1990) Malignant transformation of immortalized transgenic hepatocytes after transfection with hepatitis B virus DNA. EMBO J 9:1137–1145

Huo TI, Wang XW, Forgues M et al (2001) Hepatitis B virus X mutants derived from human hepatocellular carcinoma retain the ability to abrogate p53-induced apoptosis. Oncogene 20:3620–3628

Jackson S, Xiong Y (2009) CRL4 s: the CUL4-RING E3 ubiquitin ligases. Trends Biochem Sci 34:562–570

Keasler VV, Hodgson AJ, Madden CR et al (2007) Enhancement of hepatitis B virus replication by the regulatory X protein in vitro and in vivo. J Virol 81:2656–2662

Kim CM, Koike K, Saito I et al (1991) HBx gene of hepatitis B virus induces liver cancer in transgenic mice. Nature 351:317–320

Kim S, Park SY, Yong H et al (2008) HBV X protein targets hBubR1, which induces dysregulation of the mitotic checkpoint. Oncogene 27:3457–3464

Kumar V, Jayasuryan N, Kumar R (1996) A truncated mutant (residues 58–140) of the hepatitis B virus X protein retains transactivation function. Proc Natl Acad Sci USA 93:5647–5652

Lee TH, Elledge SJ, Butel JS (1995) Hepatitis B virus X protein interacts with a probable DNA repair protein. J Virol 69:1107–1114

Leupin O, Bontron S, Schaeffer C et al (2005) Hepatitis B virus X protein stimulates viral genome replication via a DDB1-dependent pathway distinct from that leading to cell death. J Virol 79:4238–4245

Leupin O, Bontron S, Strubin M (2003) Hepatitis B virus X protein and simian virus 5 V protein exhibit similar UV-DDB1 binding properties to mediate distinct activities. J Virol 77:6274–6283

Li SK, Ho SF, Tsui KW et al (2008) Identification of functionally important amino acid residues in the mitochondria targeting sequence of hepatitis B virus X protein. Virology 381:81–88

Li T, Robert EI, van Breugel PC et al (2010) A promiscuous alpha-helical motif anchors viral hijackers and substrate receptors to the CUL4-DDB1 ubiquitin ligase machinery. Nat Struct Mol Biol 17:105–111

Lin Y, Nomura T, Yamashita T et al (1997) The transactivation and p53-interacting functions of hepatitis B virus X protein are mutually interfering but distinct. Cancer Res 57:5137–5142

Lin-Marq N, Bontron S, Leupin O et al (2001) Hepatitis B virus X protein interferes with cell viability through interaction with the p127-kDa UV-damaged DNA-binding protein. Virology 287:266–274

Liu S, Zhang H, Gu C et al (2009) Associations between hepatitis B virus mutations and the risk of hepatocellular carcinoma: a meta-analysis. J Natl Cancer Inst 101:1066–1082

Lizzano RA, Yang B, Clippinger AJ et al (2011) The C-terminal region of the hepatitis B virus X protein is essential for its stability and function. Virus Res 155:231–239

Lucifora J, Arzberger S, Durantel D et al (2011) Hepatitis B virus X protein is essential to initiate and maintain virus replication after infection. J Hepatol 55:996–1003

Luo N, Cai Y, Zhang J et al (2012) The C-terminal region of the hepatitis B virus X protein is required for its stimulation of HBV replication in primary mouse hepatocytes. Virus Res 165:170–178

Ma NF, Lau SH, Hu L et al (2008) COOH-terminal truncated HBV X protein plays key role in hepatocarcinogenesis. Clin Cancer Res 14:5061–5068

Madden CR, Finegold MJ, Slagle BL (2001) Hepatitis B virus X protein acts as a tumor promoter in development of diethylnitrosamine-induced preneoplastic lesions. J Virol 75:3851–3858

Maguire HF, Hoeffler JP, Siddiqui A (1991) HBV X protein alters the DNA binding specificity of CREB and ATF-2 by protein-protein interactions. Science 252:842–844

Malmassari SL, Deng Q, Fontaine H et al (2007) Impact of hepatitis B virus basic core promoter mutations on T cell response to an immunodominant HBx-derived epitope. Hepatology 45:1199–1209

Martin-Lluesma S, Schaeffer C, Robert EI et al (2008) Hepatitis B virus X protein affects S phase progression leading to chromosome segregation defects by binding to damaged DNA binding protein 1. Hepatology 48:1467–1476

Mathonnet G, Lachance S, Alaoui-Jamali M et al (2004) Expression of hepatitis B virus X oncoprotein inhibits transcription-coupled nucleotide excision repair in human cells. Mutat Res 554:305–318

Matsuzaki Y, Chiba T, Hadama T et al (1997) HBV genome integration and genetic instability in HBsAg-negative and anti-HCV-positive hepatocellular carcinoma in Japan. Cancer Lett 119:53–61

Moolla N, Kew M, Arbuthnot P (2002) Regulatory elements of hepatitis B virus transcription. J Viral Hepat 9:323–331

Murakami S (1999) Hepatitis B virus X protein: structure, function and biology. Intervirology 42:81–99

Murakami Y, Saigo K, Takashima H et al (2005) Large scaled analysis of hepatitis B virus (HBV) DNA integration in HBV related hepatocellular carcinomas. Gut 54:1162–1168

Ott JJ, Stevens GA, Groeger J et al (2012) Global epidemiology of hepatitis B virus infection: new estimates of age-specific HBsAg seroprevalence and endemicity. Vaccine 30:2212–2219

Pagano JS, Blaser M, Buendia MA et al (2004) Infectious agents and cancer: criteria for a causal relation. Sem Cancer Biol 14:453–471

Papatheodoridis GV, Lampertico P, Manolakopoulos S et al (2010) Incidence of hepatocellular carcinoma in chronic hepatitis B patients receiving nucleos(t)ide therapy: a systematic review. J Hepatol 53:348–356

Parkin DM (2006) The global health burden of infection-associated cancers in the year 2002. Int J Cancer 118:3030–3044

Pollicino T, Belloni L, Raffa G et al (2006) Hepatitis B virus replication is regulated by the acetylation status of hepatitis B virus cccDNA-bound H3 and H4 histones. Gastroenterology 130:823–837

Poussin K, Dienes H, Sirma H et al (1999) Expression of mutated hepatitis B virus X genes in human hepatocellular carcinomas. Int J Cancer 80:497–505

Raimondo G, Allain JP, Brunetto MR et al (2008) Statements from the Taormina expert meeting on occult hepatitis B virus infection. J Hepatol 49:652–657

Rakotomalala L, Studach L, Wang WH et al (2008) Hepatitis B virus X protein increases the Cdt1-to-geminin ratio inducing DNA re-replication and polyploidy. J Biol Chem 283:28729–28740

Rui E, Moura PR, Goncalves Kde A et al (2005) Expression and spectroscopic analysis of a mutant hepatitis B virus onco-protein HBx without cysteine residues. J Virol Methods 126:65–74

Saigo K, Yoshida K, Ikeda R et al (2008) Integration of hepatitis B virus DNA into the myeloid/ lymphoid or mixed-lineage leukemia (MLL4) gene and rearrangements of MLL4 in human hepatocellular carcinoma. Hum Mutat 29:703–708

Scaglione SJ, Lok AS (2012) Effectiveness of hepatitis B treatment in clinical practice. Gastroenterology 142(1360–1368):e1361

Schaefer S (2005) Hepatitis B virus: significance of genotypes. J Viral Hepat 12:111–124

Sirma H, Giannini C, Poussin K et al (1999) Hepatitis B virus X mutants, present in hepatocellular carcinoma tissue abrogate both the antiproliferative and transactivation effects of HBx. Oncogene 18:4848–4859

Sitterlin D, Bergametti F, Tiollais P et al (2000a) Correct binding of viral X protein to UVDDB-p127 cellular protein is critical for efficient infection by hepatitis B viruses. Oncogene 19:4427–4431

Sitterlin D, Bergametti F, Transy C (2000b) UVDDB p127-binding modulates activities and intracellular distribution of hepatitis B virus X protein. Oncogene 19:4417–4426

Sitterlin D, Lee TH, Prigent S et al (1997) Interaction of the UV-damaged DNA-binding protein with hepatitis B virus X protein is conserved among mammalian hepadnaviruses and restricted to transactivation-proficient X-insertion mutants. J Virol 71:6194–6199

Soussan P, Garreau F, Zylberberg H et al (2000) In vivo expression of a new hepatitis B virus protein encoded by a spliced RNA. J Clin Invest 105:55–60

Studach L, Wang WH, Weber G et al (2010) Polo-like kinase 1 activated by the hepatitis B virus X protein attenuates both the DNA damage checkpoint and DNA repair resulting in partial polyploidy. J Biol Chem 285:30282–30293

Studach LL, Rakotomalala L, Wang WH et al (2009) Polo-like kinase 1 inhibition suppresses hepatitis B virus X protein-induced transformation in an in vitro model of liver cancer progression. Hepatology 50:414–423

Su Q, Schröder CH, Hofman WJ et al (1998) Expression of hepatitis B virus X protein in HBV-infected human livers and hepatocellular carcinoma. Hepatology 27:1109–1120

Summers J, Smith PM, Huang M et al (1991) Morphogenetic and regulatory effects of mutations in the envelope proteins of an avian hepadnavirus. J Virol 65:1310–1317

Sung WK, Zheng H, Li S et al (2012) Genome-wide survey of recurrent HBV integration in hepatocellular carcinoma. Nat Genet 44:765–769

Takahashi K, Akahane Y, Hino K et al (1998) Hepatitis B virus genomic sequence in the circulation of hepatocellular carcinoma patients: comparative analysis of 40 full-length isolates. Arch Virol 143:2313–2326

Tamori A, Yamanishi Y, Kawashima S et al (2005) Alteration of gene expression in human hepatocellular carcinoma with integrated hepatitis B virus DNA. Clin Cancer Res 11:5821–5826

Tang H, Delgermaa L, Huang F et al (2005) The transcriptional transactivation function of HBx protein is important for its augmentation role in hepatitis B virus replication. J Virol 79:5548–5556

Terradillos O, Billet O, Renard CA et al (1997) The hepatitis B virus X gene potentiates c-myc-induced liver oncogenesis in transgenic mice. Oncogene 14:395–404

Toh H, Hyashida H, Miyata T (1983) Sequence homology between retroviral reverse transcriptase and putative polymerases of hepatitis B virus and cauliflower mosaic virus. Nature 305:827–829

Tu H, Bonura C, Giannini C et al (2001) Biological impact of natural COOH-terminal deletions of hepatitis B virus X protein in hepatocellular carcinoma tissues. Cancer Res 61:7803–7810

Vivekanandan P, Thomas D, Torbenson M (2009) Methylation regulates hepatitis B viral protein expression. J Infect Dis 199:1286–1291

Wang HC, Huang W, Lai MD et al (2006) Hepatitis B virus pre-S mutants, endoplasmic reticulum stress and hepatocarcinogenesis. Cancer Sci 97:683–688

Wang Q, Zhang T, Ye L et al (2012) Analysis of hepatitis B virus X gene (HBx) mutants in tissues of patients suffered from hepatocellular carcinoma in China. Cancer Epidemiol 36:369–374

Wang WH, Studach LL, Andrisani OM (2011) Proteins ZNF198 and SUZ12 are down-regulated in hepatitis B virus (HBV) X protein-mediated hepatocyte transformation and in HBV replication. Hepatology 53:1137–1147

Wen Y, Golubkov VS, Strongin AY et al (2008) Interaction of hepatitis B viral oncoprotein with cellular target HBXIP dysregulates centrosome dynamics and mitotic spindle formation. J Biol Chem 283:2793–2803

Wentz MJ, Becker SA, Slagle BL (2000) Dissociation of DDB1-binding and transactivation properties of the hepatitis B virus X protein. Virus Res 68:87–92

Yeh CT, Shen CH, Tai DI et al (2000) Identification and characterization of a prevalent hepatitis B virus X protein mutant in Taiwanese patients with hepatocellular carcinoma. Oncogene 19:5213–5220

Prevention of Hepatitis B Virus Infection and Liver Cancer

Mei-Hwei Chang

Abstract

Hepatocellular carcinoma (HCC) is one of the five leading causes of cancer death in human. Hepatitis B virus (HBV) is the most common etiologic agent of HCC in the world, particularly in areas prevalent for HBV infection such as Asia, Africa, southern part of Eastern and Central Europe, and the Middle East. Risk factors of HBV-related HCC include (1) viral factors—persistent high viral replication, HBV genotype C or D, pre-S2 or core promoter mutants; (2) host factors—older age (>40 years old) at HBeAg seroconversion, male gender; (3) mother-to-infant transmission; and (4) other carcinogenic factors— smoking, habitual use of alcohol, etc. Prevention is the best way to control cancer. There are three levels of liver cancer prevention, i.e., primary prevention by HBV vaccination targeting the general population, secondary prevention by antiviral agent for high-risk subjects with chronic HBV infection, and tertiary prevention by antiviral agent to prevent recurrence for patients who have been successfully treated for liver cancer. Primary prevention by hepatitis B vaccination is most cost effective. Its cancer preventive efficacy supports it as the first successful example of cancer preventive vaccine in human. This experience can be extended to the development of other cancer preventive vaccine. Careful basic and clinical research is needed to develop ideal vaccines to induce adequate protection. Understanding the main transmission route and age at primary infection may help to set the optimal target age to start a new cancer preventive vaccination program. Besides timely HBVvaccination, the earlier administration of hepatitis B immunoglobulin immediately after birth, and even antiviral agent during the third trimester of pregnancy to block

M.-H. Chang (✉)
Department of Pediatrics, College of Medicine, National Taiwan University, Taipei, Taiwan
e-mail: changmh@ntu.edu.tw

M. H. Chang and K.-T. Jeang (eds.), *Viruses and Human Cancer*,   75
Recent Results in Cancer Research 193, DOI: 10.1007/978-3-642-38965-8_5,
© Springer-Verlag Berlin Heidelberg 2014

mother-to-infant transmission of HBV are possible strategies to enhance the prevention efficacy of HBV infection and its related liver cancer.

Keywords

Hepatocellular carcinoma · Hepatitis B virus · Mother-to-infant transmission · Hepatitis B vaccination · Primary cancer prevention · Cancer preventive vaccine

Contents

1   Introduction

Hepatocellular carcinoma (HCC) is one of the five leading cancers in the world (Parkin et al. 2001). Because of its high fatality (overall ratio of mortality to incidence of 0.93), liver cancer is the third most common cause of death from cancer worldwide. There were an estimated 694,000 deaths from liver cancer in 2008 (477,000 in men, 217,000 in women) (Source: www.globocan.iarc.fr/) (accessed March 13, 2012). Persistent infection with hepatitis B virus (HBV) or hepatitis C virus (HCV) is associated with approximately 90 % of HCC. Data from epidemiologic studies, case–control studies, animal experiments, and molecular biology studies, all support the important oncogenic role, either directly or indirectly, of HBV and HCV in HCC. As evidenced by the large population infected with HBV in the developing world, HBV remains the most prevalent oncogenic virus for HCC in humans. HBV is estimated to cause around 55–70 % of HCC worldwide, while HCV accounts for around 25 % of HCC (Bosch and Ribes 2002). Liver cirrhosis is a common precancerous lesion, accounting for approximately 80 % of patients with HCC, including children (Hsu et al. 1983). This sequela usually results from severe liver injury caused by chronic HBV or HCV infection.

Hepatitis B immunization provides the first evidence to support the success of human cancer prevention by vaccination against an oncogenic infectious agent. In-depth research in epidemiology, transmission routes, clinical long-term follow-up of the consequence of persistent HBV infection, virology, and oncogenic mechanisms of HBV provide the most helpful example in the development of effective strategies for the control of infection related cancer.

2   Disease Burden of HCC

Globally, an estimated 57 % of cases of liver cirrhosis and 78 % of cases of primary liver cancer result from hepatitis B or C virus infection. About 2 billion people have been infected with HBV worldwide, of whom more than 350 million are chronically infected, and between 500,000 and 700,000 people die annually as a result of HBV infection (http://www.who.int/immunization/topics/hepatitis/en/) (accessed January 26, 2012). The high incidence areas of HCC are mainly in developing regions, such as Eastern and Southeastern Asia, Middle and Western Africa (Ferlay et al. 2010. The geographic distribution of the mortality rates is similar to that observed for incidence. Even in the same country, different ethnic groups may have varied incidence of HCC. The annual incidence of HCC in Alaskan Eskimo males was 11.2 per 100,000, five times that of white males in the USA (Heyward et al. 1981).

The world geographic distribution of HCC overlaps well with that of the distribution for chronic HBV infection (Beasley 1982). Regions with a high prevalence of HBV infection also have high rates of HCC. HBV causes 60–80 % of the primary liver cancer, which accounts for one of the three major cancer deaths, particularly in areas highly prevalent for HBV infection, such as Asia, the Pacific Rim, and Africa (Bosch and Ribes 2002). The southern parts of Eastern and Central Europe, the Amazon basin, the Middle East, and the Indian subcontinent are also areas with high prevalence of HBV infection and HCC (Lavanchy 2004).

3   Transmission Routes of Hepatitis B Virus Infection

HBV infection is most prevalent in Asia, Africa, Southern Europe, and Latin America, where the HBsAg-positive rates in the general population ranged between 2 and 20 %. The age and source of primary HBV infection are important factors affecting the outcome of infection.

Maternal serum HBsAg and hepatitis B e antigen (HBeAg) status affect the outcome of HBV infection in their offspring. In Asia and many other endemic areas, before the era of universal HBV immunization, perinatal transmission through HBsAg carrier mothers accounts for 40–50 % of HBsAg carriers. Irrespective of the extent of HBsAg carrier rate in the population, around 85–90 % of the infants of HBeAg-seropositive carrier mothers became HBsAg carriers (Stevens et al. 1975). In endemic areas, HBV infection occurs mainly during infancy and early childhood. In contrast to the infection in adults, HBV infection during early childhood results in a much higher rate of persistent infection and long-term serious complications, such as liver cirrhosis and HCC. Taking Taiwan as an example of endemic areas, the HBsAg carrier rate is highest in chronic HBsAg carriers resulted from infections before 2 years of age in this population (Hsu et al. 1986).

Horizontal transmission, which accounts for another half of the transmission route before the HBV vaccination era, is through the use of unsterile needles or medical equipment, unsafe blood product or transfusions, unprotected sex, unsterile skin piercing, or other intrafamilial close contact. In Africa, where HBV infection is also endemic, horizontal infection during early childhood is the main route of transmission. In rural Senegal, by the age of 2 years, 25 % of children are infected, while at age 15, the infection rate rises to 80 % (Feret et al. 1987).

---

4 Chronic Hepatitis B Virus Infection and Liver Cancer

Liver injury caused by chronic HBV infection is the most important initiation event of hepatocarcinogenesis (Bruix et al. 2004). The role of HBV in tumor formation appears to be complex and may involve both direct and indirect mechanisms of carcinogenesis (Grisham 2001; Villanueva et al. 2007) (see also Chap. 4). The outcome of persistent HBV infection is affected by the interaction of host, viral, and environmental factors (Table 1).

4.1 Viral (HBV) Risk Factors for HCC

4.1.1 Seropositive HBsAg

Chronic HBV infection with persistent positive serum HBsAg is the most important determinant for HCC. A prospective general population study of 22,707 men in Taiwan showed that the incidence of HCC among subjects with chronic HBV infection is much higher than among non-HBsAg carriers during long-term follow-up. The relative risk is 66. These findings support the hypothesis that HBV has a primary role in the etiology of HCC (Beasley et al. 1981).

Table 1 Summary of risk factors for progression to HCC in HBV-infected individuals

Risk factors	High risk/low risk	References
Viral factors		
1. HBsAg	Positive/negative = 66/1	Beasley et al. (1981)
2. HBeAg in HBsAg-positive persons	Positive/negative = 60/10	Yang et al. (2002)
3. HBV DNA level	High {[>10^6]/10^5 ~ 10^6/ [10^4 ~ <10^5]}/low [<10^4]copies/ ml = 11/9/3/1	Chen et al. (2006)
4. HBV genotype	[C or D]/[A or B]	Tseng et al. (2012)
Host factors		
1. Age	>40/<40 years = 2–12/1	Chen et al. (2008), Tseng et al. (2012)
2. Age at HBeAg seroconversion	Older (>40 Years)/younger (<30 Years) = 5/1	Chen et al. (2010)
3. Ethnic group	Asian or African/others	Ferlay et al. (2010)
4. Gender	Male/female = 2–4/1	Ferlay et al. (2010), Ni et al. (1991), Schafer and Sorrell (1999)
5. Family HCC history	Positive/negative = 2–3/1	Turati et al. (2012)
6. Liver cirrhosis	Yes/no = 12/1	Yu et al. (1997)
7. Maternal HBsAg	Positive/negative = 30/1	Chang et al. (2009)
Other factors		
Smoking	Yes/no = 1–2/1	Yu et al. (1997), Jee et al. (2004)
Habitual alcohol	Yes/no = 1–2/1	Yu et al. (1997), Jee et al. (2004)

4.1.2 Active Viral Replication

HBeAg is a marker of active HBV replication. Chronic HBV-infected subjects with prolonged high HBV replication levels or positive HBeAg after 30 years of age have a higher risk of developing HCC during follow-up. Those HBsAg carriers with persistent seropositive HBeAg have 3–6 times higher risk of developing HCC than those with negative serum HBeAg (Yang et al. 2002) (Table 1). Higher HBV DNA levels predict higher rates of HCC in those with chronic HBV infection. In comparison with those with serum HBV DNA level < 10^4 copies/ml, those with greater serum HBV DNA levels [10^4 ~ <10^5], [10^5 ~ 10^6], or [>10^6] copies/ml have a higher risk of HCC [2.7, 8.9, or 10.7] during long-term follow-up (Chen et al. 2006).

4.1.3 HBV Genotype

There are at least ten genotypes of HBV identified with geographic variation.

Those with HBV genotype C or D infection have a higher risk of developing HCC than those infected with genotype A or B HBV (Tseng et al. 2012). In Alaska, those infected with genotype F have a higher risk of HCC than other genotypes (Livingston et al. 2007).

4.1.4 HBV Mutants

The presence of pre-S mutants carries a high risk of HCC in HBV carriers and was proposed to play a potential role in HBV-related hepatocarcinogenesis (Wang et al. 2006). Subjects infected with HBV core promoter mutants were reported to have a higher risk of developing HCC.

4.2 Host Factors for HCC (Table 1)

4.2.1 Age Effect

Older age (>40 years) is a risk factor for HCC development (Tseng et al. 2012; Chen et al. 2008). It is very likely due to the accumulation of genetic alterations with gain or loss of genes and liver injury with time during chronic HBV infection. HCC patients are mostly (around 80 %) anti-HBe seropositive at diagnosis (Chien et al. 1981). This implies that HCC occurs after long-term HBV infection and liver injury, and that the patients have seroconverted to anti-HBe. Chronic HBV-infected patients with delayed HBeAg seroconversion after age 40 have significantly higher risk of developing HCC (hazard ratio 5.22), in comparison with patients with HBeAg seroconversion before the age of 30 (Chen et al. 2010).

4.2.2 Male Predominance

There is a strong male predominance in HBV-related HCC, with a male to female ratio of 2–4:1, even in children (Ferlay et al. 2010; Ni et al. 1991; Schafer and Sorrell 1999). Male gender is a risk factor for the development of HCC, but the mechanisms are not fully understood. The higher activity of androgen pathway functions as a tumor-promoting factor in male hepatocarcinogenesis, and the higher activity of the estrogen pathway functions as a tumor-suppressing factor in female hepatocarcinogenesis. As both mechanisms function in a ligand-dependent manner, both the ligand and the receptor of these sex hormones are suggested to be included in assessing the relative risk of HCC patients of each gender (Yeh and Chen 2012). Additionally, the RNA-binding motif (RRM) gene on Y chromosome (RBMY), which encodes a male germ cell-specific RNA-binding protein associated with spermatogenesis, is considered as a candidate oncogene specific for male liver cancer (Tsuei et al. 2004).

4.2.3 Other Host Factors

Chronic severe liver injury caused by hepatitis virus or other agents, leading to hepatocyte transformation and finally HCC. Liver cirrhosis is a precancerous lesion for HCC (Yu et al. 1997). Cirrhotic HBV carriers have a 3–8 % annual rate of developing HCC. The risk of HCC also varies in different ethnic groups. In North America as an example, Asian, Hispanic, and African American have higher rates of HCC than non-Hispanic White people (Ferlay et al. 2010).

Those with positive HCC family history have a higher risk of HCC in comparison with those without a positive history of HCC. Familial clustering of HCC suggests the role of genetic predisposing factors in addition to the intrafamilial transmission of HBV infection (Chang et al. 1984). In a meta-analysis, based on nine case–control and four cohort studies, for a total of approximately 3,600 liver cancer cases, the pooled relative risk for family history of liver cancer was 2.50 (95 % CI, 2.06–3.03) (Turati et al. 2012) (Table 1).

4.3   Maternal Effect

Those with positive maternal serum HBsAg have a higher risk of 30 times in developing HCC than those with negative maternal HBsAg (Chang et al. 2009). HBeAg is a soluble antigen produced by HBV. It can cross the placenta barrier from the mother to the infant. Transplacental HBeAg from the mother induces a specific loss of responsiveness of helper T cells to HBeAg and HBcAg in neonates born to HBeAg-positive HBsAg carrier mothers (Hsu et al. 1992). This may help to explain why 85–90 % of the infants of HBeAg-positive carrier mothers became persistently infected (Beasley et al. 1977), while only approximately 5 % of the infants of HBeAg-negative HBsAg carrier mothers became persistently infected. The immune tolerance state persists for years to decades after neonatal HBV infection.

4.4   Environmental/Life Style Factors

Smoking, habitual alcohol drinking, and in some regions aflatoxin exposure are factors which were related to higher risk of HCC (Yu et al. 1997; Jee et al. 2004; Chen et al. 2008).

5   Prevention of Liver Cancer

The prognosis of HCC is grave, unless it is detected early and complete resection or ablation is performed. Even in such cases, *de novo* recurrence of HCC is always a problem. Prevention is thus the best way toward the control of HCC. There are three levels of liver cancer prevention, i.e., primary, secondary, and tertiary prevention of liver cancer (Fig. 1). Universal vaccination to block both mother-to-infant and horizontal transmission routes, starting from neonates to interrupt HBV infection, is the most effective and safe way to prevent HCC.

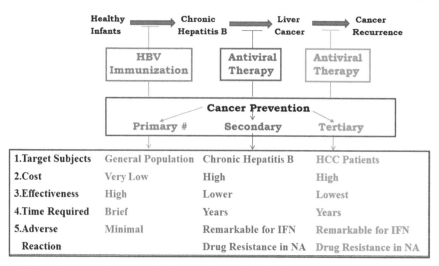

Other primary prevention: Screening blood products and avoid risky behavior, diet, or environment

Fig. 1 Strategies for primary, secondary, and tertiary prevention of liver cancer. HBV immunization is the most effective way. For persons who have been infected by hepatitis virus, antiviral therapy may delay or reduce the risk of developing HCC in a minor degree. The effect of other strategies such as chemoprevention and avoidance of risky behavior is still not confirmed and is under investigation. *HBV* hepatitis B virus; *IFN* interferon; *NA* nucleos(t)ide analog

Antiviral therapy for hepatitis B is aimed at the normalization of the liver enzymes, HBeAg clearance, reduction in HBV DNA levels as well as inflammation and fibrosis in the liver. Studies have shown that a finite course of conventional interferon-α (IFN) therapy may provide long-term benefit for achieving a cumulative response as well as reducing the progression of fibrosis and the development of cirrhosis and/or HCC. Long-term therapy with nucleos(t)ide analogs may also improve fibrosis or reverse advanced fibrosis as well as reduce disease progression and possibly the development of HCC. However, current antiviral therapies and immune modulating agents do not reach a high sustained response rate. Even in those with sustained response, HBV cannot be eradicated from the hosts in the majority of treated cases because it is difficult to eliminate cccDNA by this treatment. Drug resistance after prolonged use of NUC is another problem. Antiviral therapy, therefore, may have some but limited beneficial effect in preventing HBV-related complications including HCC.

A prospective randomized controlled trial of antiviral therapy was conducted in patients with HBV-related cirrhosis. HCC was diagnosed in 3.9 % of lamivudine-treated patients and in 7.4 % of placebo controls after a median follow-up of 32 months ($p = 0.047$) (Liaw et al. 2004). Another retrospective study revealed that the cumulative incidence of HCC at the end of 15 years (median 6.8 years) follow-up in the 233 interferon (IFN)-treated patients and 233 controls was 2.7 versus 12.5 % ($p = 0.011$). Yet, significant reduction in HCC was only observed in patients with preexisting cirrhosis and HBeAg seroconverters (Lin et al. 2007).

The outcome with pegylated IFN (PEG-IFN) and newer nucleoside/nucleotide may be better because of more effective viral suppression effect, and/or a low risk of resistance. However, the treatment outcomes still need to be improved, and more effective, safe, and affordable anti-HBV agents/strategies are needed (Liaw 2011).

For HCC patients who have been treated successfully by surgery, liver transplantation, or local therapy, tertiary prevention of HCC using antiviral therapy against HBV or HCV may potentially prevent late tumor recurrence (Braitenstein et al. 2009). Yet, further study is needed to confirm its efficacy.

Other strategies to prevent HCC such as chemoprevention of high-risk subjects (Jacobson et al. 1997; Egner et al. 2001), prevention of high-risk behavior, changes in environment and/or diet, and liver transplantation for precancerous lesion (e.g., liver cirrhosis) may also be helpful to prevent liver cancer. Yet, more evidence is needed to support their efficacy.

5.1 Primary Prevention of Hepatitis B Virus Infection by Immunization

5.1.1 Universal Hepatitis B Vaccination in Infancy

Currently, there are mainly three strategies of universal immunization programs in the world, depending on the resources and prevalence of HBV infection (Table 2). In countries with adequate resources, such as the USA, pregnant women are screened for HBsAg but not HBeAg. It is recommended that every infant receive three doses of HBV vaccine. In addition, infants of all HBsAg-positive mothers, regardless of HBeAg status, also receive HBIG within 24 h after birth

Table 2 Current pregnant women screening and universal infant hepatitis B virus (HBV) immunoprophylaxis strategies in different countries and proposed surveillance program for high-risk children with breakthrough infection linked to the specific strategies

Strategy type	Pregnant women screening		Neonatal immunization#		
	HBsAg	HBeAg	HBV vaccine	HBIG to children of HBsAg (+)/HBeAg(−) mothers	HBIG to children of HBsAg (+)/HBeAg (+) mothers
I	Yes	No	Yes	Yes	Yes
II*	Yes	Yes	Yes	No	Yes
III	No	No	Yes	No	No

#Examples of applied countries: Strategy type I: USA, Italy, Korea; Strategy type II: Taiwan, Singapore; Strategy type III: Thailand
*In Strategy type II, either simultaneous or sequential HBsAg and HBeAg tests can be applied. All pregnant women are screened for HBsAg and HBeAg at the same time; or all pregnant women are screened for HBsAg, while HBeAg is tested only in those positive for HBsAg; the former strategy is time saving and the latter is budget saving

(Shepard et al. 2006). This strategy saves the cost and the procedure of maternal HBeAg screening but increases the cost of HBIG, which is very expensive.

The first universal hepatitis B vaccination program in the world was launched in Taiwan since July 1984 (Chen et al. 1987). Pregnant women were screened for both serum HBsAg and HBeAg (Strategy II).

To save the cost of screening and HBIG, some countries with intermediate/low prevalence of chronic HBV infection or inadequate resources do not screen pregnant women, and all infants receive three doses of HBV vaccines without HBIG. Using this strategy, the cost of maternal screening and HBIG can be saved. The efficacy of preventing the infants from chronic infection seems satisfactory (Poovorawan et al. 1989).

5.2 Effect of HBV Vaccination on the Reduction in HBV Infection and Related Complications

HBV vaccine has been part of the WHO global immunization, resulting in major declines in acute and chronic HBV infection. Approximately 90 % of the incidence of chronic HBV infection in children has been reduced remarkably in areas where universal HBV vaccination in infancy has been successfully introduced. After the universal vaccination program of HBV, the rate of chronic HBV infection was reduced to one-tenth of that before the vaccination program in the vaccinated infants worldwide. Fulminant or acute hepatitis also has been reduced.

Serial epidemiologic surveys of serum HBV markers were conducted in Taiwan (Hsu et al. 1986; Chen et al. 1996; Ni et al. 2001, 2007). The HBsAg carrier rate decreased significantly from around 10 % before the vaccination program to 0.6–0.7 % afterward in vaccinated children younger than 20 years of age. A similar effect has also been observed in many other countries (Whittle et al. 2002; Jang et al. 2001), where universal vaccination programs have been successfully conducted. Universal HBV vaccination in infancy is more effective than selective immunization for high-risk groups.

The HBV vaccination program has indeed reduced both the perinatal and horizontal transmission of HBV worldwide (Da Villa et al. 1995; Whittle et al. 2002). In the reports from many countries, such as Gambia and Korea, universal vaccination programs have been equally successful. The hepatitis B carrier rate has fallen from 5 to 10 % to less than 1 %, demonstrating that universal vaccination is more effective than selective immunization for high-risk groups (Montesano 2011).

6 Effect of Liver Cancer Prevention by Immunization against Hepatitis B Virus Infection (Table 3)

Current therapies for HCC are not satisfactory. Even with early detection and therapy for HCC, recurrence or newly developed HCC is often a troublesome problem during long-term follow-up. Therefore, vaccination is the best way to

Table 3 Incidence rates of HCC among children 6–19 years old and born before or after universal HBV vaccination program

Age at diagnosis, year	HBV vaccination	Years	Hepatocellular carcinoma		
			No. of HCCs	Incidence rate (per 10^5 person-years)	Rate ratio (95 % CI)
Taiwan		Birth year			
6–19	No	1963–1984	444	0.57	1 (referent)
	Yes	1984–1998	64	0.17	0.30 (0.18–0.42)
Khon Kaen	(Thailand)	Birth year			
>5–18	No	Before 1990	15	0.097	1 (referent)
	Yes	After 1990	3	0.024	0.25
Alaska	(USA)	Diagnostic year			
<20	No	1969–1984		0.7–2.6	
	Yes	1984–1988		2.9	
	Yes	1989–2008		0.0–1.4	

References
1. Taiwan-Chang et al. (2009); 2. Thailand-Wichajarn et al. (2008); 3. Alaska-McMahon et al. (2011)

prevent HBV infection and HCC. Since it usually takes 40 years or longer for HCC to develop after HBV infection in most occasions, we may expect to see the reduction in HCC in adults 40 years after the implementation of the universal HBV vaccination program. Studies on the changes of the incidence of HCC in children born before and after universal HBV vaccination program in a hyperendemic area may facilitate our understanding of the effect of HCC prevention by HBV immunization, if the HCC in children is similar to the HCC in adults.

HCC in children is closely related to HBV infection, and the characteristics are similar to HCC in adults (Chang et al. 1989). In comparison with most other parts of the world, Taiwan has a high prevalence of HBV infection and HCC in children. Children with HCC in Taiwan are nearly 100 % HBsAg seropositive, and most (86 %) of them are HBeAg negative. HCC children are mostly (94 %) having positive maternal HBsAg. Most (80 %) of the non-tumor portion has liver cirrhosis. Integration of HBV genome into host genome was demonstrated in the HCC tissues in children (Chang et al. 1991). The histological features of HCC in children are very similar to that in adults.

The reduction in HBV infection after the launch of universal hepatitis B vaccination program in July 1984 in Taiwan has had a dramatic effect on the reduction in HCC incidence in children. The annual incidence of HCC in children aged 6–14 was reduced to one-fourth from 0.52 to 0.54 per 100,000 children born before July 1984 to 0.13 to 0.20 per 100,000 born after July 1984 (Chang et al. 1997, 2000, 2005).

Approximately 90 % of the mothers of the HCC children with known serum HBsAg status were positive for HBsAg. This provides strong evidence of perinatal transmission of maternal HBV as the main route of HBV transmission in HCC children born after the immunization era and was not effectively eliminated by the HBV immunization program (Chang et al. 2009).

The cancer prevention effect by the universal HBV vaccine program has extended further beyond childhood after 20 years of the program (Table 3). HCC incidence was statistically significantly lower among children and adolescents aged 6–19 years in vaccinated compared with unvaccinated birth cohorts.

The incidence of HCC is significantly lower in Thai children who received hepatitis B vaccine at birth (year of birth after 1990) than unvaccinated children. Cases of liver tumors in children under 18 years old diagnosed during 1985–2007 of Khon Kaen region, Thailand. The age-standardized incidence rates (ASRs) for liver cancer in children >10 years of age of non-vaccinated and vaccinated children were 0.88 and 0.07 per million, respectively ($p = 0.039$) (Wichajarn et al. 2008).

Alaska Native people experience the highest rates of acute and chronic HBV infection and HCC in the USA. Universal newborn HBV vaccination coupled with mass screening and immunization of susceptible Alaska Natives has eliminated HCC among Alaska Native children. The incidence of HCC in persons <20 years decreased from 3/100,000 in 1984–1988 to zero in 1995–1999, and no HCC cases have occurred since 1999 (McMahon et al. 2011).

7 Problems to be Solved in Liver Cancer Prevention

The risk of developing HCC for vaccinated cohorts was statistically significantly associated with incomplete HBV vaccination, prenatal maternal HBsAg or HBeAg seropositivity (Chang et al. 2005, 2009). Failure to prevent HCC results mostly from unsuccessful control of HBV infection by highly infectious mothers. To eradicate HBV infection and its related diseases, we have to overcome the difficulties that hinder the success of universal HBV vaccination.

7.1 Low Coverage Rate

7.1.1 Inadequate Resources in Developing Countries

Failure to attract national government fund delays the integration of HBV vaccination into the EPI program, such as in some countries in Southeast Asia and Africa. Even with integration into the EPI program, the coverage rate is still inadequate. One main reason is that parents still have to pay for the HBV vaccines in those countries.

Since 1991, WHO has called for all countries to add hepatitis B vaccine to their national immunization programs. Yet, many low-income countries do not use the vaccine. In 1992, WHO further recommended that all countries with a high burden of HBV-related diseases should introduce hepatitis B vaccine in their routine infant immunization programs by 1995 and that all countries do so by 1997 (Kane 1996). In 1996, an additional target was added, that is, an 80 % reduction in the incidence of new hepatitis B carriers among children worldwide by 2001 (Kane and Brooks 2002). This is particularly urgent in areas where HBV infection and HCC are prevalent.

How to reduce the cost of the vaccine and to increase funds for HBV vaccination to help children of endemic areas with poor economic conditions are important issues to solve for the eradication of HBV infection and its related liver cancer. The Global Alliance for Vaccines and Immunization (GAVI) was established in 1999. GAVI has contributed very much in helping the developing countries to increase the coverage of HBV vaccination.

Increasing uptake of HBV vaccine was noted globally year by year. According to the recent report of WHO, hepatitis B vaccine for infants was introduced nationwide in 179 countries of the 193 WHO member states (including in parts of India and the Sudan) by the end of 2010. Global hepatitis B vaccine coverage is estimated at 75 % and is as high as 91 % in the Western Pacific and 89 % in the Americas. However, the coverage in the Southeast Asia region only reached 52 % in 2010 (World Health Organization 2012).

7.1.2 Poor Compliance Caused by Anxiety to the Adverse Effects of Vaccination or Ignorance

In countries with adequate resources, the ignorance of the parents/guardians or an anti-vaccine mentality drives some of the people to refuse vaccination. Opposition to vaccination may be reduced by clarification of the vaccine-related side effects. For instance, although there is a lack of supporting evidence, the correlation of central nervous system demyelinating diseases and hepatitis B vaccine has been suspected. Clarification of this question may help to eliminate opposition to vaccination, which hampers the effort of HBV vaccination and, hence, the goal of eradication of HBV and its related liver diseases (Halsey et al. 1999). Education and propagation of the benefits of HBV vaccination will enhance the motivation of the public and the governments to accept HBV vaccination, even in low prevalence areas.

7.2 Breakthrough Infection in Vaccinees

Causes of breakthrough infection or non-responders include high maternal viral load (Lee et al. 1986), intrauterine infection (Tang et al. 1998; Lin et al. 1987), surface gene mutants (Hsu et al. 1999, 2004, 2010), poor compliance, genetic hyporesponsiveness, and immune-compromised status. A positive maternal HBsAg serostatus was found in 89 % of the HBsAg-seropositive subjects born after the launch of the HBV vaccination program in Taiwan (Ni et al. 2007).

Maternal transmission is the primary reason for breakthrough HBV infection and is the challenge that needs to be addressed in future vaccination programs.

7.2.1 Mother-to-Infant Transmission: High Maternal Viral Load and Positive Serum HBeAg and HBsAg

Risk factors of immunoprophylaxis failure include a high level of maternal HBV DNA, uterine contraction and placental leakage during the process of delivery, and low level of maternal anti-HBc (Lin et al. 1991; Chang et al. 1996).

Mother-to-infant transmission is the major cause of HBV infection among immunized children. Among 2,356 Taiwan children born to HBsAg-positive mothers, identified through prenatal maternal screens, children born to HBeAg-positive mothers are at greatest risk for chronic HBV infection (9.26 %), despite HBIG injection <24 h after birth and full course of HBV vaccination in infancy (Chen et al. 2012). Intrauterine HBV infection, though infrequent, is a possible reason.

Although immunoprophylaxis for HBV infection is very successful, still around 2.4 % of infants of HBeAg-positive mothers already had detectable HBsAg in the serum at birth or shortly after birth (Tang et al. 1998) and persisted to 12 months of age and later. They become HBsAg carriers despite complete immunoprophylaxis.

7.2.2 Hepatitis B Surface Gene Mutants

The rate of HBsAg gene mutants in HBsAg carriers born after the vaccination program is increasing with time. The rate of HBV surface gene mutation was 7.8, 17.8, 28.1, and 23.8 %, respectively, among those seropositive for HBV DNA, at before, and 5, 10, and 15 years after the launch of the HBV vaccination program (Hsu et al. 1999, 2010). Fortunately, it has remained stationary (22.6 %) at 20 years after the vaccination program. HBV vaccine covering surface gene mutant proteins is still not urgently needed for routine HBV immunization at present, but careful and continuous monitoring for the surface gene mutants is needed.

7.2.3 Immunocompromised Host, Hyporesponders, or Non-responders

Before receiving immunosuppressant or organ transplantation, hepatitis B markers and anti-HBs need to be monitored routinely. Hepatitis B vaccination should be given to those with inadequate anti-HB levels. A double dose of HBV vaccine can be given to hyporesponders to enhance the vaccine response. Further development of a better vaccine is needed for non-responders to conventional HBV vaccine.

8   Strategies Toward a Successful Control of HBV-Related HCC

Primary prevention by universal vaccination is most cost effective toward a successful control of HBV infection and its complications. Yet currently, there are several problems that remain to be solved. The most important strategy is to provide effective primary prevention to every infant for a better control of HBV infection globally, including further increasing the world coverage rates of HBV vaccine, and better methods to act against breakthrough HBV infection/vaccine non-responsiveness. It is extremely important to find ways to reduce the cost of HBV vaccines and to increase funding for HBV vaccination of children living in developing countries endemic of HBV infection. It is particularly urgent in areas where HBV infection and HCC are prevalent.

Increasing efforts are required to eliminate acute and chronic hepatitis B. Due to the competition of other new vaccines, HBV has not captured sufficient attention from policymakers, advocacy groups, or the general public. This is a major challenge for the future (Van Herck et al. 2008). It is very important to persuade and support the policy makers of countries that still have no universal HBV vaccination program to establish a program, and to encourage the countries which already have a program to increase the coverage rates. A comprehensive public health prevention program should include prevention, detection, and control of HBV infections and its related complications, and evaluation of the effectiveness of prevention activities (Lavanchy 2008).

8.1   Prevention of Breakthrough HBV infection

Further investigation into the mechanisms of breakthrough HBV infection or non-responders is crucial for setting effective strategies to prevent breakthrough infection of HBV. Current HBV vaccine induces good immune response and protection against HBV infection in most vaccines. Yet, approximately 10 % breakthrough infection rate occurs in high-risk infants of HBsAg carrier mothers with positive HBeAg and/or high viral load.

8.1.1   Better Vaccine

Whether the development of new HBV vaccines against surface antigen gene mutants and better vaccines for immune-compromised individuals may further reduce the incidence of new HBV infections requires further investigation.

8.1.2   Treating High-Risk Pregnant Mothers

Nucleoside analog treatment during pregnancy was used in the attempt to prevent perinatal transmission of HBV infection. A pilot study included 8 highly viremic (HBV DNA $> = 1.2 \times 10^9$ copies/mL) mothers treated with lamivudine per day during the last month of pregnancy. At 12 months old, 12.5 % of infants in the lamivudine group and 28 % in the control group were still HBsAg and HBV DNA positive (van Zonneveld et al. 2003). Another clinical trials using lamivudine 100 mg per day was given from 34 weeks of gestation to 4 weeks after delivery for HBsAg seropositive highly viremic mothers (Xu et al. 2009) and demonstrated a reduction in HBsAg-seropositive rate in the infants of the treated group (18 %) in comparison with infants of the control group (39 %) at week 52.

A recent study recruited mothers with positive HBeAg and HBV DNA> 1.0×10^7 copies/ml. The incidence of perinatal transmission was lower in the infants that completed follow-up born to the telbivudine-treated mothers than to the controls (0 vs. 8 %; $p = 0.002$) (Han et al. 2011). Further studies to clarify the benefit and efficacy of nucleoside/nucleotide analogs in the prevention of intrauterine/ perinatal infection are needed.

8.2   Screening High-Risk Subjects and Secondary Prevention of HBV-Related HCC

HBsAg carriers are at high risk for HCC. Screening for serum HBsAg is the first step to early detect the high-risk persons as the target for HCC screening. With limited resources, the priority target subjects to be screened are illustrated in Fig. 2. Subjects with an HBsAg carrier mother or HBsAg carrier family member(s)

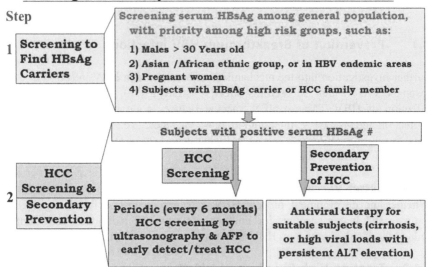

Fig. 2 Screening and secondary prevention of hepatocellular carcinoma. HBsAg carriers are at high risk of developing HCC. So, the first step is screening to find HBsAg-positive persons. # Subjects with positive HBsAg, particularly those with special high risk of HCC, i.e., males >40 years, positive HCC family history, cirrhosis, high viral load with persistent abnormal ALT levels, are the priority target groups to receive periodic HCC screening and secondary prevention of HCC

are at higher risk of chronic HBV infection and HCC. They should be screened for HBsAg, and if positive, further screened for HCC. Screening HBsAg among pregnant women is helpful to interrupt mother-to-infant transmission. Furthermore, those females with positive HBsAg can be followed up regularly to screen for or secondarily prevent HCC.

The HCC risk is higher in HBsAg carriers who are males, over 40 years old, Asian or African ethnic groups, with liver cirrhosis, a family history of HCC, and high HBV DNA >10,000 copies/mL (Table 1). For the HBV carriers at high risk for HCC, periodic (every 6 month) screening of HCC by ultrasonography and alpha-fetoprotein (AFP) is recommended. For those who are living in areas where ultrasound is not readily available, periodic screening with AFP should be considered (Bruix and Sherman 2005).

Secondary prevention of HCC can be considered in high HCC risk patients with chronic HBV infection, such as those with liver cirrhosis, or with high HBV DNA levels (>10,000 copies/mL) and persistent or intermittent abnormal ALT levels.

9   Future Prospects and Implications in Other Cancer Prevention

Prevention is the best way to control cancer. Prevention of liver cancer by hepatitis B vaccination is the first successful example of cancer preventive vaccine in human. With the universal hepatitis B vaccination program starting from neonates in most countries in the world, HBV infection and its complications will be further reduced in this century. It is expected that an effective decline in the incidence of HCC in adults will be achieved in the near future. Furthermore, the impact of HBV vaccination on the control of hepatitis B and its related diseases can be extrapolated to other infectious agent-related cancers, such as human papillomavirus and cervical cancer, Epstein-Barr virus (EBV) and endemic Burkitt's lymphoma, as well as nasopharyngeal carcinoma, human T cell virus type 1 (HTLV-1) and adult T cell leukemia/lymphoma (ATL), and *Helicobacter pylori* and mucosa-associated lymphoid tissue lymphoma (MALToma) or gastric cancer. Besides vaccination, the addition of hepatitis B immunoglobulin immediately after birth, and even antiviral agent during the third trimester of pregnancy to block mother-to-infant transmission of HBV are existing or possible emerging strategies to enhance the prevention efficacy of HBV infection and its related liver cancer.

Acknowledgments The work was supported by grants from the National Health Research Institute, Taiwan.

References

Beasley RP (1982) Hepatitis B virus as the etiologic agent in hepatocellular carcinoma: epidemiologic considerations. Hepatology 2:21s–26s

Beasley RP, Hwang LY, Lin CC et al (1981) Hepatocellular carcinoma and hepatitis B virus. A prospective study of 22707 men in Taiwan. Lancet 2:1129–1133

Beasley RP, Trepo C, Stevens CE et al (1977) The e antigen and vertical transmission of hepatitis B surface antigen. Amer J Epidemiol 105:94–98

Braitenstein S, Dimitroulis D, Petrowsky H et al (2009) Systematic review and meta-analysis of interferon after curative treatment of hepatocellular carcinoma in patients with viral hepatitis. Br J Surg 96:975–981

Bosch FX, Ribes J (2002) The epidemiology of primary liver cancer: global epidemiology. In: Tabor E (ed) Viruses and liver cancer. Elsevier, Amsterdam, pp 1–16

Bruix J, Boix L, Sala M et al (2004) Focus on hepatocellular carcinoma. Cancer Cell 5:215–219

Bruix J, Sherman M (2005) Management of hepatocellular carcinoma AASLD Guideline recommendations for HCC screening. Hepatology 42:1208–1236

Chang MH, Chen CJ, Lai MS et al (1997) Universal hepatitis B vaccination in Taiwan and the incidence of hepatocellular carcinoma in children. N Engl J Med 336:1855–1859

Chang MH, Chen DS, Hsu HC et al (1989) Maternal transmission of hepatitis B virus in childhood hepatocellular carcinoma. Cancer 64:2377–2380

Chang MH, Chen PJ, Chen JY et al (1991) Hepatitis B virus integration in hepatitis B virus-related hepatocellular carcinoma in childhood. Heptology 13:316–320

Chang MH, Chen TH, Hsu HM et al (2005) Problems in the prevention of childhood hepatocellular carcinoma in the era of universal hepatitis B immunization. Clin Cancer Res 11:7953–7957

Chang MH, Hsu HC, Lee CY et al (1984) Fraternal hepatocellular carcinoma in young children in two families. Cancer 53:1807–1810

Chang MH, Hsu HY, Huang LM et al (1996) The role of transplacental hepatitis B core antibody in the mother-to-infant transmission of hepatitis B virus. J Hepatol 24:674–679

Chang MH, Shau WY, Chen CJ et al (2000) The effect of universal hepatitis B vaccination on hepatocellular carcinoma rates in boys and girls. JAMA 284:3040–3042

Chang MH, You SL, Chen CJ et al (2009) Decreased incidence of hepatocellular carcinoma in hepatitis B vaccinees: a 20-year follow-up study. J Natl Cancer Inst 101:1348–1355

Chen CJ, Yang HI, Su J et al (2006) Risk of hepatocellular carcinoma across a biological gradient of serum hepatitis B virus DNA level. JAMA 295:65–73

Chen CL, Yang HI, Yang WS et al (2008) Metabolic factors and risk of hepatocellular carcinoma by chronic hepatitis B/C infection: a follow-up study in Taiwan. Gastroenterology 135:111–121

Chen DS, Hsu NHM, Sung JL et al (1987) A mass vaccination program in Taiwan against hepatitis B virus infection in infants of hepatitis B surface antigen-carrier mothers. JAMA 257:2597–2603

Chen HL, Chang MH, Ni YH et al (1996) Seroepidemiology of hepatitis B virus infection in children: ten years after a hepatitis B mass vaccination program in Taiwan. JAMA 276:906–908

Chen HL, Lin LH, Hu FC et al (2012) Effects of maternal screening and universal immunization to prevent mother- to-infant transmission of HBV. Gastroenterology 142:773–781

Chen YC, Chu CM, Liaw YF (2010) Age- specific prognosis following spontaneous hepatitis B e antigen seroconversion in chronic hepatitis B. Hepatology 51:435–444

Chien MC, Tong MJ, Lo KJ et al (1981) Hepatitis B viral markers in patients with primary hepatocellular carcinoma in Taiwan. J Natl Cancer Inst 66:475–479

Da Villa G, Picciottoc L, Elia S et al (1995) Hepatitis B vaccination: universal vaccination of newborn babies and children at 12 years of age versus high risk groups: a comparison in the field. Vaccine 13:1240–1243

Egner PA, Wang JB, Shu YR et al (2001) Chlorophyline intervention reduces aflatoxin-DNA adducts in individuals at high risk for liver cancer. Proc Natl Acad Sci USA 98:14601–14606

Feret E, Larouze B, Diop B et al (1987) Epidemiology of hepatitis B virus infection in the rural community of tip, Senegal. Am J Epidemiol 125:140–149

Ferlay J, Parkin DM, Curado MP, et al (2010) Cancer incidence in five continents, Volumes I to IX: IARC CancerBase No. 9 [Internet]. Lyon, France: International Agency for Research on Cancer; 2010. [http://ci5.iarc.fr, accessed 26 Jan 2012]

Grisham JW (2001) Molecular genetic alterations in primary hepatocellular neoplasms. In: Coleman WB, Tsongalis GJ (eds) The molecular basis of human cancer. Humana Press, Totowa, pp 269–346

Han GR, Cao MK, Zhao W et al (2011) A prospective and open-label study for the efficacy and safety of telbivudine in pregnancy for the prevention of perinatal transmission of hepatitis B virus infection. J Hepatol 55:1215–1221

Halsey NA, Duclos P, van Damme P et al (1999) Hepatitis B vaccine and central nervous system demyelinating diseases. Pediatrc Infect Dis J 18:23–24

Heyward WL, Lanier AP, Bender TR et al (1981) Primary hepatocellular carcinoma in Alaskan natives, 1969–1979. Int J Cancer 28:47–50

Hsu HY, Chang MH, Ni YH, Chen HL (2004) Survey of hepatitis B surface variant infection in children 15 years after nationwide vaccination program in Taiwan. Gut 53:1499–503

Hsu HC, Lin WS, Tsai MJ (1983) Hepatitis B surface antigen and hepatocellular carcinoma in Taiwan. With special reference to types and localization of HBsAg in the tumor cells. Cancer 52:1825–1832

Hsu HY, Chang MH, Chen DS et al (1986) Baseline seroepidemiology of hepatitis B virus infection in children in Taipei, 1984: a study just before mass hepatitis B vaccination program in Taiwan. J Med Virol 18:301–307

Hsu HY, Chang MH, Hsieh KH et al (1992) Cellular immune response to hepatitis B core antigen in maternal-infant transmission of hepatitis B virus. Hepatology 15:770–776

Hsu HY, Chang MH, Liaw SH et al (1999) Changes of hepatitis B surface variants in carrier children before and after universal vaccination in Taiwan. Hepatology 30:1312–1317

Hsu HY, Chang MH, Ni YH et al (2010) Twenty-year trends in the emergence of hepatitis B surface antigen variants in children and adolescents after universal vaccination in Taiwan. J Infect Dis 201:1192–1200

Jacobson LP, Zhang BC, Shu YR et al (1997) Oltipratz chemoprevention trial in Qidong, People's Republic of China: study design and clinical outcomes. Cancer Epidemiol Biomarkers Prev 6:257–265

Jang MK, Lee JY, Lee JH et al (2001) Seroepidemiology of HBV infection in South Korea, 1995 through 1999. Korean J Intern Med 16:153–159

Jee SH, Ohrr H, Sull JW et al (2004) Cigarette smoking, alcohol drinking, hepatitis B, and risk for hepatocellular carcinoma in Korea. J Natl Cancer Inst 96:1851–1855

Kane MA (1996) Global status of hepatitis B immunization. Lancet 348:696

Kane MA, Brooks A (2002) New immunization initiatives and progress toward the global control of hepatitis B. Curr Opin Infect Dis 15:465–469

Lavanchy D (2004) Hepatitis B virus epidemiology, disease burden, treatment, and current and emerging prevention and control measures. J Viral Hepatitis 11:97–107

Lavanchy D (2008) Chronic viral hepatitis as a public health issue in the world. Best Pract Res Clin Gastroenterol 22:991–1008

Lee SD, Lo KJ, Wu JC et al (1986) Prevention of maternal-infant hepatitis B virus transmission by immunization: the role of serum hepatitis B virus DNA. Hepatology 6(3):369–73

Liaw YF, Sung JJ, Chow WC et al (2004) Lamivudine for patients with chronic hepatitis B and advanced liver disease. N Engl J Med 351:1521–1531

Liaw YF (2011) Impact of hepatitis B therapy on the long-term outcome of liver disease. Liver Int Suppl 1:117–121

Lin HH, Chang MH, Chen DS et al (1991) Early predictor of the efficacy of immunoprophylaxis against perinatal hepatitis B transmission: analysis of prophylaxis failure. Vaccine 9:457–460

Lin HH, Lee TY, Chen DS et al (1987) Transplacental leakage of HBeAg-positive maternal blood as the most likely route in causing intrauterine infection with hepatitis intrauterine infection with hepatitis B virus. J Pediatr 111:877–881

Lin SM, Yu ML, Lee CM et al (2007) Interferon therapy in HBeAg positive chronic hepatitis reduces progression to cirrhosis and hepatocellular carcinoma. J Hepatol 46:45–52

Livingston SE, Simonetti J, McMahon B et al (2007) Hepatitis B virus genotypes in Alaska Native people with hepatocellular carcinoma: preponderance of genotype F. J Infect Dis 195(5–11):1

McMahon BJ, Bulkow LR, Singleton RJ et al (2011) Elimination of hepatocellular carcinoma and acute hepatitis B in children 25 years after a hepatitis B newborn and catch-up immunization program. Hepatology 54:801–807

Montesano R (2011) Preventing primary liver cancer: the HBV vaccination project in the Gambia (West Africa). Environ Health 10(Suppl 1):S6

Ni YH, Chang MH, Hsu HY et al (1991) Hepatocellular carcinoma in childhood. Clinical manifestations and prognosis. Cancer 68:1737–1741

Ni YH, Chang MH, Huang LM et al (2001) Hepatitis B virus infection in children and adolescents in an hyperendemic area: 15 years after universal hepatitis B vaccination. Ann Intern Med 135:796–800

Ni YH, Huang LM, Chang MH et al (2007) Two decades of universal hepatitis B vaccination in Taiwan: impact and implication for future strategies. Gastroenterology 132:1287–1293

Parkin DM, Bray F, Ferlay J et al (2001) Estimating the world cancer burden: Globocan 2000. Int J Cancer 94:153–156

Poovorawan Y, Sanpavat S, Pongpunlert W et al (1989) Protective efficacy of a recombinant DNA hepatitis B vaccine in neonates of HBe antigen-positive mothers. JAMA 261:3278–3281

Schafer DF, Sorrell MF (1999) Hepatocellular carcinoma. Lancet 353:1253–1257

Shepard CW, Simard EP, Finelli L et al (2006) Hepatitis B virus infection: epidemiology and vaccination. Epidemiol Rev 28:112–125

Stevens CE, Beasley RP, Tsui J et al (1975) Vertical transmission of hepatitis B antigen in Taiwan. N Engl J Med 292:771–774

Tang JR, Hsu HY, Lin HH et al (1998) Hepatitis B surface antigenemia at birth: a long-term follow-up study. J Pediatr 133:374–377

Tseng TC, Liu CJ, Yang HC et al (2012) High levels of hepatitis B surface antigen increase risk of hepatocellular carcinoma in patients with low HBV load. Gastroenterology 142:1140–1149

Tsuei DJ, Hsu HC, Lee PH et al (2004) RBMY, a male germ cell-specific RNA-binding protein, activated in human liver cancers and transforms rodent fibroblasts. Oncogene 29(23):5815–5822

Turati F, Edefonti V, Talamini R et al (2012) Family history of liver cancer and hepatocellular carcinoma. Hepatology 55:1416–1425

Van Herck K, Vorsters A, Van Damme P (2008) Prevention of viral hepatitis (B and C) reassessed. Best Pract Res Clin Gastroenterol 22:1009–1029

van Zonneveld M, van Nunen AB, Niesters HG et al (2003) Lamivudine treatment during pregnancy to prevent perinatal transmission of hepatitis B virus infection. J Viral Hepat 10:294–297

Villanueva A, Newell P, Chiang DY et al (2007) Genomics and signaling pathways in hepatocellular carcinoma. Semin Liver Dis 27:55–76

Wang HC, Huang W, Lai MD et al (2006) Hepatitis B virus pre-S mutants, endoplasmic reticulum stress and hepatocarcinogenesis. Cancer Sci 97:683–688

Whittle H, Jaffar S, Wansbrough M et al (2002) Observational study of vaccine efficacy 14 years after trial of hepatitis B vaccination in Gambian children. BMJ 325:569–573

Wichajarn K, Kosalaraksa P, Wiangnon S (2008) Incidence of hepatocellular carcinoma in children in Khon Kaen before and after national hepatitis B vaccine program. Asian Pac J Cancer Prev 9:507–509

World Health Organization (2012) Data, statistics and graphics. Global routine vaccination coverage, (http://www.who.int/immunization_monitoring/data/en/, accessed Mar 2012)

Xu WM, Cui YT, Wang L et al (2009) Lamivudine in late pregnancy to prevent perinatal transmission of hepatitis B virus infection: a multicentre, randomized, double-blind, placebo-controlled study. J Viral Hepat 16:94–103

Yang HI, Lu SN, Liaw YF et al (2002) Hepatitis B e antigen and the risks of hepatocellular carcinoma. N Engl J Med 347:168–174

Yeh SH, Chen PJ (2012) Gender disparity of hepatocellular carcinoma: the roles of sex hormones. Oncology 78(suppl 1):172–179

Yu MW, Hsu FC, Sheen IS et al (1997) Prospective study of hepatocellular carcinoma and liver cirrhosis in asymptomatic chronic hepatitis B virus carriers. Am J Epidemiol 145:1039–1047

The Oncogenic Role of Hepatitis C Virus

Kazuhiko Koike

Abstract

Persistent infection with hepatitis C virus (HCV) is a major risk toward development of hepatocellular carcinoma (HCC). However, it remains controversial in the pathogenesis of HCC associated with HCV whether the virus plays a direct or an indirect role. The observation that chronic hepatitis C patients with sustained high levels of serum alanine aminotransferase are prone to develop HCC suggests the significance of inflammation in hepatocarcinogenesis in hepatitis C. However, the rare development of HCC in patients with autoimmune hepatitis, which is accompanied by robust inflammation, even after the progress into cirrhosis, implies a possibility of the direct role of HCV in HCC development. What is the role of HCV, a simple plus-stranded RNA virus, whose genome is never integrated into the host genome, in hepatocarcinogenesis? The studies using transgenic mouse and cultured cell models, in which the HCV proteins are expressed, indicate the direct pathogenicity of HCV, including oncogenic activities. In particular, the core protein of HCV induces overproduction of oxidative stress by impairing the mitochondrial electron transfer system, through insulting the function of molecular chaperon, prohibitin. HCV also modulates the intracellular signaling pathways including mitogen-activated protein kinase, leading to the acquisition of growth advantage by hepatocytes. In addition, HCV induces disorders in lipid and glucose metabolisms, thereby accelerating the progression of liver fibrosis and HCC development. These results would provide a clue for further

K. Koike (✉)
Department of Gastroenterology, Graduate School of Medicine,
The University of Tokyo, Tokyo, Japan
e-mail: kkoike-tky@umin.ac.jp

M. H. Chang and K.-T. Jeang (eds.), *Viruses and Human Cancer*,
Recent Results in Cancer Research 193, DOI: 10.1007/978-3-642-38965-8_6,
© Springer-Verlag Berlin Heidelberg 2014

understanding of the role of HCV in pathogenesis of persistent HCV infection including hepatocarcinogenesis.

Keywords
Hepatitis C · Hepatocellular carcinoma · Core protein · Oxidative stress · Lipid metabolism · Insulin resistance

Contents

1   Introduction

Worldwide, approximately 170 million people are persistently infected with hepatitis C virus (HCV), which induces a spectrum of chronic liver diseases, from chronic hepatitis, cirrhosis, eventually, to hepatocellular carcinoma (HCC) (Saito et al. 1990). HCV has been given an increasing attention because of its wide and deep penetration in the community, tied with a very high incidence of HCC in persistent HCV infection. Once liver cirrhosis is established in hosts persistently infected with HCV, HCC develops at a yearly rate of approximately 7 % (Ikeda et al. 1998), resulting in the development of HCC in nearly 90 % of HCV-associated cirrhotic patients in 15 years. In addition, the outstanding features in the mode of hepatocarcinogenesis in HCV infection, i.e., development of HCC in a multicentric fashion and in a very high incidence, are not common in other malignancies except for hereditary cancers such as familial polyposis of the colon. The knowledge on the mechanism underlying HCC development in persistent HCV infection, therefore, is imminently required for the prevention of HCC.

2   Hepatitis C Virus and Viral Proteins

The hepatitis C virus is an enveloped RNA virus belonging to the family *Flaviviridae*, and it contains a positive-sense, single-stranded RNA genome of approximately 9,600 nucleotides (nt) within the nucleocapsid (Houghton et al. 1991). The genome consists of a large open reading frame (ORF) encoding a

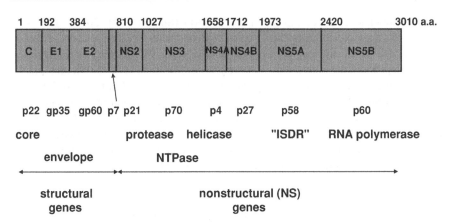

Fig. 1 The structure of hepatitis C virus genome. The HCV genome RNA encodes a polyprotein of 3,010 a.a., which is processed into structural and non-structural proteins by the cellular or viral proteases. One of the structural proteins, the core protein, shows a versatile character in experiments both in vitro and in vivo. ISDR, interferon sensitivity–determining region

polyprotein of approximately 3,010 amino acids (aa) (Fig. 1). The ORF is contiguous to highly conserved untranslated regions (UTRs) at both the 5' and 3' termini. The complete 5' UTR consists of 341 nt and acts as an internal ribosomal entry site. Such feature leads to the translation of the RNA genome using a cap-independent mechanism, rather than ribosome scanning from the 5' end of a capped molecule.

The polyprotein is processed by both the cellular and viral proteases to generate the viral gene products, which are subdivided into structural and non-structural proteins. The structural proteins, which are encoded by the NH$_2$-terminal quarter of the genome, include the core protein and the envelope proteins, E1 and E2. The E2 has an alternative form, E2-p7, though it is not clear whether or not the p7 composes the viral particle. The NS2, NS3, NS4A, NS4B, NS5A, and NS5B are the non-structural proteins that are coded in the remaining portion of the polyprotein. These include serine protease (NS3/4A), NTPase/helicase (NS3), and RNA-dependent RNA polymerase (NS5B).

The core protein of HCV occupies residues 1–191 of the precursor polyprotein and is cleaved between the core and E1 protein by host signal peptidase. The C-terminal membrane anchor of the core protein is further processed by host signal peptide peptidase (Moradpour et al. 2007). The mature core protein is estimated to consist of 177–179 amino acids and shares high homology among HCV genotypes. The HCV core protein possesses the hydrophilic N-terminal region "domain 1" (residues 1–117) followed by a hydrophobic region called "domain 2," which is located from residue 118–170. The domain 1 is rich in basic residues and is implicated in RNA binding and homo-oligomerization. The amphipathic helices I and II spanning from residue 119–136 and residue 148–164, respectively, in domain 2 are involved in the association of HCV core protein with lipid (Boulant et al. 2006). In addition, the region spanning from residue 112–152 is associated with membranes of the endoplasmic reticulum and mitochondria (Suzuki et al. 2005). The core protein is also localized into the nucleus (Miyamoto et al. 2007; Shirakura et al. 2007) and binds to the nuclear proteasome activator PA28γ/REG γ, resulting in PA28γ-dependent degradation of the core protein (Moriishi et al. 2003). Autophagy is involved in the degradation of cellular organelles and the elimination of invasive microorganisms. Disruption of autophagy often leads to several protein deposition diseases. Recently, it has been shown that replication of HCV RNA induces autophagy in a strain-dependent manner, suggesting that HCV harnesses autophagy to circumvent cell death, and dysfunction of autophagy flux may participate in the genotype-specific pathogenesis of HCV (Taguwa et al. 2011).

3   Possible Role of HCV in Hepatocarcinogenesis

The mechanism underlying hepatocarcinogenesis in HCV infection is not fully understood yet, despite the fact that nearly 80 % of patients with HCC in Japan and 30 % of those in the world (Perz et al. 2006) are persistently infected with HCV (Kiyosawa et al. 1990; Saito et al. 1990; Yotsuyanagi et al. 2000). These lines of evidence prompted us to seek for determining the role of HCV in hepatocarcinogenesis. Inflammation induced by HCV should be considered in hepatitis viral infection: necrosis of hepatocytes due to chronic inflammation followed by regeneration enhances genetic aberrations in host cells, the accumulation of which culminates in HCC. This theory presupposes an indirect involvement of HCV in HCC via hepatic inflammation. However, this context leaves us with a serious question: Can inflammation alone result in the development of HCC in such a high incidence (90 % in 15 years) or multicentric nature in HCV infection?

The other role of HCV would have to be weighed against a rare occurrence of HCC in patients with autoimmune hepatitis in which severe inflammation in the liver persists, even after the development of cirrhosis. These backgrounds and reasonings lead to a possible activity of viral proteins for inducing neoplasia. This possibility has been evaluated by introducing genes of HCV into hepatocytes in culture with little success. One of the difficulties in using cultured cells is the carcinogenic capacity of HCV, if any, which would be weak and would take a long time to manifest itself. Actually, it takes 30–40 years for HCC to develop in individuals infected with HCV. On the basis of these viewpoints, investigation was started on carcinogenesis in chronic hepatitis C, in vivo, using transgenic mouse technology.

4   HCV Shows an In Vivo Oncogenic Activity in Mouse Studies

One of the major issues regarding the pathogenesis of HCV-associated liver lesion is whether the HCV proteins have direct effects on pathological phenotypes. Although several strategies have been used to characterize the hepatitis C viral

Table 1 Consequences of the expression of HCV proteins in mice

HCV gene	Genotype	Promoter	Protein expression	Phenotypes	References
Core	1b	HBV	Similar to patients	Steatosis, HCC, insulin resistance, oxidative stress	Moriya et al. (1997, 1998) Moriishi et al. (2003, 2007) Shintani et al. (2004) Miyamoto et al. (2007)
Core	1b	EF-1a	Similar to patients	Steatosis, adenoma, HCC, oxidative stress	Machida et al. (2006)
E1-E2	1b	HBV	Abundant	None in the liver	Koike et al. (1995, 1997)
Core-E1-E2	1b	Albumin	Similar to patients	Steatosis, HCC, oxidative stress	Lerat 2003
Core-E1-E2	1a	CMV	Similar to patients	Steatosis, HCC	Naas et al. (2005)
Structural proteins	1b	MHC	Low in the liver	Hepatitis	Honda et al. (1999)
Entire Polyprotein	1b	Albumin	Only mRNA detectable	Steatosis, HCC	Lerat 2003
Entire Polyprotein	1a	A1-antitrypsin		Steatosis, intrahepatic T-cell recruitment	Alonzi et al. (2004)
NS3/4A	1a	MUP		None (modulation of immunity)	Frelin et al. (2006)
NS5A	1a	ApoE		None (resistance to TNF)	Majumder et al. (2002)

HBV, hepatitis B virus; *EF*, elongation factor; *MUP*, major urinary protein; *Alb*, albumin; *CMV*, cytomegalovirus; *MHC*, major histocompatibility complex; *AT*, antitrypsin; *apo E*, apolipoprotein E

proteins, the relationship between the protein expression and disease phenotype has not been clarified. For this purpose, several lines of mice have been established which were transgenic for the HCV cDNA (Table 1). They include the ones carrying the entire coding region of HCV genome (Lerat et al. 2002), the core region only (Machida et al. 2006; Moriya et al. 1997), the envelope region only (Koike et al. 1995; Pasquinelli et al. 1997), the core and envelope regions (Lerat et al. 2002; Naas et al. 2005), and the core to NS2 regions (Wakita et al. 1998). Although detection of mRNA from the NS regions of the HCV cDNA has been reported (Honda et al. 1999; Lerat et al. 2002), the detection of HCV NS proteins in the transgenic mouse liver has not been successful. The reason for this failure in detecting NS proteins is unclear, but the expression of the NS enzymes may be harmful to mouse development and may allow the establishment of only low-expression mice.

We have engineered transgenic mouse lines carrying the HCV genome where by introducing the genes from the cDNA of the HCV genome of genotype 1b (Moriya et al. 1997, 1998). Established are four different kinds of transgenic mouse lines, which carry the core gene, envelope genes, the entire nonstructural (NS) genes, and NS5A gene, respectively, under the same transcriptional regulatory element. Among these mouse lines, only the transgenic mice carrying the core gene developed HCC in two independent lineages (Moriya et al. 1998). The envelope gene transgenic mice do not develop HCC, despite high expression levels of both E1 and E2 proteins (Koike et al. 1995, 1997), and the transgenic mice carrying the entire NS genes or NS5A gene have developed no HCC.

The core gene transgenic mice, early in life, develop hepatic steatosis, which is one of the histologic characteristics of chronic hepatitis C, along with lymphoid follicle formation and bile duct damages. Thus, the core gene transgenic mouse model well reproduces the feature of chronic hepatitis C. Of note, any pictures of significant inflammation are not observed in the liver of this animal model. Late in life, these transgenic mice develop HCC. Notably, the development of steatosis and HCC has been reproduced by other HCV transgenic mouse lines, which harbor the entire HCV genome or structural genes including the core gene (Lerat et al. 2002; Machida et al. 2006; Naas et al. 2005). These outcomes indicate that the core protein *per se* of HCV has an oncogenic potential when expressed in vivo.

5   HCV Augments Oxidative Stress Production and Modulates Intracellular Signaling

There is a notable feature in the localization of the core protein in hepatocytes; while the core protein predominantly exists in the cytoplasm associated with lipid droplets, it is also present in the mitochondria and nuclei (Moriya et al. 1998). On the basis of this finding, the pathways related to these two organelles, the mitochondria and nuclei, were thoroughly investigated.

One effect of the core protein is an increased production of oxidative stress in the liver. We would like to draw particular attention to the fact that the production of oxidative stress is increased in the core gene transgenic mouse model in the absence of inflammation in the liver. The overproduction of oxidative stress results in the generation of deletions in the mitochondrial and nuclear DNA, an indicator of genetic damage (Moriya et al. 2001a).

Augmentation of oxidative stress is implicated in the pathogenesis of liver disease in HCV infection as shown by a number of clinical and basic studies (Farinati et al. 1995). Reactive oxygen species (ROS) are endogenous oxygen-containing molecules formed as normal products during aerobic metabolism. ROS can induce genetic mutations as well as chromosomal alterations and thus contribute to cancer development in multistep carcinogenesis (Fujita et al. 2008; Kato et al. 2001). Oxidative stress has been shown to be more augmented in hepatitis C than in other types of hepatitis such as hepatitis B (Farinati et al. 1995). Thus, a major role in the pathogenesis of HCV-associated liver disease has been attributed to oxidative stress augmentation, but little has been known about the mechanism of increased oxidative stress in HCV infection. Hence, it is an important issue to understand the mechanism of oxidative stress augmentation, both on generation and scavenging of ROS, which may allow us to develop new tools of therapies for chronic hepatitis C.

Other pathways in hepatocarcinogenesis would be the alteration of the expression of cellular genes and modulation of intracellular signaling pathways. For example, tumor necrosis factor (TNF)-α and interleukin-1β have been found transcriptionally activated (Tsutsumi et al. 2002). The mitogen-activated protein kinase (MAPK) cascade is also activated in the liver of the core gene transgenic mouse model. The MAPK pathway, which consists of three routes, c-Jun N-terminal kinase (JNK), p38, and extracellular signal-regulated kinase (ERK), is involved in numerous cellular events including cell proliferation. In the liver of the core gene transgenic mouse model prior to HCC development, only the JNK route is activated. In the downstream of the JNK activation, transcription factor–activating protein (AP)-1 activation is markedly enhanced (Tsutsumi et al. 2002, 2003). At far downstream, the levels of cyclin D1 and cyclin-dependent kinase (CDK) 4 are increased. Thus, the HCV core protein modulates the intracellular signaling pathways and gives advantage for cell proliferation to hepatocytes. In addition, HCV core protein suppresses the expression of suppressor of cytokine signaling (SOCS)-1, a negative regulator of cytokine signaling pathway, which may work as a tumor suppressor gene (Miyoshi et al. 2005).

Such an effect of the core protein on the MAPK pathway, combined with that on oxidative stress, may explain the extremely high incidence of HCC development in chronic hepatitis C (Fig. 2).

---

6   Mitochondria as Origin of ROS Production in HCV Infection

What is the origin for the increase in oxidative stress in the liver of hepatitis C patients? The core protein is mostly localized to the endoplasmic reticulum (ER), but it is also localized to the mitochondria in cultured cells and transgenic mice (Moriya et al. 1998; Suzuki et al. 2005). In addition, the double structure of mitochondrial membranes is disrupted in hepatocytes of core gene transgenic mice. Evidence suggests that the core protein modulates some mitochondrial functions, including fatty acid β-oxidation, the impairment of which may induce lipid abnormalities and hepatic steatosis. In addition, the mitochondrion is an important source of ROS. In livers of transgenic mice harboring the core gene, increased ROS production has been observed (Moriya et al. 2001a). A recent study found, by the proteomic profiling of biopsy specimens, that impairment in key mitochondrial processes, including fatty acid oxidation and oxidative

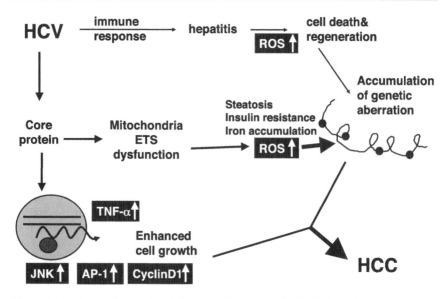

Fig. 2 Molecular pathogenesis of hepatocarcinogenesis in HCV infection. Induction of oxidative stress together with hepatic steatosis by the HCV core protein would play a pivotal role in the development of HCC. Alterations in cellular gene expressions, such as TNF-α, and those in the intracellular signaling pathways including JNK would be co-accelerators to hepatocarcinogenesis in HCV infection. ROS, reactive oxygen species; HCC, hepatocellular carcinoma; TNF-α, tumor necrosis factor-α; JNK, c-Jun N-terminal kinase; AP-1, activating protein-1; ETS, electron transfer system

phosphorylation, and in the response to oxidative stress occurs in HCV-infected human liver with advanced fibrosis (Diamond et al. 2007). Therefore, it is probable that the HCV core protein affects mitochondrial functions, since such pathogenesis is observed in HCV core transgenic mice, cultured cells expressing the core protein (Korenaga et al. 2005), and HCV-infected patients.

The recent progress in proteomics has opened new avenues for disease-related biomarker discovery. We performed a two-dimensional polyacrylamide gel electrophoresis (2D-PAGE) of mitochondria isolated from HepG2 cells stably expressing the HCV core protein and identified several proteins of different expressions when compared with control HepG2 cells. Among upregulated proteins in the core-expressing cells, we focused on prohibitin, which functions as a mitochondrial protein chaperone, and found that the core protein interacts with prohibitin and represses the interaction between prohibitin and subunit proteins of cytochrome C oxidase (COX), which leads to a decrease in the expression level of the proteins and in COX activity.

Prohibitin, a mitochondrial protein chaperone, was identified as an upregulated protein in core-expressing cells. Prohibitin is a ubiquitously expressed and highly conserved protein that was originally determined to play a predominant role in inhibiting cell cycle progression and cellular proliferation by attenuating DNA synthesis (Mishra et al. 2005). It exists in the nucleus and interacts with transcription factors that are vital in cell cycle progression. In core-expressing cells, prohibitin was also detected in the nucleus, and its expression level was also higher than that in control cells. Mitochondrial prohibitin acts as a protein chaperone by stabilizing newly synthesized mitochondrial translation products through direct interaction (Nijtmans et al. 2000). We examined the interaction between prohibitin and mitochondria-encoded subunit II of COX and found a suppressed interaction between these proteins in core-expressing cells. In addition, there are several studies that showed the association of prohibitin with the assembly of mitochondrial respiratory complex I as well as complex IV (COX) (Nijtmans et al. 2000). Complex I also consists of both nuclear and mitochondrial DNA-encoded subunits; therefore, it is probable that the assembly and function of complex I are impaired by the core protein. With respect to the complex I function, we previously found a decreased complex I activity in core-expressing cells. Other groups have also shown that complex I activity is decreased in cultured cells (Piccoli et al. 2007). From these findings, the interaction between prohibitin and the core protein may impair the function of complex I as well as complex IV, leading to an increase in ROS production. In fact, the suppression of the prohibitin function is shown to result in an increased production of ROS (Theiss et al. 2007), a phenomenon observed in core-expressing cells used in this study as well as in the liver of core gene transgenic mice (Moriya et al. 2001a). Very interestingly, the liver-specific deletion of prohibitin resulted in the morphological abnormality and HCC (Ko et al. 2010).

This is a new mechanism for ROS overproduction in viral infection in that HCV induces mitochondrial dysfunction through the inhibition of chaperone function in the mitochondria (Tsutsumi et al. 2009).

7 HCV not only Induces ROS But Attenuates Some Antioxidant System

As discussed above, chronic hepatitis C is characterized by its prominent augmentation of oxidative stress. Related to this, iron accumulation in the liver has been shown to aggravate the oxidative stress as shown by the increase in the amount of DNA adducts in the liver (Farinati et al. 1995). Iron is accumulated in the liver of the HCV core gene transgenic mice (Moriya et al. 2010). The accumulation of iron observed in the liver of the core gene transgenic mice fed with normal chow corroborates well with the observation in chronic hepatitis C patients (Farinati et al. 1995; Fujita et al. 2008). Then, the impact of iron overloading on the oxidant/antioxidant system was examined using this mouse model and cultured cells. Iron overloading caused the induction of ROS as well as antioxidants. However, some of the key antioxidant enzymes, including HO-1 and NADH dehydrogenase, and quinone 1 (NDQ-1), were not augmented sufficiently by iron overloading, while other antioxidant enzymes such as catalase and GST were augmented more strongly in the iron-overloaded core gene transgenic mice than in the iron-overloaded control or non-iron-overloaded core gene transgenic mice. The attenuation of iron-induced augmentation of HO-1 was also confirmed in HepG2 cells expressing the core protein. HO-1 catalyzes the initial and rate-limiting reaction in heme catabolism and cleaves pro-oxidant heme to form biliverdin, which is converted to bilirubin in mammals, both of which have been known to have very strong antioxidant activities (Stocker et al. 1987). In addition, HO-1 has been also suggested to be a central antioxidant under the condition of glutathione depletion. Thus, HO-1 is an essential protective endogenous mechanism against oxidative stress, particularly, in the case of iron overload. Therefore, it is probable that the attenuation of HO-1 and NQO-1 would hamper the antioxidant system and lead to a robust production of oxidative stress in HCV infection.

Thus, HCV infection not only induces ROS but also hampers the antioxidant activation in the liver, thereby exacerbating oxidative stress that would facilitate hepatocarcinogenesis. Aggravation of oxidative stress by overloading of iron was also shown using other transgenic mouse lines carrying HCV genome (Nishina et al. 2008).

8   Metabolic Changes in HCV Infection: Co-factor for Liver Disease Progression

Steatosis is frequently observed in chronic hepatitis C patients and significantly associated with accelerated progression rate of fibrosis of the liver (Powell et al. 2005). The composition of fatty acids that are accumulated in the liver of core gene transgenic mice is different from that in fatty liver due to simple obesity. Carbon 18 mono-unsaturated fatty acids (C18:1) such as oleic or vaccenic acid, which favor the proliferation of cancer cells (Kudo et al. 2011), are significantly increased. This is also the case in the comparison of liver tissues from hepatitis C patients and simple fatty liver patients due to obesity (Moriya et al. 2001b).

The mechanism of steatogenesis in hepatitis C was investigated using this mouse model. At least three pathways are involved in the development of steatosis. One is the frequent presence of insulin resistance in hepatitis C patients as well as in the core gene transgenic mice, which occurs through the inhibition of tyrosine phosphorylation of insulin receptor substrate (IRS)-1 (Shintani et al. 2004). Insulin resistance increases the peripheral release and hepatic uptake of fatty acids, resulting in the accumulation of lipid in the liver. The second pathway is the suppression of the activity of microsomal triglyceride transfer protein (MTP) by HCV core protein (Perlemuter et al. 2002). This inhibits the secretion of very-low-density protein (VLDL) from the liver, yielding an increase in triglycerides in the liver. The last one involves the sterol regulatory element–binding protein (SREBP)-1c, which regulates the production of triglycerides and phospholipids. In HCV core gene transgenic mice, SREBP-1c is upregulated, neither SREBP-2 nor SREBP-1a (Moriishi et al. 2007). This corroborates the results of in vitro studies

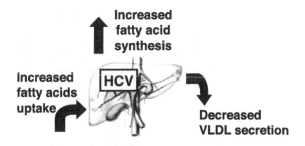

Fig. 3 HCV induces steatosis in the liver by involving three pathways of lipid metabolism. First, HCV core protein induces insulin resistance, leading to the increase in peripheral release and hepatic uptake of fatty acids. Second, HCV core protein suppresses the activity of MTP, inhibiting the secretion VLDL from the liver, yielding an increase in triglycerides in the liver. Lastly, a transcription factor, SREBP-1c, is upregulated by HCV core protein, resulting in an increased production of triglycerides. Thus, the involvement of three pathways easily leads to the development of hepatic steatosis in hepatitis C patients. MTP, microsomal triglyceride transfer protein; VLDL, very-low-density protein; SREBP, sterol regulatory element–binding protein

(Kim et al. 2007; Waris et al. 2007) and a chimpanzee study (Su et al. 2002). Thus, the involvement of three pathways would easily lead to the development of hepatic steatosis in hepatitis C patients (Fig. 3). The presence of steatosis exacerbates the production of ROS and accelerates the progression of liver disease in hepatitis C.

9   Conclusion

The results of HCV mouse studies indicate a carcinogenic activity of the HCV core protein in vivo; thus, HCV would have an oncogenic potential in the liver. In research studies of carcinogenesis, it has been established that the accumulation of a complete set of cellular genetic aberrations is necessary for the development of neoplasia such as colorectal cancer (Kinzler and Vogelstein 1996). They have deduced that mutations in the APC gene for inactivation, those in K-*ras* for activation, and those in the p53 gene for inactivation accumulate, which cooperate toward the development of colorectal cancer. Their theory has been extended to the carcinogenesis of other cancers as well, called "Vogelstein-type" carcinogenesis.

On the basis of the results we obtained for the induction of HCC by the HCV core protein, we would like to present a different mechanism for the hepatocarcinogenesis in HCV infection. We do allow multistages in the induction of all cancers; it would be mandatory for hepatocarcinogenesis that many mutations accumulate in hepatocytes. Some of these steps, however, may be skipped in the development of HCC in HCV infection to which the core protein would contribute. The overall effects achieved by the expression of the viral protein would be the induction of HCC, even in the absence of a complete set of genetic aberrations, required for carcinogenesis (Fig. 4). By considering such a "non-Vogelstein-type" process for the induction of HCC, a reasonable explanation may be given for

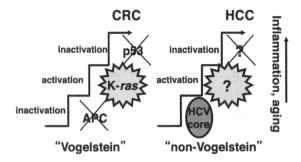

Fig. 4 The role of HCV in hepatocarcinogenesis. Multiple steps are required in the induction of all cancers; it would be mandatory for hepatocarcinogenesis that genetic mutations accumulate in hepatocytes. However, in HCV infection, some of these steps may be skipped in the development of HCC in the presence of the core protein. The effect achieved by the core protein would be one step up in the stairway to HCC, even in the absence of a complete set of genetic aberrations required for carcinogenesis. By considering such a "non-Vogelstein-type" process for the induction of HCC, a reasonable explanation would be given for many unusual modes of hepatocarcinogenesis in hepatitis C, such as a very high incidence and multicentric nature of HCC development. CRC, colorectal cancer; HCC, hepatocellular carcinoma; APC, adenomatous polyposis coli

unusual events happening in HCV carriers (Koike 2005). Now it does not seem so difficult as before to determine why HCC develops in persistent HCV infection at an outstandingly high incidence. Our theory may also give an account of the non-metastatic and multicentric *de novo* occurrence characteristics of HCC, which would be the result of persistent HCV infection.

References

Alonzi T, Agrati C, Costabile B, Cicchini C, Amicone L, Cavallari C, Rocca CD, Folgori A, Fipaldini C, Poccia F, Monica NL, Tripodi M (2004) Steatosis and intrahepatic lymphocyte recruitment in hepatitis C virus transgenic mice. J Gen Virol 85:1509–1520

Boulant S, Montserret R, Hope RG, Ratinier M, Targett-Adams P, Lavergne JP et al (2006) Structural determinants that target the hepatitis C virus core protein to lipid droplets. J Biol Chem 281:22236–22247

Diamond DL, Jacobs JM, Paeper B, Proll SC, Gritsenko MA, Carithers RL Jr et al (2007) Proteomic profiling of human liver biopsies: hepatitis C virus-induced fibrosis and mitochondrial dysfunction. Hepatology 46:649–657

Farinati F, Cardin R, De Maria N, Della Libera G, Marafin C, Lecis E, Burra P, Floreani A, Cecchetto A, Naccarato R (1995) Iron storage, lipid peroxidation and glutathione turnover in chronic anti-HCV positive hepatitis. J Hepatol 22:449–456

Frelin L, Brenndörfer ED, Ahlén G, Weiland M, Hultgren C, Alheim M, Glaumann H, Rozell B, Milich DR, Bode JG, Sällberg M (2006) The hepatitis C virus and immune evasion: non-structural 3/4A transgenic mice are resistant to lethal tumour necrosis factor alpha mediated liver disease. Gut 55:1475–1483

Fujita N, Sugimoto R, Ma N, Tanaka H, Iwasa M, Kobayashi Y, Kawanishi S, Watanabe S, Kaito M, Takei Y (2008) Comparison of hepatic oxidative DNA damage in patients with chronic hepatitis B and C. J Viral Hepat 15:498–507

Honda A, Arai Y, Hirota N, Sato T, Ikegaki J, Koizumi T, Hatano M, Kohara M, Moriyama T, Imawari M, Shimotohno K, Tokuhisa T (1999) Hepatitis C virus structural proteins induce liver cell injury in transgenic mice. J Med Virol 59:281–289

Houghton M, Weiner A, Han J, Kuo G, Choo QL (1991) Molecular biology of hepatitis C viruses. Implications for diagnosis, development and control of viral diseases. Hepatology 14:381–388

Ikeda K, Saitoh S, Suzuki Y, Kobayashi M, Tsubota A, Koida I et al (1998) Disease progression and hepatocellular carcinogenesis in patients with chronic viral hepatitis: a prospective observation of 2215 patients. J Hepatol 28:930–938

Kato J, Kobune M, Nakamura T, Kuroiwa G, Takada K, Takimoto R, Sato Y, Fujikawa K, Takahashi M, Takayama T, Ikeda T, Niitsu Y (2001) Normalization of elevated hepatic 8-hydroxy-2'-deoxyguanosine levels in chronic hepatitis C patients by phlebotomy and low iron diet. Cancer Res 61:8697–8702

Kim KH, Hong SP, Kim K, Park MJ, Kim KJ, Cheong J (2007) HCV core protein induces hepatic lipid accumulation by activating SREBP1 and PPARgamma. Biochem Biophys Res Commun 55:883–888

Kinzler KW, Vogelstein B (1996) Lessons from hereditary colorectal cancer. Cell 87:159–170

Kiyosawa K, Sodeyama T, Tanaka E, Gibo Y, Yoshizawa K, Nakano Y et al (1990) Interrelationship of blood transfusion, non-A, non-B hepatitis and hepatocellular carcinoma: analysis by detection of antibody to hepatitis C virus. Hepatology 12:671–675

Ko KS, Tomasi ML, Iglesias-Ara A, French BA, French SW, Ramani K, Lozano JJ, Oh P, He L, Stiles BL, Li TW, Yang H, Martínez-Chantar ML, Mato JM, Lu SC (2010) Liver-specific deletion of prohibitin 1 results in spontaneous liver injury, fibrosis, and hepatocellular carcinoma in mice. Hepatology 52:2096–2108

Koike K (2005) Molecular basis of hepatitis C virus-associated hepatocarcinogenesis: lessons from animal model studies. Clin Gastroenterol Hepatol 3:S132–S135

Koike K, Moriya K, Ishibashi K, Matsuura Y, Suzuki T, Saito I et al (1995) Expression of hepatitis C virus envelope proteins in transgenic mice. J Gen Virol 76:3031–3038

Koike K, Moriya K, Yotsuyanagi H, Shintani Y, Fujie H, Ishibashi K et al (1997) Sialadenitis resembling Sjögren's syndrome in mice transgenic for hepatitis C virus envelope genes. Proc Natl Acad Sci USA 94:233–236

Korenaga M, Wang T, Li Y, Showalter LA, Chan T, Sun J, Weinman SA (2005) Hepatitis C virus core protein inhibits mitochondrial electron transport and increases reactive oxygen species (ROS) production. J Biol Chem 280:37481–37488

Kudo Y, Tanaka Y, Tateishi K, Yamamoto K, Yamamoto S, Mohri D, Isomura Y, Seto M, Nakagawa H, Asaoka Y, Tada M, Ohta M, Ijichi H, Hirata Y, Otsuka M, Ikenoue T, Maeda S, Shiina S, Yoshida H, Nakajima O, Kanai F, Omata M, Koike K (2011) Altered composition of fatty acids exacerbates hepatotumorigenesis during activation of the phosphatidylinositol 3-kinase pathway. J Hepatol 55:1400–1408

Lerat H, Honda M, Beard MR, Loesch K, Sun J, Yang Y et al (2002) Steatosis and liver cancer in transgenic mice expressing the structural and nonstructural proteins of hepatitis C virus. Gastroenterology 122:352–365

Machida K, Cheng KT, Lai CK, Jeng KS, Sung VM, Lai MM (2006) Hepatitis C virus triggers mitochondrial permeability transition with production of reactive oxygen species, leading to DNA damage and STAT3 activation. J Virol 80:7199–7207

Majumder M, Ghosh AK, Steele R, Zhou XY, Phillips NJ, Ray R, Ray RB (2002) Hepatitis C virus NS5A protein impairs TNF-mediated hepatic apoptosis, but not by an anti-FAS antibody, in transgenic mice. Virology 294:94–105

Mishra S, Murphy LC, Nyomba BL, Murphy LJ (2005) Prohibitin: a potential target for new therapeutics. Trends Mol Med 11:192–197

Miyamoto H, Moriishi K, Moriya K, Murata S, Tanaka K, Suzuki T et al (2007) Hepatitis C virus core protein induces insulin resistance through a PA28γ-dependent pathway. J Virol 81:1727–1735

Miyoshi H, Fujie H, Shintani Y, Tsutsumi T, Shinzawa S, Makuuchi M, Kokudo N, Matsuura Y, Suzuki T, Miyamura T, Moriya K, Koike K (2005) Hepatitis C virus core protein exerts an inhibitory effect on suppressor of cytokine signaling (SOCS)-1 gene expression. J Hepatol 43:757–763

Moradpour D, Penin F, Rice CM (2007) Replication of hepatitis C virus. Nat Rev Microbiol 5:453–463

Moriishi K, Okabayashi T, Nakai K, Moriya K, Koike K, Murata S et al (2003) Proteasome activator PA28 gamma-dependent nuclear retention and degradation of hepatitis C virus core protein. J Virol 77:10237–10249

Moriishi K, Mochizuki R, Moriya K, Miyamoto H, Mori Y, Abe T et al (2007) Critical role of PA28g in hepatitis C virus-associated steatogenesis and hepatocarcinogenesis. Proc Natl Acad Sci USA 104:1661–1666

Moriya K, Yotsuyanagi H, Shintani Y, Fujie H, Ishibashi K, Matsuura Y et al (1997) Hepatitis C virus core protein induces hepatic steatosis in transgenic mice. J Gen Virol 78:1527–1531

Moriya K, Fujie H, Shintani Y, Yotsuyanagi H, Tsutsumi T, Matsuura Y et al (1998) Hepatitis C virus core protein induces hepatocellular carcinoma in transgenic mice. Nat Med 4:1065–1068

Moriya K, Nakagawa K, Santa T, Shintani Y, Fujie H, Miyoshi H et al (2001a) Oxidative stress in the absence of inflammation in a mouse model for hepatitis C virus-associated hepatocarcinogenesis. Cancer Res 61:4365–4370

Moriya K, Todoroki T, Tsutsumi T, Fujie H, Shintani Y, Miyoshi H et al (2001b) Increase in the concentration of carbon 18 monounsaturated fatty acids in the liver with hepatitis C: analysis in transgenic mice and humans. Biophys Biochem Res Commun 281:1207–1212

Moriya K, Miyoshi H, Shinzawa S, Tsutsumi T, Fujie H, Goto K, Shintani Y, Yotsuyanagi H, Koike K (2010) Hepatitis C virus core protein compromises iron-induced activation of antioxidants in mice and HepG2 cells. J Med Virol 82:776–792

Naas T, Ghorbani M, Alvarez-Maya I, Lapner M, Kothary R, De Repentigny Y et al (2005) Characterization of liver histopathology in a transgenic mouse model expressing genotype 1a hepatitis C virus core and envelope proteins 1 and 2. J Gen Virol 86:2185–2196

Nijtmans LG, de Jong L, Artal Sanz M, Coates PJ, Berden JA, Back JW et al (2000) Prohibitins act as a membrane-bound chaperone for the stabilization of mitochondrial proteins. EMBO J 19:2444–2451

Nishina S, Hino K, Korenaga M, Vecchi C, Pietrangelo A, Mizukami Y, Furutani T, Sakai A, Okuda M, Hidaka I, Okita K, Sakaida I (2008) Hepatitis C virus-induced reactive oxygen species raise hepatic iron level in mice by reducing hepcidin transcription. Gastroenterology 134:226–238

Pasquinelli C, Shoenberger JM, Chung J et al (1997) Hepatitis C virus core and E2 protein expression in transgenic mice. Hepatology 25:719–727

Perlemuter G, Sabile A, Letteron P, Vona G, Topilco A, Koike K et al (2002) Hepatitis C virus core protein inhibits microsomal triglyceride transfer protein activity and very low density lipoprotein secretion: a model of viral-related steatosis. FASEB J 16:185–194

Perz JF, Armstrong GL, Farrington LA, Hutin YJ, Bell BP (2006) The contributions of hepatitis B virus and hepatitis C virus infections to cirrhosis and primary liver cancer worldwide. J Hepatol 45:529–538

Piccoli C, Scrima R, Quarato G, D'Aprile A, Ripoli M, Lecce L et al (2007) Hepatitis C virus protein expression causes calcium-mediated mitochondrial bioenergetic dysfunction and nitro-oxidative stress. Hepatology 46:58–65

Powell EE, Jonsson JR, Clouston AD (2005) Steatosis: co-factor in other liver diseases. Hepatology 42:5–13

Saito I, Miyamura T, Ohbayashi A, Harada H, Katayama T, Kikuchi S et al (1990) Hepatitis C virus infection is associated with the development of hepatocellular carcinoma. Proc Natl Acad Sci USA 87:6547–6549

Shintani Y, Fujie H, Miyoshi H, Tsutsumi T, Kimura S, Moriya K et al (2004) Hepatitis C virus and diabetes: direct involvement of the virus in the development of insulin resistance. Gastroenterology 126:840–848

Shirakura M, Murakami K, Ichimura T, Suzuki R, Shimoji T, Fukuda K et al (2007) E6AP ubiquitin ligase mediates ubiquitylation and degradation of hepatitis C virus core protein. J Virol 81:1174–1185

Suzuki R, Sakamoto S, Tsutsumi T, Rikimaru A, Tanaka K, Shimoike T et al (2005) Molecular determinants for subcellular localization of hepatitis C virus core protein. J Virol 79:1271–1281

Stocker R, Yamamoto Y, McDonagh AF, Glazer AN, Ames BN (1987) Bilirubin is an antioxidant of possible physiological importance. Science 235:1043–1046

Su AI, Pezacki JP, Wodicka L, Brideau AD, Supekova L, Thimme R et al (2002) Genomic analysis of the host response to hepatitis C virus infection. Proc Natl Acad Sci USA 99:15669–15674

Taguwa S, Kambara H, Fujita N, Noda T, Yoshimori T, Koike K, Moriishi K, Matsuura Y (2011) Dysfunction of autophagy participates in vacuole formation and cell death in cells replicating hepatitis C virus. J Virol 85:13185–13194

Theiss AL, Idell RD, Srinivasan S, Klapproth JM, Jones DP, Merlin D et al (2007) Prohibitin protects against oxidative stress in intestinal epithelial cells. FASEB J 21:197–206

Tsutsumi T, Suzuki T, Moriya K, Yotsuyanagi H, Shintani Y, Fujie H et al (2002) Intrahepatic cytokine expression and AP-1 activation in mice transgenic for hepatitis C virus core protein. Virology 304:415–424

Tsutsumi T, Suzuki T, Moriya K, Shintani Y, Fujie H, Miyoshi H et al (2003) Hepatitis C virus core protein activates ERK and p38 MAPK in cooperation with ethanol in transgenic mice. Hepatology 38:820–828

Tsutsumi T, Matsuda M, Aizaki H, Moriya K, Miyoshi H, Fujie H, Shintani Y, Yotsuyanagi H, Miyamura T, Suzuki T, Koike K (2009) Proteomics analysis of mitochondrial proteins reveals overexpression of a mitochondrial protein chaperone, prohibitin, in cells expressing hepatitis C virus core protein. Hepatology 50:378–386

Wakita T, Taya C, Katsume A et al (1998) Efficient conditional transgene expression in hepatitis C virus cDNA transgenic mice mediated by the Cre/loxP system. J Biol Chem 273:9001–9006

Waris G, Felmlee DJ, Negro F, Siddiqui A (2007) Hepatitis C virus induces proteolytic cleavage of sterol regulatory element binding proteins and stimulates their phosphorylation via oxidative stress. J Virol 81:8122–8130

Yotsuyanagi H, Shintani Y, Moriya K, Fujie H, Tsutsumi T, Kato T et al (2000) Virological analysis of non-B, non-C hepatocellular carcinoma in Japan: frequent involvement of hepatitis B virus. J Infect Dis 181:1920–1928

Prevention of Hepatitis C Virus Infection and Liver Cancer

E. J. Lim and J. Torresi

Abstract

Hepatocellular carcinoma (HCC) is the fifth most prevalent cancer and the third leading cause of cancer-related death, and its incidence is increasing. The majority of HCC cases are associated with chronic viral hepatitis. With over 170 million individuals chronically infected with hepatitis C virus (HCV) worldwide, HCV is currently a serious global health concern, leading to chronic hepatitis, cirrhosis and HCC, thereby causing significant morbidity and mortality. With the incidence of HCV infection increasing, the problem of HCV-associated HCC is expected to worsen as well, with the majority of HCCs developing in the setting of cirrhosis. Thus, it is imperative to provide antiviral therapy to infected individuals prior to the development of established cirrhosis in order to reduce the risk of subsequent HCC. Indeed, the successful eradication of HCV is associated with clinical and histological improvement as well as a greatly reduced risk of subsequent HCC development. Even after the development of cirrhosis, successful viral clearance is still associated with reduced HCC risk. Current standard of care antiviral treatment consists of pegylated interferon-α and ribavirin, but viral clearance rates are suboptimal with this regimen, especially in difficult to treat cohorts. However, there is a myriad of different classes of HCV-specific direct-acting antiviral agents currently in development, which can be used in combination with one another

J. Torresi
Department of Infectious Diseases, Austin Hospital, Heidelberg, Victoria 3084, Australia

E. J. Lim (✉)
Department of Gastroenterology, Austin Hospital, Heidelberg, Victoria 3084, Australia
e-mail: josepht@unimelb.edu.au

M. H. Chang and K.-T. Jeang (eds.), *Viruses and Human Cancer*,
Recent Results in Cancer Research 193, DOI: 10.1007/978-3-642-38965-8_7,
© Springer-Verlag Berlin Heidelberg 2014

or with standard of care treatment to improve HCV cure rates. Preventative and therapeutic vaccines against HCV remain an area of ongoing research with good progress towards developing an effective vaccine in the future.

Keywords
Hepatitis C · Hepatocellular carcinoma · Cirrhosis · Antiviral treatment · Direct-acting antiviral agents · Vaccines

Contents

1   Introduction

Hepatocellular carcinoma (HCC) is the fifth most prevalent cancer and the third leading cause of cancer-related death worldwide (Parkin 2001). The incidence of HCC is increasing (El-Serag and Rudolph 2007), and currently around 80 % of HCC cases worldwide are associated with chronic viral hepatitis (Thomas and Zhu 2005). An estimated one-third of HCC cases in the USA are attributable to hepatitis C (HCV), making HCV the leading risk factor for HCC in this country (National institutes of health consensus development conference statement: Management of hepatitis C 2002). Although a strong association exists between chronic HCV and the development of HCC, the precise mechanisms by which HCV infection ultimately results in HCC are uncertain. There is, however, good evidence to suggest that eradication of the virus in both cirrhotic and non-cirrhotic HCV-infected patients reduces the subsequent risk of developing of HCC.

2   Hepatitis C Infection

There are currently over 170 million individuals chronically infected with HCV worldwide, and this is associated with an estimated 476,000 deaths annually (Dore et al. 2003). HCV is transmitted via infected blood. Currently in Western countries, acquisition of HCV occurs primarily through intravenous drug use and tattoos (Razali et al. 2007), whereas in Asia and Africa, infection mainly occurs through the use of contaminated blood products and medical instruments. In contrast to these modes, HCV transmission via sexual and perinatal routes is infrequent (Razali et al. 2007).

Following infection with HCV, up to 85 % of patients are unable to clear the virus, resulting in chronic infection which may ultimately progress to cirrhosis in approximately 20 % of individuals (National institutes of health consensus development conference statement: Management of hepatitis c 2002). The natural history of HCV infection is shown in Fig. 1. The majority of chronically infected patients are asymptomatic for several decades, thus delaying both the diagnosis and treatment of this disease. It is often not until the development of complications of cirrhosis such as hepatic decompensation and HCC that these patients present to medical care practitioners. The impact of chronic hepatitis C is highlighted by the fact that HCV-related liver failure is now the commonest indication for liver transplantation in the USA, Europe and Australia (Davis et al. 2003; Law et al. 2003).

HCV belongs to the genus *Hepacivirus* in the *Flaviviridae* family (Forns and Bukh 1999). It has a single-stranded linear RNA genome of approximately 9,600 nucleotides that encodes a large polyprotein of approximately 3,000 amino acids (Bartenschlager et al. 2011). The structure of the HCV genome and functions of the various

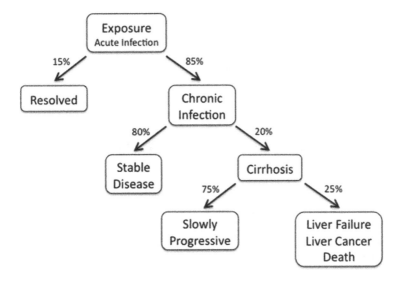

Fig. 1 Natural history of HCV infection. Adapted from Alter et al. (1995)

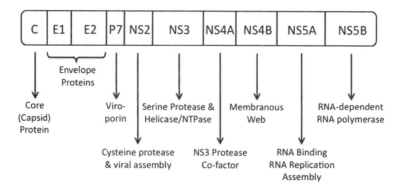

Fig. 2 HCV genome structure and functions of viral proteins. Adapted from Bartenschlager et al. (2011)

viral proteins are shown in Fig. 2. HCV exists as six major genotypes with genotype 1 being the dominant genotype in the USA, Europe and Australia (Simmonds et al. 2005). Genotype 3 is more prevalent in India and Southeast Asia, whereas genotype 4 is more commonly seen in the Middle East and Africa (Kamal and Nasser 2008). Each genotype contains multiple subtypes that are detected by viral sequencing and that are identified with lower case letters (a, b, c, etc.). HCV genotype influences treatment response and also the severity of liver disease. For example, chronic infection with genotype 1b may be associated with the development of more advanced fibrosis, cirrhosis and HCC. In fact, a recent meta-analysis found that genotype 1b HCV infection was associated with a doubling of the risk of HCC development compared to infection with other genotypes (Raimondi et al. 2009).

Like other RNA viruses, the RNA polymerase protein of HCV lacks proof-reading ability, and as a consequence, replication of the viral genome is error-prone resulting in a high mutation rate. The result is great genetic heterogeneity that leads to the evolution of diverse viral quasispecies. This viral diversity interferes with the development of effective host humoral immune responses against the virus, thereby promoting viral persistence within infected individuals (Forns and Bukh 1999).

3   Hepatitis C and Associated Risk Factors for HCC Development

HCV is recognized as a major cause of HCC globally. In a large population-based prospective study, infection with HCV conferred a 20-fold increased risk of developing HCC compared to HCV-negative individuals (Sun et al. 2003). This strong association between chronic HCV and HCC has been noted since the early 1980s, when the virus was known as non-A, non-B hepatitis (Kiyosawa et al. 1984). Almost all HCV-related HCCs occur in the setting of established cirrhosis, with cirrhosis itself being a strong independent risk factor for developing HCC.

In the setting of HCV-induced cirrhosis, the incidence of HCC is between 2 and 8 % per year (Bruix et al. 2005). Consequently, HCC develops many years (often 2–3 decades) after initial HCV infection.

However, not all HCV-related HCCs occur in patients with pre-existing cirrhosis. In a large prospective study, about 17 % of HCV-positive patients with HCC were not cirrhotic, but were noted to have at least an Ishak fibrosis score of 3 or more on serial liver biopsies (Lok et al. 2009), indicating that even in the absence of cirrhosis, HCC may develop in HCV-infected individuals with established chronic hepatitis and advanced liver fibrosis.

Risk factors that may contribute to the progression of HCV-associated liver disease leading to cirrhosis and HCC include concurrent alcohol consumption, older age at time of infection, male gender, co-infection with HIV or hepatitis B, immunosuppression, associated insulin resistance or non-alcoholic steatohepatitis, and a higher degree of inflammation and fibrosis on liver biopsy (Chen and Morgan 2006).

Significant alcohol intake of >40 g alcohol/day in women and >60 g of alcohol/day in men for more than 5 years is associated with a two to threefold increased risk of cirrhosis and decompensated liver disease in HCV-infected individuals (Wiley et al. 1998). Furthermore, the risk of developing HCC is doubled in HCV-infected individuals who consume >60 g of alcohol/day compared to those consuming <60 g/day (Donato et al. 2002). Also, the presence of chronic HCV infection has been associated with more advanced liver disease and increased mortality in alcoholic individuals compared to alcoholic patients with chronic hepatitis B infection (Mendenhall et al. 1991).

Age of infection is also an independent risk factor for the development of more severe liver disease in chronic HCV. After controlling for duration of HCV infection, patients who acquire HCV infection at an older age (>40 years old) are significantly more likely to progress to advanced liver fibrosis than individuals infected at a younger age (Poynard et al. 1997). The incidence of HCC is up to 29 times higher for individuals who become infected with HCV after 39 years of age compared to those infected before the age of 19 years (Pradat et al. 2007).

In the setting of HIV–HCV co-infection, a low CD4 count is associated with higher HCV viral loads as well as an accelerated progression to cirrhosis (Di Martino et al. 2001). Other causes of immunosuppression, such as organ transplantation, have also been associated with more rapid liver fibrosis progression (Berenguer et al. 2000). Finally, individuals with HCV–HBV co-infection may also be at a higher risk of HCC development (Cho et al. 2011).

How HCV infection results in the development of HCC is not entirely clear. There is evidence to suggest that HCV may interact with various intracellular signal transduction pathways or effect epigenetic changes, thereby altering hepatocyte physiology to directly promote malignant transformation. For example, HCV core protein has been noted to interact with the mitogen-activated protein (MAP) kinase signalling pathway, thereby promoting cell proliferation (Hayashi et al. 2000). Also, it has been shown that the tumour suppressor gene p16INK4A in tumour tissue resected from the livers of HCV-infected patients with HCCs is hypermethylated, and this results in the inactivation of p16INK4A, a feature not seen in non-HCV-associated HCCs (Li et al. 2004). Secondly, immune-mediated liver inflammation and the promotion of apoptosis of HCV-infected hepatocytes result in a compensatory stimulation of cell proliferation to replace dead hepatocytes. The increased cell turnover permits the accumulation of genetic mutations within hepatocytes, and this together with the surrounding inflammatory liver milieu promotes HCC development (Hino et al. 2002).

4   Prevention of HCC in Patients With Hepatitis C–induced Cirrhosis

The incidence of HCV-related HCC continues to rise worldwide because of the increasing number of individuals with chronic HCV infection, the presence of associated co-morbidities and the longer survival of patients with advanced liver disease as a result of improved management of the complications of liver failure. HCC develops at an annual rate of about 3 % in patients with HCV-related cirrhosis, with an 11.5 % four-year risk of HCC (Serfaty et al. 1998), hence emphasizing the importance of HCC screening in this population. The successful clearance of HCV with interferon therapy in patients with cirrhosis is associated with a threefold reduction in the risk of subsequent HCC development (Bruno et al. 2007; Singal et al. 2010b), indicating that despite the presence of cirrhosis, successful antiviral therapy of HCV-infected patients will still reduce the future risk of HCC. However, as the risk of HCC is not completely eliminated despite achieving viral eradication, ongoing HCC screening is still indicated in patients with cirrhosis. In contrast, there appears to be no benefit of interferon therapy on reducing HCC risk if viral eradication has not been successful. In patients with HCV-induced cirrhosis who have not achieve viral clearance with standard antiviral therapy, the provision of long-term half-dose maintenance pegylated interferon may reduce the incidence of HCC development (Lok et al. 2011). In this study, the beneficial effect in reducing HCC incidence was not seen in HCV-infected patients who had advanced hepatic fibrosis but not cirrhosis. However, in a meta-analysis of four studies with 1,152 non-responders, there was no statistically significant reduction in HCC risk in patients maintained on low-dose interferon compared with those who did not receive maintenance therapy (Singal et al. 2010b).

A meta-analysis of patients with HCV-induced cirrhosis who developed HCC has shown that after curative treatment for HCC via local ablative therapy or surgical resection, successful eradication of HCV with antiviral therapy was associated with a reduced risk of HCC recurrence from 61 to 35 % (Singal et al. 2010a). Furthermore, successful treatment with pegylated interferon and ribavirin was associated with improved hepatic functional reserve and increased survival (96 % vs. 61 % at 3 years) in this cohort (Ishikawa et al. 2012). In patients undergoing liver transplantation for HCV-associated HCC, interferon therapy for recurrent HCV post-liver transplant was found to decrease subsequent HCC recurrence from 27 to 4 % (Kohli et al. 2012). These studies indicate that antiviral therapy is important in the secondary prevention of HCC in HCV-induced cirrhosis.

5   Prevention of Cirrhosis and HCC in Patients with Hepatitis C–Induced Chronic Hepatitis

The potential long-term benefits of successful antiviral therapy of HCV-infected patients with chronic hepatitis include the normalization of serum transaminase levels a reduction in hepatic necroinflammation and fibrosis, improvement in health-related quality of life and the reduction in HCC risk, all of which enhance patient survival (Patel et al. 2006). We know that HCC primarily develops in HCV-infected patients with cirrhosis. In order to reduce the risk of developing HCV-related HCC, the aim would be to provide treatment to eradicate HCV prior to the development of cirrhosis. Indeed, studies have shown that successful viral clearance with antiviral therapy results in clinical and histological improvement in the vast majority of patients (Marcellin et al. 1997), with an associated reduction in the risk of subsequent development of cirrhosis and HCC (Pradat et al. 2007).

In a large multi-centre European study of largely non-cirrhotic (89 %) HCV-infected patients, successful viral eradication was associated with the progression to cirrhosis in only 2.3 % of patients with no patients developing HCC, whereas a failure to achieve viral clearance was associated with progression to cirrhosis in 20 % of patients and development of HCC in 4.2 % (Pradat et al. 2007). In another study of HCV-infected patients, the majority of whom (90 %) did not have cirrhosis, successful viral clearance with interferon therapy was associated with a reduction in the risk of HCC development from 2.31 to 0.24 %/year (Maruoka et al. 2012). This study also showed that patients who failed to achieve viral eradication despite antiviral therapy had the same risk of subsequent HCC development as patients who did not receive antiviral therapy (Maruoka et al. 2012).

Even in non-cirrhotic patients who do not achieve viral clearance with interferon therapy, the provision of long-term interferon monotherapy appears to reduce the risk of subsequent development of HCV-related HCC. In a retrospective study, patients with genotype 1 HCV infection who failed previous therapy with interferon, 93 % of whom did not have cirrhosis at the time of treatment, receiving a further 48 weeks or more of interferon monotherapy was associated with a reduction in the 10-year incidence of HCC from 16.4 to 11.5 % (Takeyasu et al. 2012).

6   Antiviral Treatment for Hepatitis C

A vaccine for the prevention of HCV is not yet available, and therefore, the only effective means to prevent the development of liver cirrhosis and HCC is with antiviral therapy. The aim of such treatment is to clear the virus, thereby halting the progression of liver injury and fibrosis. The currently available standard of care in the treatment for chronic HCV infection comprises the combination of pegylated interferon alpha and ribavirin. However, the advent of direct-acting antiviral

(DAA) drugs has vastly changed the landscape of antiviral therapy for chronic HCV and may provide a more effective approach to the long-term prevention of HCV-associated HCC.

6.1 Standard of Care Therapy

The current standard of care for the treatment for HCV infection consists of a combination of subcutaneous pegylated interferon alpha given weekly and daily oral ribavirin (Fried et al. 2002). The two forms of pegylated interferon currently approved for HCV treatment are pegylated interferon alpha-2a (Pegasys, Hoffmann-La Roche, Nutley, NJ) and pegylated interferon alpha-2b (Pegintron, Merck Sharp and Dohme, Whitehouse Station, NJ), with no statistically significant difference in efficacy seen between the two forms (McHutchison et al. 2009b). Interferon alpha exerts its immunoregulatory properties by binding to interferon alpha 1 and 2 cell surface receptors to induce JAK-STAT signal transduction. Upon ligand binding, receptor-associated Janus kinase (JAK) tyrosine kinases are activated, leading to the phosphorylation and activation of members of the signal transducers and activators of transcription (STAT) family of transcription factors (Kisseleva et al. 2002). These transcription factors then translocate to the cell nucleus where they transcribe several genes involved in cell cycle regulation, apoptosis and promotion of an antiviral state within hepatocytes (Thomas et al. 2003). Besides the direct effect against virally-infected host cells, interferon alpha is also involved in modulating the immune system by enhancing the CD8+ cytotoxic T cell response against infected hepatocytes, as well as promoting the proliferation of B cells to augment the production of antibodies against HCV (Thomas et al. 2003).

Ribavirin (Copegus, Hoffmann-La Roche, Nutley, NJ; Rebetol, Merck Sharp and Dohme, Whitehouse Station, NJ) is a synthetic guanosine nucleoside analogue which has in vitro activity against a range of RNA and DNA viruses (Patterson and Fernandezlarsson 1990). When used as monotherapy in patients with chronic HCV, ribavirin was noted to reduce serum aminotransferase levels without having an antiviral effect (Dibisceglie et al. 1995). The precise mechanism by which ribavirin produces its antiviral effect when combined with pegylated interferon is unknown; however, there are a number of hypotheses. Ribavirin is administered as an orally active pro-drug, which is metabolized within the body into a form that resembles purine RNA nucleotides, thus enabling the incorporation of ribavirin into the viral genome during RNA replication. This induces hypermutation of the viral RNA genome resulting in viral lethality (Crotty et al. 2002). Ribavirin also inhibits the viral RNA-dependent RNA polymerase that is essential for HCV replication and it is also an inosine monophosphate dehydrogenase inhibitor. Inhibition of this intracellular enzyme leads to the depletion of intracellular GTP levels, thus decreasing viral protein synthesis and reducing RNA genome replication (Streeter et al. 1973).

6.2 Predictors of Pegylated Interferon/Ribavirin Antiviral Response

A successful response to treatment, which effectively equates to permanent viral eradication, is defined as the absence of HCV RNA in the serum 24 weeks after the completion of antiviral therapy and is referred to as a sustained virological response (SVR) (Pradat et al. 2007). An undetectable HCV RNA in the serum at week 4 of treatment, known as a rapid virological response (RVR), is predictive of a highly favourable outcome, with over 86 % of patients achieving an RVR subsequently attaining an SVR (Yu et al. 2007). If viral eradication is achieved prior to the development of cirrhosis, and no other hepatotoxic factors were present, hepatic fibrosis is expected to improve and the patient's risk of developing HCC returns to that of the baseline population.

There are a number of host and viral factors that can predict antiviral treatment response prior to commencing therapy, with the most reliable of these being the HCV genotype. Patients infected with genotypes 2 and 3 attain much higher viral clearance rates than genotype 1–infected patients (Hadziyannis et al. 2004).

Single nucleotide polymorphisms at different loci within the IL28B gene on chromosome 19, which encodes interferon lambda 3, have also recently been shown to be strongly associated with response to interferon therapy, especially for individuals infected with genotype 1 HCV. Compared to patients possessing CT and TT polymorphisms at the rs12979860 locus of this gene, patients with a CC polymorphism are twice as likely to achieve viral clearance with pegylated interferon/ribavirin treatment (Ge et al. 2009).

In addition to the HCV genotype and IL28B gene polymorphisms, a low pre-treatment HCV viral load (<600,000 IU/mL) also predicts for a favourable outcome to antiviral treatment (Fried et al. 2002). Other factors that predict a poor response to treatment with pegylated interferon/ribavirin include vitamin D deficiency, the presence of insulin resistance or a body mass index of more than 30 kg/m^2 (Bressler et al. 2003; Romero-Gomez et al. 2005), and the presence of cirrhosis or decompensated liver disease (Fried et al. 2002; Manns et al. 2001).

Although pegylated interferon and ribavirin treatment offers permanent eradication of the virus from chronically infected individuals, it is extremely expensive and is associated with a significant number of side effects. Furthermore, overall cure rates using the standard regimen are disappointing in patients infected with genotype 1 in whom SVR rates of 40–50 % are achieved compared to 70–80 % in patients with genotypes 2 and 3, and approximately 70 % in patients with genotype 4 (Chevaliez and Pawlotsky 2007; Khuroo et al. 2004).

6.3 Direct-Acting Antiviral Agents

Significant advances have recently been made in the treatment for chronic HCV infection. The development of direct-acting antiviral agents (DAAs), which target specific HCV proteins crucial for the replication cycle of HCV, have resulted in

Table 1 HCV direct-acting antiviral agents

DAA	Type of inhibitor	Study phase	Genotypic activity	Barrier to resistance	Company
Telaprevir	NS3/4A protease (1st generation, 1st wave)	Phase III	1	Low	Vertex
Boceprevir	NS3/4A protease (1st generation, 1st wave)	Phase III	1	Low	Merck Sharp and Dohme
Danoprevir	NS3/4A Protease (1st generation, 2nd wave)	Phase IIb	1	Low	Roche/ Genentech
Asunaprevir	NS3/4A protease (1st generation, 2nd wave)	Phase III	1	Low	Bristol-Myers Squibb
Vaniprevir	NS3/4A protease (1st generation, 2nd wave)	Phase IIb	1	Low	Merck Sharp and Dohme
MK-5172	NS3/4A protease (2nd generation)	Phase II	1, 2, 3	Moderate	Merck Sharp and Dohme
Mericitabine	NS5B polymerase (nucleoside)	Phase IIb	1, 2, 3	High	Roche/ Genentech
Sofosbuvir	NS5B polymerase (nucleoside)	Phase III	1, 2, 3	High	Pharmasset/ Gilead
BI207127	NS5B polymerase (non-nucleoside)	Phase IIb	1	Low	Boehringer Ingelheim
VX-222	NS5B polymerase (non-nucleoside)	Phase II	1	Low	Vertex
ABT-072	NS5B polymerase (non-nucleoside)	Phase II	1	Low	Abbott
Daclatasvir	NS5A	Phase IIb	1	Low	Bristol-Myers Squibb

significant improvements in achieving SVR. This has been most beneficial for patients infected with HCV genotype 1, patients with poor pre-treatment prognostic factors, and non-responders and relapsers to pegylated interferon and ribavirin therapy. Some of the DAAs that are currently in use or in development are listed in Table 1.

Protease inhibitors: The first NS3/4A protease inhibitors to be licensed include telaprevir (Incivek, Vertex) and boceprevir (Victrelis, Merck Sharp and Dohme). Both are only effective in patients infected with genotype 1 HCV. The addition of these protease inhibitors to standard of care therapy significantly improves SVR rates in both naïve patients and those who have failed to achieve an SVR with previous therapy.

In a large phase III, randomized, double-blind, placebo-controlled trial of treatment-naïve patients infected with genotype 1 HCV, telaprevir was administered for 12 weeks together with pegylated interferon and ribavirin, after which pegylated interferon and ribavirin were continued for a further 12–36 weeks

(Jacobson et al. 2011). Telaprevir increased SVR rates from 44 to 75 % compared to standard therapy alone. Even for patients with high pre-treatment viral loads, SVR was improved from 36 to 74 % with the addition of telaprevir (Jacobson et al. 2011). In another large phase III study, after a 4-week "lead-in" phase of pegylated interferon and ribavirin, boceprevir was added and therapy with all three drugs continued for a total of 28–48 weeks (24–44 weeks of boceprevir) (Poordad et al. 2011). The addition of boceprevir enhanced SVR rates from 38 to 66 % compared to pegylated interferon and ribavirin alone. In both studies, the SVR rates in treatment-naïve cirrhotic patients were also improved from 33 to 62 % with the addition of telaprevir (Jacobson et al. 2011), and 38–52 % with the addition of boceprevir (Poordad et al. 2011).

For patients who experienced virological relapse after becoming HCV RNA PCR negative with previous pegylated interferon and ribavirin, the addition of telaprevir for the first 12 weeks of a 48-week course of pegylated interferon and ribavirin treatment improved SVR rates from 24 % to over 80 % compared to pegylated interferon and ribavirin alone for 48 weeks (Zeuzem et al. 2011a). The addition of boceprevir to pegylated interferon and ribavirin for 44 weeks after the 4-week "lead-in" phase also resulted in an improvement in SVR from 29 to 75 % compared to pegylated interferon and ribavirin alone for 48 weeks (Bacon et al. 2011). For treatment-experienced cirrhotic patients, the results were also encouraging for relapsers, with SVR rates increasing from 13 to 84 % with telaprevir (Zeuzem et al. 2011a), and from 20 to 83 % with boceprevir (Bacon et al. 2011).

Common side effects of telaprevir therapy include skin rash, anaemia and gastrointestinal symptoms, while common side effects of boceprevir include anaemia and dysgeusia (Jacobson et al. 2011; Poordad et al. 2011). Ribavirin remains crucial to antiviral therapy with the trials employing telaprevir and boceprevir to date showing that even in the era of potent DAAs, SVR and virological relapse rates were suboptimal without the use of full-dose ribavirin (Kwo et al. 2010; McHutchison et al. 2009a). Although potent inhibition of HCV replication is achieved with these agents, the rapid development of drug-resistant HCV variants has also been noted, with the potential to confer cross-resistance to other protease inhibitors (Halfon and Locarnini 2011). As such, these agents should not be used as monotherapy.

In addition to telaprevir and boceprevir, a number of other NS3/4A protease inhibitors, such as danoprevir (Roche/Genentech), asunaprevir (Bristol-Myers Squibb) and vaniprevir (Merck Sharp and Dohme) have now entered into clinical trials. These second wave, first generation protease inhibitors have potent antiviral activity, but also have improved pharmacokinetics and side effect profiles compared to telaprevir and boceprevir. Like telaprevir and boceprevir, these agents suffer from a low genetic barrier to resistance and should be used in combination with pegylated interferon and ribavirin (Gane et al. 2011a). The second generation protease inhibitor MK-5172 (Merck Sharp and Dohme) inhibits the NS3/4A protease of HCV genotypes 1, 2 and 3 in vitro and is also active against telaprevir- and boceprevir-resistant HCV. MK-5172 has been shown to have potent antiviral activity clinically against both genotypes 1 and 3 HCV, was well tolerated and was not associated with viral rebounds indicating an improved resistance profile (Brainard et al. 2010).

Polymerase inhibitors: Agents that inhibit the active site of the HCV RNA-dependant RNA polymerase (NS5B) are also currently in development. There are two classes of polymerase inhibitors: nucleoside and non-nucleoside inhibitors.

Nucleoside NS5B inhibitors: These nucleoside inhibitors are incorporated by the NS5B polymerase into the elongating HCV genomic RNA causing premature chain termination. Nucleoside inhibitors have a high barrier to drug resistance and appear to act across different genotypes because the active site of the polymerase is highly conserved across the HCV genotypes (Buhler and Bartenschlager 2012). Mericitabine (Roche/Genentech), which is metabolized to a pyrimidine (cytosine) analogue, has been shown to increase SVR rates when combined with pegylated interferon and ribavirin, without the development of resistant variants (Pockros et al. 2011). Another nucleoside inhibitor sofosbuvir (Pharmasset/Gilead), a pyrimidine (uridine) analogue, has been shown to have potent efficacy against genotypes 1, 2 and 3 HCV (Lalezari et al. 2011; Nelson et al. 2011).

Non-nucleoside NS5B inhibitors: These NS5B inhibitors bind to 1 of 4 allosteric sites on the viral polymerase away from the catalytic site and induce conformational changes within the polymerase that results in the loss of enzymatic activity and subsequent inhibition of viral replication. Because these allosteric binding sites are genotype specific, all non-nucleoside NS5B inhibitors developed to date are only active against genotype 1 HCV. Some of these agents currently in development include BI207127 (Boehringer Ingelheim), VX-222 (Vertex) and ABT-072 (Abbott). Non-nucleoside NS5B inhibitors have a low genetic barrier to resistance and antiviral resistant variants are rapidly selected if these drugs are used alone (Lagace et al. 2010).

NS5A inhibitors: NS5A inhibitors bind to domain I of the NS5A protein, but the precise mechanism by which these drugs inhibit HCV replication remains unknown. Daclatasvir (Bristol-Myers Squibb) has been shown to potently inhibit HCV replication at low concentrations in phase I trials, but rapidly selects for antiviral resistance when used as monotherapy (Gao et al. 2010). In combination with pegylated interferon and ribavirin, daclatasvir results in viral clearance in about 90 % of treatment-naïve patients infected with HCV genotype 1 (Pol et al. 2011).

At present, almost all of the DAAs developed to date have to be used in combination with pegylated interferon and ribavirin in order to prevent the rapid selection of drug-resistant HCV variants. Hence, therapy with these DAAs results in further adverse effects in addition to those already experienced by using interferon and ribavirin alone. However, with the development of more potent DAAs, it is envisaged that interferon-free regimens using combinations of different classes of DAAs could result in successful HCV eradication without the development of antiviral resistance. Recently, it has been reported that the combination of the nucleoside polymerase inhibitor RG7128 and the protease inhibitor danoprevir, without pegylated interferon or ribavirin, was able to rapidly reduce HCV viral load by 5 $\log_{10}$ IU/mL without the development of drug-resistant HCV (Gane et al. 2010). In a subsequent study, the nucleoside polymerase inhibitor sofosbuvir in combination with ribavirin without pegylated interferon produced viral clearance in 100 % of patients with genotypes 2 and 3 HCV after only 12 weeks of therapy (Gane et al. 2011b). Combinations of DAAs may also be effective in patients who are unlikely to respond to treatment with pegylated interferon and ribavirin. In one study of patients with HCV genotype 1 who were null responders (less than 2 log IU/mL reduction in HCV viral load after 12 weeks of pegylated interferon and ribavirin) to previous treatment with pegylated interferon and ribavirin, 24 weeks of dual therapy with the NS5A inhibitor daclatasvir and the NS3/4A protease inhibitor asunaprevir produced a viral clearance rate of 90 % (Chayama et al. 2012).

6.4 Host-Targeted Agents

6.4.1 Interferon Lambda

Interferon lambda binds to a different cell surface receptor than interferon alpha, but shares the same intracellular JAK-STAT signalling pathway to produce its antiviral action against HCV. Unlike interferon alpha receptors, which are found ubiquitously in most tissues, interferon lambda-1 receptors are expressed abundantly on hepatocytes but not in many other cell types, thus allowing for a more targeted action within the liver (Sommereyns et al. 2008). In a study comparing pegylated interferon lambda and ribavirin with pegylated interferon alpha and ribavirin, interferon lambda produced higher RVR and complete EVR rates than interferon alpha and had significantly less adverse effects (Zeuzem et al. 2011b), indicating that interferon lambda may be a more effective and better-tolerated interferon for use in future DAA combinations.

6.4.2 Cyclophilin Inhibitors

The cyclophilins are a family of proteins found abundantly within all cells and perform a diverse array of functions including acting as protein chaperones to ensure correct protein folding, modulate protein function and intracellular signalling events (Wang and Heitman 2005). The cyclophilins are crucial for HCV viral replication, with NS5A known to interact with cyclophilin A (Hanoulle et al. 2009). Inhibition of cyclophilin A by cyclosporine was noted to inhibit HCV replication, but treatment with cyclosporine also results in the inhibition of calcineurin which is immunosuppressive. Subsequently, non-immunosuppressive cyclophilin inhibitors that do not interact with calcineurin have been developed. Alisporivir (Debiopharm/Novartis) interferes with the interaction between NS5A and cyclophilin A and suppresses viral load in patients infected with HCV genotypes 1–4 (Flisiak et al. 2009). The development of alisporivir, however, was recently suspended following reports of pancreatitis in association with its use in patients with chronic HCV.

6.5   Therapeutic Vaccines

The function of an effective therapeutic vaccine would be to prime the immune system of infected individuals to produce HCV-specific T cell-mediated responses in order to improve SVR rates when antiviral therapy is employed (Torresi et al. 2011). Although the results of therapeutic vaccines to date have been encouraging with regard to viral load reduction in inoculated individuals, they have had limited success in clearing HCV when used as monotherapy. As such, their current role is primarily as an adjunct to standard of care antiviral therapy. This treatment strategy would be beneficial in patients with poor pre-treatment prognostic markers or who have failed previous antiviral therapy.

Vaccines aimed at stimulating T cell responses against HCV are currently being studied in clinical trials. These include HCV vaccines based on synthetic peptide antigens (IC41, Intercell AG), tarmogens (GI-5005a, Globeimmune), modified HCV DNA (ChronVac-C, Tripep AB, Sweden) and modified vaccinia Ankara virus (TG4040, Transgene). Other strategies utilizing recombinant adenoviruses, and virus-like particles, are currently in the pre-clinical phase of development (Torresi et al. 2011). Some of the HCV vaccines currently in development are listed in Table 2.

Table 2 HCV preventive and therapeutic vaccines

Vaccine	Immunogenicity	Challenge inoculum	Outcome	Study phase	Company
IC41	Induce HCV-specific T cell responses	HCV peptide vaccine with polyarginine	Completed	Phase II	Intercell AG
GI-5005a	Reduction in ALT compared with placebo and reduction in viral load to −1.4 log	Inactivated recombinant Saccharomyces cerevisiae expressing NS3-Core fusion protein	Ongoing	Phase I	GlobeImmune
ChronVac-C	Safe, immunogenic and transient effects on viral load	DNA-based vaccine in combination with electroporation	Recruiting patients	Phase IIa	Tripep AB
TG4040	T cell responses and viral load reduction up to 1.5 log observed	MVA virus expressing non-structural proteins (NS3, NS4 and NS5B)	Recruiting patients	Phase I	Transgene
E1/E2 vaccine		Recombinant E1 and E2 proteins	Data not yet published	Phase I	Chiron Corp

Adapted from Torresi et al. (2011)

7 Prevention of the Acquisition of HCV in High-Risk Patients

7.1 Preventative Vaccines

Despite HCV being discovered over 20 years ago, there is still no effective vaccine to prevent HCV infection, and the development of a preventative vaccine remains an area of intense research (Torresi et al. 2011). We know that individuals who spontaneously clear HCV infection develop a strong and broadly cross-reactive CD4+ and CD8+ T cell responses against HCV core and non-structural proteins NS3, NS4 and NS5 (Lauer et al. 2004) as well as the production of cross-reactive neutralizing antibodies (Pestka et al. 2007). As such, prospective preventative vaccines would need to generate all these responses against the various genotypes of HCV. Few vaccine strategies other than live attenuated viruses or virus-like particles are likely to fulfil these criteria.

One vaccine in development that incorporates recombinant HCV E1 and E2 envelope glycoproteins (Chiron Corp) has been reported to protect chimpanzees against HCV genotype 1 infection (Vajdy et al. 2006). Another vaccine candidate has utilized a recombinant HCV core protein, although this vaccine is still in a relatively early stage of development (Drane et al. 2009).

7.2 Public Health Measures

As a safe and effective preventative vaccine remains elusive, strategies to reduce HCV transmission among individuals at high risk of acquiring the virus should be employed. In particular, harm reduction measures to reduce unsafe injecting practices among intravenous drug users, such as behavioural interventions, the access to sterile needles and syringes, and the management of substance abuse, have been shown to reduce the risk of HCV infection by about 75 % (Hagan et al. 2011). The universal screening of blood donors is also important to prevent transmission via contaminated blood products, as is adhering to universal precautions and strict needle-stick protocols within health care facilities.

8 Conclusions

HCV is currently a serious global health concern, chronically infecting about 3 % of the world's population, leading to chronic hepatitis, cirrhosis, liver failure and HCC, thereby causing significant morbidity and mortality. With the incidence of HCV infection increasing, the problem of HCV-associated HCC is expected to worsen as well, with the majority of HCCs developing in the setting of cirrhosis. Thus, it is imperative to provide antiviral therapy to infected individuals prior to the development of established cirrhosis in order to reduce the risk of subsequent

HCC. Even after the development of cirrhosis, successful HCV clearance is still associated with reduced HCC risk. There is a myriad of different classes of HCV-specific DAAs currently in development, which can be used in combination with one another or with current standard of care treatment to improve HCV cure rates. Preventative and therapeutic vaccines against HCV remain an area of ongoing research and hopefully an effective vaccine will be available in the future.

References

National institutes of health consensus development conference statement: Management of hepatitis c (2002) (June 10–12, 2002). Gastroenterology, 123(6), 2082–2099

Alter MJ (1995) Epidemiology of hepatitis c in the west. Semin Liver Dis 15(1):5–14

Bacon BR, Gordon SC, Lawitz E, Marcellin P, Vierling JM, Zeuzem S, Poordad F, Goodman ZD, Sings HL, Boparai N, Burroughs M, Brass CA, Albrecht JK, Esteban R, Investigators HR (2011) Boceprevir for previously treated chronic HCV genotype 1 infection. N Engl J Med 364(13):1207–1217

Bartenschlager R, Penin F, Lohmann V, Andre P (2011) Assembly of infectious hepatitis c virus particles. Trends Microbiol 19(2):95–103

Berenguer M, Ferrell L, Watson J, Prieto M, Kim M, Rayon M, Cordoba J, Herola A, Ascher N, Mir J, Berenguer J, Wright TL (2000) HCV-related fibrosis progression following liver transplantation: Increase in recent years. J Hepatol 32(4):673–684

Brainard DM, Petry A, Van Dyck K, Nachbar RB, De Lepeleire IM, Caro L, Stone JA, Sun P, Uhle M, Wagner FD, O'mara E, Wagner JA (2010) Safety and antiviral activity of MK-5172, a novel HCV N53/4a protease inhibitor with potent activity against known resistance mutants, in genotype 1 and 3 HCV-infected patients. Hepatology 52(4):706A–707A

Bressler BL, Guindi M, Tomlinson G, Heathcote J (2003) High body mass index is an independent risk factor for nonresponse to antiviral treatment in chronic hepatitis c. Hepatology (Baltimore, Md.), vol 38(3) pp 639–44

Bruix J, Sherman M, Practice Guidelines Committee A, A. F. T. S. O. L. D. (2005) Management of hepatocellular carcinoma. Hepatology (Baltimore, Md.) 42(5), 1208–1236

Bruno S, Stroffolini T, Colombo M, Bollani S, Benvegno L, Mazzella G, Ascione A, Santantonio T, Piccinino F, Andreone P, Mangia A, Gaeta GB, Persico M, Fagiuoli S, Almasio PL, Italian Association of the Study of the Liver D (2007) Sustained virological response to interferon-alpha is associated with improved outcome in HCV-related cirrhosis: A retrospective study. Hepatology (Baltimore, Md.), 45(3), 579–587

Buhler S, Bartenschlager R (2012) New targets for antiviral therapy of chronic hepatitis c. Liver international: official J Int Assoc Liver, 32 Suppl 1, pp 9–16

Chayama K, Takahashi S, Toyota J, Karino Y, Ikeda K, Ishikawa H, Watanabe H, Mcphee F, Hughes E Kumada H (2012) Dual therapy with the nonstructural protein 5a inhibitor, daclatasvir, and the nonstructural protein 3 protease inhibitor, asunaprevir, in hepatitis c virus genotype 1b-infected null responders. Hepatology (Baltimore, Md.) 55(3), 742–748

Chen SL, Morgan TR (2006) The natural history of hepatitis c virus (HCV) infection. Int J Med Sci 3(2):47–52

Chevaliez S, Pawlotsky J-M (2007) Hepatitis c virus: virology, diagnosis and management of antiviral therapy. World J Gastroenterology:WJG 13(17):2461–2466

Cho LY, Yang JJ, Ko K-P, Park B, Shin A, Lim MK, Oh J-K, Park S, Kim YJ, Shin H-R, Yoo K-Y, Park SK (2011) Coinfection of hepatitis b and c viruses and risk of hepatocellular carcinoma: systematic review and meta-analysis. Int J Cancer J Int du Cancer 128(1):176–184

Crotty S, Cameron C, Andino R (2002) Ribavirin's antiviral mechanism of action: Lethal mutagenesis? J Molecular Med-Jmm 80(2):86–95

Davis GL, Albright JE, Cook SF, Rosenberg DM (2003) Projecting future complications of chronic hepatitis c in the united states. Liver Transpl 9(4):331–338

Di Martino V, Rufat P, Boyer N, Renard P, Degos F, Martinot-Peignoux M, Matheron S, Le Moing V, Vachon F, Degott C, Valla D, Marcellin P (2001) The influence of human immunodeficiency virus coinfection on chronic hepatitis c in injection drug users: A long-term retrospective cohort study. Hepatology (Baltimore, Md.) 34(6), 1193–1199

Dibisceglie AM, Conjeevaram HS, Fried MW, Sallie R, Park Y, Yurdaydin C, Swain M, Kleiner DE, Mahaney K, Hoofnagle JH, Wright D (1995) Ribavirin as therapy for chronic hepatitis-c—a randomized, double-blind, placebo-controlled trial. Ann Intern Med 123(12), 897

Donato F, Tagger A, Gelatti U, Parrinello G, Boffetta P, Albertini A, Decarli A, Trevisi P, Ribero ML, Martelli C, Porru S, Nardi G (2002) Alcohol and hepatocellular carcinoma: the effect of lifetime intake and hepatitis virus infections in men and women. Am J Epidemiol 155(4):323–331

Dore GJ, Law M, Macdonald M, Kaldor JM (2003) Epidemiology of hepatitis c virus infection in Australia. J Clin Virol 26(2):171–184

Drane D, Maraskovsky E, Gibson R, Mitchell S, Barnden M, Moskwa A, Shaw D, Gervase B, Coates S, Houghton M, Basser R (2009) Priming of CD4 + and CD8 + t cell responses using a HCV core ISCOMATRIX vaccine: a phase I study in healthy volunteers. Human vaccines 5(3):151–157

El-Serag HB, Rudolph KL (2007) Hepatocellular carcinoma: epidemiology and molecular carcinogenesis. Gastroenterology 132(7):2557–2576

Flisiak R, Feinman SV, Jablkowski M, Horban A, Kryczka W, Pawlowska M, Heathcote JE, Mazzella G, Vandelli C, Nicolas-Metral V, Grosgurin P, Liz JS, Scalfaro P, Porchet H, Crabbe R (2009) The cyclophilin inhibitor Debio 025 combined with PEG IFnalpha2a significantly reduces viral load in treatment-naive hepatitis c patients. Hepatology (Baltimore, Md.) 49(5), 1460–1468

Forns X, Bukh J (1999) The molecular biology of hepatitis c virus. Genotypes and quasispecies. Clinics in liver disease 3(4), 693–716, vii

Fried MW, Shiffman ML, Reddy KR, Smith C, Marinos G, Goncales FL Jr, Haussinger D, Diago M, Carosi G, Dhumeaux D, Craxi A, Lin A, Hoffman J, Yu J (2002) Peginterferon alfa-2a plus ribavirin for chronic hepatitis c virus infection. The New England J Med 347(13):975–982

Gane EJ, Roberts SK, Stedman CAM, Angus PW, Ritchie B, Elston R, Ipe D, Morcos PN, Baher L, Najera I, Chu T, Lopatin U, Berrey MM, Bradford W, Laughlin M, Shulman NS, Smith PF (2010) Oral combination therapy with a nucleoside polymerase inhibitor (rg7128) and danoprevir for chronic hepatitis c genotype 1 infection (inform-1): a randomised, double-blind, placebo-controlled, dose-escalation trial. Lancet 376(9751):1467–1475

Gane EJ, Rouzier R, Stedman C, Wiercinska-Drapalo A, Horban A, Chang L, Zhang Y, Sampeur P, Najera I, Smith P, Shulman NS, Tran JQ (2011a) Antiviral activity, safety, and pharmacokinetics of danoprevir/ritonavir plus peg-IFN alpha-2a/RBV in hepatitis c patients. J Hepatol 55(5):972–979

Gane EJ, Stedman CA, Hyland RH, Sorensen RD, Symonds WT, Hindes R, Berrey MM (2011b) Once daily psi-7977 plus RBV: Pegylated interferon-alfa not required for complete rapid viral response in treatment-naive patients with HCV GT2 or GT3. Hepatology, 54(377A–377A)

Gao M, Nettles RE, Belema M, Snyder LB, Nguyen VN, Fridell RA, Serrano-Wu MH, Langley DR, Sun J-H, O'boyle DR 2nd, Lemm JA, Wang C, Knipe JO, Chien C, Colonno RJ, Grasela DM, Meanwell NA, Hamann LG (2010) Chemical genetics strategy identifies an HCV NS5a inhibitor with a potent clinical effect. Nature 465(7294):96–100

Ge D, Fellay J, Thompson AJ, Simon JS, Shianna KV, Urban TJ, Heinzen EL, Qiu P, Bertelsen AH, Muir AJ, Sulkowski M, Mchutchison JG, Goldstein DB (2009) Genetic variation in IL28b predicts hepatitis c treatment-induced viral clearance. Nature 461(7262):399–401

Hadziyannis SJ, Sette H, Jr, Morgan TR, Balan V, Diago M, Marcellin P, Ramadori G, Bodenheimer H, Jr, Bernstein D, Rizzetto M, Zeuzem S, Pockros PJ, Lin A, Ackrill AM &

Group, P. I. S. (2004) Peginterferon-alpha2a and ribavirin combination therapy in chronic hepatitis c: A randomized study of treatment duration and ribavirin dose. Anna Intern Med 140(5), 346–55

Hagan H, Pouget ER, Des Jarlais DC (2011) A systematic review and meta-analysis of interventions to prevent hepatitis c virus infection in people who inject drugs. J Infect Dis 204(1):74–83

Halfon P, Locarnini S (2011) Hepatitis c virus resistance to protease inhibitors. J Hepatol 55(1):192–206

Hanoulle X, Badillo A, Wieruszeski J-M, Verdegem D, Landrieu I, Bartenschlager R, Penin F, Lippens G (2009) Hepatitis c virus NS5a protein is a substrate for the peptidyl-prolyl cis/trans isomerase activity of cyclophilins a and b. J Biol Chem 284(20):13589–13601

Hayashi J, Aoki H, Kajino K, Moriyama M, Arakawa Y, Hino O (2000) Hepatitis C virus core protein activates the MAPK/ERK cascade synergistically with tumor promoter TPA, but not with epidermal growth factor or transforming growth factor alpha. Hepatology (Baltimore, Md.) 32(5), 958–961

Hino O, Kajino K, Umeda T, Arakawa Y (2002) Understanding the hypercarcinogenic state in chronic hepatitis: a clue to the prevention of human hepatocellular carcinoma. J Gastroenterol 37(11):883–887

Ishikawa T, Higuchi K, Kubota T, Seki K-I, Honma T, Yoshida T, Kamimura T (2012) Combination peg-IFN a-2b/ribavirin therapy following treatment of hepatitis c virus-associated hepatocellular carcinoma is capable of improving hepatic functional reserve and survival. Hepatogastroenterology 59(114):529–532

Jacobson IM, Mchutchison JG, Dusheiko G, Di Bisceglie AM, Reddy KR, Bzowej NH, Marcellin P, Muir AJ, Ferenci P, Flisiak R, George J, Rizzetto M, Shouval D, Sola R, Terg RA, Yoshida EM, Adda N, Bengtsson L, Sankoh AJ, Kieffer TL, George S, Kauffman RS, Zeuzem S, Team AS (2011) Telaprevir for previously untreated chronic hepatitis c virus infection. The New England J Med 364(25):2405–2416

Kamal SM, Nasser IA (2008) Hepatitis c genotype 4: What we know and what we don't yet know. Hepatology (Baltimore, Md.), 47(4), 1371–1383

Khuroo MS, Khuroo MS, Dahab ST (2004) Meta-analysis: a randomized trial of peginterferon plus ribavirin for the initial treatment of chronic hepatitis c genotype 4. Aliment Pharmacol Ther 20(9):931–938

Kisseleva T, Bhattacharya S, Braunstein J, Schindler CW (2002) Signaling through the JAK/STAT pathway, recent advances and future challenges. Gene 285(1–2):1–24

Kiyosawa K, Akahane Y, Nagata A, Furuta S (1984) Hepatocellular carcinoma after non-a, non-b posttransfusion hepatitis. Am J Gastroenterology 79(10):777–781

Kohli V, Singhal A, Elliott L, Jalil S (2012) Antiviral therapy for recurrent hepatitis c reduces recurrence of hepatocellular carcinoma following liver transplantation. Transplant Int: official Journal of the European Society for Organ Transplantation 25(2):192–200

Kwo PY, Lawitz EJ, Mccone J, Schiff ER, Vierling JM, Pound D, Davis MN, Galati JS, Gordon SC, Ravendhran N, Rossaro L, Anderson FH, Jacobson IM, Rubin R, Koury K, Pedicone LD, Brass CA, Chaudhri E, Albrecht JK (2010) Efficacy of boceprevir, an NS3 protease inhibitor, in combination with peginterferon alfa-2b and ribavirin in treatment-naive patients with genotype 1 hepatitis c infection (sprint-1): an open-label, randomised, multicentre phase 2 trial. Lancet 376(9742):705–716

Lagace L, Cartier M, Laflamme G, Lawetz C, Marquis M, Triki I, Bernard M-J, Bethell R, Larrey DG, Lueth S, Trepo C, Stern JO, Boecher WO, Steffgen J, Kukolj G (2010) Genotypic and phenotypic analysis of the NS5b polymerase region from viral isolates of HCV chronically infected patients treated with bi 207127 for 5-days monotherapy. Hepatology 52(4):1205A–1206A

Lalezari J, Lawitz E, Rodriguez-Torres M, Sheikh A, Freilich B, Nelson DR, Hassanein T, Mader M, Albanis E, Symonds W, Berrey MM (2011) Once daily psi-7977 plus pegIFN/RBV in a phase 2b trial: Rapid virologic suppression in treatment-naive patients with HCV GT2/GT3. J Hepatol 54(S28–S28)

Lauer GM, Barnes E, Lucas M, Timm J, Ouchi K, Kim AY, Day CL, Robbins GK, Casson DR, Reiser M, Dusheiko G, Allen TM, Chung RT, Walker BD, Klenerman P (2004) High resolution analysis of cellular immune responses in resolved and persistent hepatitis c virus infection. Gastroenterology 127(3):924–936

Law MG, Dore GJ, Bath N, Thompson S, Crofts N, Dolan K, Giles W, Gow P, Kaldor J, Loveday S, Powell E, Spencer J, Wodak A (2003) Modelling hepatitis c virus incidence, prevalence and long-term sequelae in Australia. Int J Epidemiol 32(5):717–724

Li X, Hui A-M, Sun L, Hasegawa K, Torzilli G, Minagawa M, Takayama T, Makuuchi M (2004) P16INK4a hypermethylation is associated with hepatitis virus infection, age, and gender in hepatocellular carcinoma. Clinical Cancer Res: Official J Am Assoc Cancer Res 10(22):7484–7489

Lok AS, Everhart JE, Wright EC, Di Bisceglie AM, Kim H.-Y, Sterling RK, Everson GT, Lindsay KL, Lee WM, Bonkovsky H. L, Dienstag JL, Ghany MG, Morishima C, Morgan TR & Group H.-C. T (2011) Maintenance peginterferon therapy and other factors associated with hepatocellular carcinoma in patients with advanced hepatitis c. Gastroenterology, 140(3), 840–849; quiz e12

Lok AS, Seeff LB, Morgan TR, Di Bisceglie AM, Sterling RK, Curto TM, Everson GT, Lindsay KL, Lee WM, Bonkovsky HL, Dienstag JL, Ghany MG, Morishima C, Goodman Z D & Group, H.-C. T. (2009) Incidence of hepatocellular carcinoma and associated risk factors in hepatitis c-related advanced liver disease. Gastroenterology, 136(1), 138–148

Manns MP, Mchutchison JG, Gordon SC, Rustgi VK, Shiffman M, Reindollar R, Goodman ZD, Koury K, Ling MH, Albrecht JK, & Int Hepatitis Interventional, T (2001) Peginterferon alfa-2b plus ribavirin compared with interferon alfa-2b plus ribavirin for initial treatment of chronic hepatitis c: A randomised trial. Lancet, 358(9286), 958–965

Marcellin P, Boyer N, Gervais A, Martinot M, Pouteau M, Castelnau C, Kilani A, Areias J, Auperin A, Benhamou JP, Degott C, Erlinger S (1997) Long-term histologic improvement and loss of detectable intrahepatic HCV RNA in patients with chronic hepatitis c and sustained response to interferon-alpha therapy. Ann Intern Med 127(10):875–881

Maruoka D, Imazeki F, Arai M, Kanda T, Fujiwara K, Yokosuka O (2012) Long-term cohort study of chronic hepatitis c according to interferon efficacy. J Gastroenterol Hepatol 27(2):291–299

Mchutchison JG, Everson GT, Gordon SC, Jacobson IM, Sulkowski M, Kauffman R, Mcnair L, Alam J, Muir AJ (2009a) Telaprevir with peginterferon and ribavirin for chronic HCV genotype 1 infection (vol 360, pg 1827, 2009). New England J Med 361(15), 1516–1516

Mchutchison JG, Lawitz EJ, Shiffman ML, Muir AJ, Galler GW, Mccone J, Nyberg LM, Lee WM, Ghalib RH, Schiff ER, Galati JS, Bacon BR, Davis MN, Mukhopadhyay P, Koury K, Noviello S, Pedicone LD, Brass CA, Albrecht JK, Sulkowski MS, Team IS (2009b) Peginterferon alfa-2b or alfa-2a with ribavirin for treatment of hepatitis c infection. The New England J Med 361(6):580–593

Mendenhall CL, Seeff L, Diehl AM, Ghosn SJ, French S W, Gartside PS, Rouster SD, Buskell-Bales Z, Grossman CJ, Roselle GA (1991) Antibodies to hepatitis b virus and hepatitis c virus in alcoholic hepatitis and cirrhosis: Their prevalence and clinical relevance. The VA cooperative study group (no. 119). Hepatology (Baltimore, Md.) 14(4 Pt 1), 581–589

Nelson DR, Lalezari J, Lawitz E, Hassanein T, Kowdley K, Poordad F, Sheikh A, Afdhal N, Bernstein D, Dejesus E, Freilich B, Dieterich D, Jacobson I, Jensen D, Abrams GA, Darling J, Rodriguez-Torres M, Reddy R, Sulkowski M, Bzowej N, Demicco M, Strohecker J, Hyland R, Mader M, Albanis E, Symonds WT, Berrey MM (2011) Once daily psi-7977 plus peg-IFN/RBV in HCV GT1: 98% rapid virologic response, complete early virologic response: The proton study. J Hepatology, 54(S544-S544)

Parkin DM (2001) Global cancer statistics in the year. Lancet Oncol 2(9):533–543

Patel K, Muir AJ, Mchutchison JG (2006) Diagnosis and treatment of chronic hepatitis c infection. BMJ (Clinical research ed.) 332(7548):1013–1017

Patterson JL, Fernandezlarsson R (1990) Molecular mechanisms of action of ribavirin. Rev Infect Dis 12(6):1139–1146

Pestka JM, Zeisel MB, Blaser E, Schurmann P, Bartosch B, Cosset F.-L, Patel AH, Meisel H, Baumert J, Viazov S, Rispeter K, Blum HE, Roggendorf M, Baumert TF (2007) Rapid induction of virus-neutralizing antibodies and viral clearance in a single-source outbreak of hepatitis C. Proc Nat Acad Sci United States Am 104(14), 6025–6030

Pockros P, Jensen D, Tsai N, Taylor RM, Ramji A, Cooper C, Dickson R, Tice A, Stancic S, Ipe D, Thommes JA, Vierling JM (2011) First svr data with the nucleoside analogue polymerase inhibitor mericitabine (rg7128) combined with peginterferon/ribavirin in treatment-naive HCV G1/4 patients: Interim analysis from the jump-c trial. J Hepatology, 54(S538–S538)

Pol S, Ghalib RH, Rustgi V, Martorell C, Everson G, Tatum H, Hezode C, Lim J, Bronowicki J.-P, Abrams G, Brau N, Morris D, Thuluvath P, Reindollar R, Yin P, Diva U, Hindes R, Mcphee F, Gao M, Thiry A, Schnittman S, Hughes E (2011) First report of SVR12 for a NS5a replication complex inhibitor, BMS-790052 in combination with peg-IFN-alfa-2a and RBV: Phase 2a trial in treatment-naive HCV-genotype 1 subjects. 46th annual meeting of the European association for the study of the liver (easl 2011). J Hepatol 54, Suppl. 1 (S544–S545)

Poordad F, Mccone J Jr, Bacon BR, Bruno S, Manns MP, Sulkowski MS, Jacobson IM, Reddy KR, Goodman ZD, Boparai N, Dinubile MJ, Sniukiene V, Brass CA, Albrecht JK, Bronowicki J-P, Investigators S (2011) Boceprevir for untreated chronic HCV genotype 1 infection. N Engl J Med 364(13):1195–1206

Poynard T, Bedossa P, Opolon P (1997) Natural history of liver fibrosis progression in patients with chronic hepatitis c. The obsvirc, metavir, clinivir, and dosvirc groups. Lancet 349(9055):825–832

Pradat P, Tillmann HL, Sauleda S, Braconier JH, Saracco G, Thursz M, Goldin R, Winkler R, Alberti A, Esteban JI, Hadziyannis S, Rizzetto M, Thomas H, Manns MP, Trepo C, Grp H (2007) Long-term follow-up of the hepatitis C HENCORE cohort: Response to therapy and occurrence of liver-related complications. J Viral Hepatitis 14(8):556–563

Raimondi S, Bruno S, Mondelli MU, Maisonneuve P (2009) Hepatitis c virus genotype 1b as a risk factor for hepatocellular carcinoma development: a meta-analysis. J Hepatol 50(6):1142–1154

Razali K, Thein HH, Bell J, Cooper-Stanbury M, Dolan K, Dore G, George J, Kaldor J, Karvelas M, Li J, Maher L, Mcgregor S, Hellard M, Poeder F, Quaine J, Stewart K, Tyrrell H, Weltman M, Westcott O, Wodak A, Law M (2007) Modelling the hepatitis c virus epidemic in Australia. Drug Alcohol Depend 91(2–3):228–235

Romero-Gomez M, Del Mar Viloria M, Andrade RJ, Salmeron J, Diago M, Fernandez-Rodriguez CM, Corpas R, Cruz M, Grande L, Vazquez L, Munoz-De-Rueda P, Lopez-Serrano P, Gila A, Gutierrez ML, Perez C, Ruiz-Extremera A, Suarez E, Castillo J (2005) Insulin resistance impairs sustained response rate to peginterferon plus ribavirin in chronic hepatitis c patients. Gastroenterology 128(3):636–641

Serfaty L, Aumaitre H, Chazouilleres O, Bonnand AM, Rosmorduc O, Poupon RE, Poupon R (1998) Determinants of outcome of compensated hepatitis c virus-related cirrhosis. Hepatology (Baltimore, Md.) 27(5), 1435–1440

Simmonds P, Bukh J, Combet C, Deleage G, Enomoto N., Feinstone S, Halfon P, Inchauspe G, Kuiken C, Maertens G, Mizokami M, Murphy DG, Okamoto H, Pawlotsky J.-M, Penin F, Sablon E, Shin-I T, Stuyver LJ, Thiel H.-J, Viazov S, Weiner AJ, Widell A (2005) Consensus proposals for a unified system of nomenclature of hepatitis c virus genotypes. Hepatology (Baltimore, Md.) 42(4), 962–973

Singal AK, Freeman DH Jr, Anand BS (2010a) Meta-analysis: Interferon improves outcomes following ablation or resection of hepatocellular carcinoma. Aliment Pharmacol Ther 32(7):851–858

Singal AK, Singh A, Jaganmohan S, Guturu P, Mummadi R, Kuo Y-F, Sood GK (2010b) Antiviral therapy reduces risk of hepatocellular carcinoma in patients with hepatitis c virus-related cirrhosis. Clinical Gastroenterology Hepatology:Official Clinical Prac J Am Gastro-enterological Assoc 8(2):192–199

Sommereyns C, Paul S, Staeheli P, Michiels T (2008) IFN-lambda (IFN-lambda) is expressed in a tissue-dependent fashion and primarily acts on epithelial cells in vivo. PLoS Pathog 4(3):e1000017

Streeter DG, Witkowsk Jt, Khare GP, Sidwell RW, Bauer R J, Robins RK, Simon LN (1973) Mechanism of action of 1-beta-d-ribofuranosyl-1,2,4-triazole-3-carboxamide (virazole)—new broad-spectrum antiviral agent. Proc Nat Acad Sci United States Am 70(4), 1174–1178

Sun C-A, Wu D-M, Lin C–C, Lu S-N, You S-L, Wang L-Y, Wu M-H, Chen C-J (2003) Incidence and cofactors of hepatitis c virus-related hepatocellular carcinoma: a prospective study of 12,008 men in Taiwan. Am J Epidemiol 157(8):674–682

Takeyasu M, Akuta N, Suzuki F, Seko Y, Kawamura Y, Sezaki H, Suzuki Y, Hosaka T, Kobayashi M, Kobayashi M, Arase Y, Ikeda K, Kumada H (2012) Long-term interferon monotherapy reduces the risk of HCV-associated hepatocellular carcinoma. J Med Virol 84(8):1199–1207

Thomas H, Foster G, Platis D (2003) Mechanisms of action of interferon and nucleoside analogues. J Hepatology, 39 Suppl 1(S93-8)

Thomas MB, Zhu AX (2005) Hepatocellular carcinoma: the need for progress. J Clinical Oncology: Official J Am Society Clinical Oncology 23(13):2892–2899

Torresi J, Johnson D, Wedemeyer H (2011) Progress in the development of preventive and therapeutic vaccines for hepatitis c virus. J Hepatol 54(6):1273–1285

Vajdy M, Selby M, Medina-Selby A, Coit D, Hall J, Tandeske L, Chien D, Hu C, Rosa D, Singh M, Kazzaz J, Nguyen S, Coates S, Ng P, Abrignani S, Lin Y-L, Houghton M, O'hagan DT (2006) Hepatitis c virus polyprotein vaccine formulations capable of inducing broad antibody and cellular immune responses. J Gen Virol 87(Pt 8):2253–2262

Wang P, Heitman J (2005) The cyclophilins. Genome Biol 6(7):226

Wiley TE, Mccarthy M, Breidi L, Layden TJ (1998) Impact of alcohol on the histological and clinical progression of hepatitis c infection. Hepatology (Baltimore, Md.) 28(3), 805–9

Yu J-W, Wang G-Q, Sun L-J, Li X-G, Li S-C (2007) Predictive value of rapid virological response and early virological response on sustained virological response in HCV patients treated with pegylated interferon alpha-2a and ribavirin. J Gastroenterol Hepatol 22(6):832–836

Zeuzem S, Andreone P, Pol S, Lawitz E, Diago M, Roberts S, Focaccia R, Younossi Z, Foster GR, Horban A, Ferenci P, Nevens F, Muellhaupt B, Pockros P, Terg R, Shouval D, Van Hoek B, Weiland O, Van Heeswijk R, De Meyer S, Luo D, Boogaerts G, Polo R, Picchio G, Beumont M, Team RS (2011a) Telaprevir for retreatment of HCV infection. N Engl J Med 364(25):2417–2428

Zeuzem S, Arora S, Bacon B, Box T, Charlton M, Diago M, Dieterich D, Esteban Mur R, Everson GT, Fallón M, Ferenci P, Flisiak R, George J, Ghalib R, Gitlin N, Gladysz A, Gordon S, Greenbloom S, Hassanein T, Jacobson I, Jeffers L, Kowdley K, Lawitz E, Lee S, Leggett B, Lueth S, Nelson D, Pockros P, Rodriguez-Torres M, Rustgi V, Serfaty L, Sherman M, Shiffman M, Sola R, Sulkowski M, Vargas H, Vierling J, Yoffe B, Ishak L, Fontana D, Xu D, Lester J, Gray T, Horga A, Hillson J, Ramos E, Lopez-Talavera JC, Muir A (2011b) Pegylated interferon-lambda (pegIFN-l) shows superior viral response with improved safety and tolerability versus pegIFNa-2a in HCV patients (G1/2/3/4): Emerge phase IIb through week 12. J Hepatol, 54 Suppl. 1(S538–S539)

The Role of Human Papillomaviruses in Oncogenesis

Kristen K. Mighty and Laimonis A. Laimins

Abstract

Human papillomaviruses (HPVs) are the causative agents of cervical and other anogenital as well as oral cancers. Approximately fifty percent of virally induced cancers in the USA are associated with HPV infections. HPVs infect stratified epithelia and link productive replication with differentiation. The viral oncoproteins, E6, E7, and E5, play important roles in regulating viral functions during the viral life cycle and also contribute to the development of cancers. p53 and Rb are two major targets of the E6 and E7 oncoproteins, but additional cellular proteins also play important roles. E5 plays an auxiliary role in contributing to the development of cancers. This review will discuss the various targets of these viral proteins and what roles they play in viral pathogenesis.

Contents

K. K. Mighty · L. A. Laimins (✉)
Department of Microbiology-Immunology, Feinberg School of Medicine,
Northwestern University, 303 E, Chicago Avenue, Morton 6-681, Chicago, IL 60611, USA
e-mail: l-laimins@northwestern.edu

M. H. Chang and K.-T. Jeang (eds.), *Viruses and Human Cancer*,
Recent Results in Cancer Research 193, DOI: 10.1007/978-3-642-38965-8_8,
© Springer-Verlag Berlin Heidelberg 2014

1 Introduction

Human papillomaviruses (HPVs) are small, double-stranded DNA viruses that contain circular genomes of approximately 8 kilobases (kb) in length and encode approximately eight major open reading frames (ORFs) (Howley and Lowy 2007). These viruses have a tropism for squamous epithelial tissues of the feet, hands, and anogenital tracts where they propagate via an unusual life cycle utilizing the differentiation program of the host cell (zur Hausen and de Villiers 1994). There is vast diversity among the HPVs with almost 200 different types identified to date. Even though the viruses are genetically distinct, they all infect epithelial tissues, only differing in the preferential target body location for infection (zur Hausen and de Villiers 1994). About 40 types of HPVs exhibit tropism for the genital tract, while the remaining HPVs specifically infect cutaneous tissues (Howley and Lowy 2007; Moody and Laimins 2010).

Genital HPVs can be further categorized into two main groups, high-risk (HR) and low-risk (LR) types, which is determined by the propensity of the infection to progress to malignancy. HR genital HPVs, including types 16, 18, 31, 33, 45, and 5, are frequently associated with cervical carcinomas (zur Hausen and de Villiers 1994). HPV DNA is found in over 99 % of cervical cancers, while HPV DNA exists as extrachromosomal elements, or episomes, in precancerous lesions. Although up to 50 % of HPV 16-positive and most HPV 18-positive carcinomas maintain HPV DNA as episomes (Parkin et al. 2000). While at least 10 HPV types can contribute to development of cervical cancer, three types are the major contributors: HPV 16 is found in about 50 % of cervical cancers, HPV 18 in approximately 25 %, and HPV 31 in about 10 % of cases (Fehrmann and Laimins 2003; Longworth and Laimins 2004a). Conversely, LR genital HPVs are rarely associated with malignancies and instead primarily cause benign genital warts. LR genital HPVs include types 6, 11, 42, 43, and 44, of which types 6 and 11 are responsible for about 90 % of all genital warts (Lorincz et al. 1992).

High-risk HPVs (HR-HPVs) are the causative agents of most cervical cancers accounting for up to 5 % of all human cancers (Stanley 2010). Cervical cancer is the second most common cancer in women worldwide with over 500,000 new cancer cases diagnosed each year and is the third leading cancer killer, causing nearly 300,000 deaths in women annually (Ferlay et al. 2008). The use of the *Papanicolaou smear* (Pap smear) has reduced the incidence of cervical cancers by over 80 % in the USA in the past 50 years (Moody and Laimins 2010). Additionally, the Federal Drug and Administration (FDA) recently approved the use of two prophylactic polyclonal vaccines targeted against the most common HPV types associated with cervical cancers and genital warts. The use of these vaccines is approved for males and females between the ages of 10 and 25. Gardasil® is the quadrivalent vaccine (HPV 6, 11, 16, 18) developed by Merck, and Cervarix® is the bivalent (HPV 16, 18) vaccine developed by GlaxoSmithKline. Both of these vaccines are effective in preventing initial HPV infection (Markowitz et al. 2007). These vaccines are recommended for individuals, who have never been sexually active since they can only prevent new infections, and not treat existing infections or used as a cancer treatment. Immunization with one of these vaccines along with annual Pap smear screening is the most effective prevention strategy.

HPV infections are sexually transmitted and rank as the most common sexually transmitted viral infections (Markowitz et al. 2007). Recent analyses indicate that approximately 20 million Americans are currently infected with HPV. Further studies by the American Social Health Association estimate that 75 % of sexually active individuals between the ages of 15 and 49 have been infected with HPV at some point in their lives. It is also estimated that more than half of infection-related cancers in females are attributed to HPV (zur Hausen 2009). Together, these findings demonstrate the major public health impact HPV infection and cervical cancer has in the USA and globally. This exemplifies the need for continued research for prevention and treatment of HPV infections and HPV-related cancers.

Infection with genital HPVs in most individuals can last for up to 2 years due to the ability of the virus to evade the innate immune surveillance, and thus delaying onset of an adaptive immune response. This immune evasion is due in part to HPVs not being cytolytic and the fact that viral proteins are expressed at very low levels (Bodily and Laimins 2011). In addition, HPV-induced innate immune evasion results in a delayed onset of the adaptive immune response, facilitating the persistence and productive replication of the virus. Eventually, most people are able to mount an effective cell-mediated immune response, which clears the viral infection (Stanley 2008). However, up to 20 % of women fail to clear the infection and are at high risk of developing cervical carcinoma (Bodily and Laimins 2011). Although persistent HR-HPV infection is the single most important contributing factor to the development of cervical cancer, other risk factors such as immuno-suppression, cigarette smoking, and infection with human immunodeficiency virus (HIV) also contribute to progression to malignancy (Markowitz et al. 2007; Bodily and Laimins 2011; zur Hausen 1996).

The development of cervical carcinoma is not a rapid event as it typically occurs over a period of several decades. In the cases where initial infection is not cleared by immune surveillance, precancerous lesions can typically develop within a decade and cervical cancer within several decades. To understand how viral infection can progress to cervical cancer, it is first important to examine the unusual life cycle of the virus and genomic arrangement.

1.1   The HPV Life Cycle

The HPV life cycle is intimately associated with the differentiation program of epithelial cells (Fig. 1). To understand the unusual life cycle of HPV, it is first important to understand the normal epithelial differentiation program. In normal epithelial cells, the only actively dividing cells are present in the basal layers of the stratified epithelium, which consist of transit amplifying (TA) cells and stem cells.

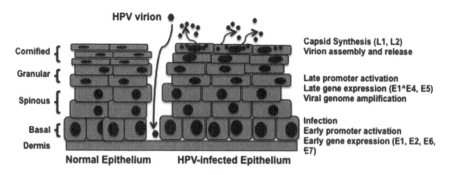

Fig. 1 Differentiation-dependent HPV Life Cycle. The HPV life cycle is intimately associated with the differentiation program of epithelial cells. HPV infects actively dividing cells in basal layer of the epithelium via a micro-abrasion. Following entry, viral gene expression is activated and episomal HPV DNA is maintained at approximately 50–100 copies per cell. HPV oncoproteins then enable the infected cells that exit the basal layer to remain active in the cell cycle. Once the infected cells begin to differentiate, the late promoter is activated, which results in the onset of the productive phase of the viral life cycle. During this phase, viral DNA amplification occurs and viral protein expression is increased. Finally, synthesis of viral capsids and packaging occurs in the uppermost differentiated layer of the epithelium, followed by release of the progeny virions

TA cells are defined as cells that are proliferating and can terminally differentiate. In contrast, stem cells have the potential to proliferate indefinitely, but divide infrequently in order to replenish the TA cell pool. Once cells in the basal layer divide and one daughter cell migrates suprabasally and differentiates, it loses its ability to remain active in the cell cycle. The differentiating cells go through a number of events including changes in gene expression and desquamation (Bodily and Laimins 2011).

Upon infection, the virus establishes its double-stranded DNA genome in the nuclei of infected host cells (Howley and Lowy 2007). HPV gains entry to cells in the basal layer of the epithelium that become exposed through micro-abrasions (Moody and Laimins 2010). Infection of the basal layer allows the virus to establish a persistent infection, as the basal cells are the only cells of the epithelium undergoing active replication (Moody and Laimins 2010). Since the HPV genomes are only 8 kb in size, they do not encode viral polymerases or other enzymes required for viral replication. The virus must therefore rely on host cell replication machinery to facilitate viral DNA synthesis (Moody and Laimins 2010). Following entry, viral genomes are established in the nucleus as extra-chromosomal plasmids, or episomes. In the infected basal cells, early viral gene expression is activated, and genome copy numbers are maintained at approximately 20–100 copies per cell (Moody and Laimins 2010). As HPV-infected basal cells divide, one of the infected daughter cells remains in the basal layer. The other daughter cell migrates away from the basal layer and begins to differentiate, resulting in the activation of the late viral promoter. This results in the onset of the productive phase of the life cycle, which includes viral DNA amplification, with copy number increasing to over 100 copies of HPV DNA per cell, and the onset of capsid gene expression. Finally, synthesis of viral capsids and packaging of viral genomes occur in the uppermost differentiated layer of the epithelium, ultimately resulting in the release of the progeny virions.

The signals that control the induction of late viral events in the life cycle are not well characterized, but studies have shown that HPV oncoproteins enable infected cells in the suprabasal layer to remain active in the cell cycle and to reenter S phase or arrest in G2/M to allow for viral amplification. This alteration of cell cycle control is essential for activation of the productive phase of the life (Moody and Laimins 2010). Furthermore, studies have indicated that viral proteins E6, E7, E1^E4, and E5 are needed for this activation, and these activities will be briefly summarized below.

1.2   The HPV Genome

The small, double-stranded DNA genome of all HPVs is approximately 8 kb in size. On average, HPVs encode eight major ORFs which are expressed from polycistronic mRNAs transcribed from a single DNA strand (Howley and Lowy 2007) (Fig. 2). The early proteins, E1, E2, E6, and E7, are expressed early in infection in undifferentiated cells and have drastically different functions. Sequences within the upstream regulatory region (URR) located in the non-coding region of the genome are responsible for regulation of viral transcription and replication. Expression of HPV gene products is directed from two different promoters, the early promoter and the late promoter (Moody and Laimins 2010). The early promoter, termed P_{97} in HPV 31, is located upstream of the E6 ORF and directs expression of early (E) gene products in undifferentiated cells. Early proteins include E1, E2, E6, E7, E1^E4, and E5. Translation of HPV messages occurs by a leaky scanning mechanism resulting in high levels of E6 and E7, but low levels of E1^E4 and E5 protein synthesis. The E1 and E2 proteins function in replication and transcription control, while E1^E4 modulates late viral functions. The late promoter, P_{742} in HPV 31, directs the expression of late (L) gene products and is located within the E7 ORF. Importantly, P_{742} is activated upon epithelial cell differentiation. Late proteins include L1 and L2 as well as E1^E4, and E5, and these are all expressed from P_{742} (Moody and Laimins 2010).

1.3   Oncoproteins: E5, E6, and E7

The E6 and E7 proteins are expressed upon initial HPV infection of host keratinocytes, while E5 is primarily expressed in the late phase of the life cycle. Although these proteins all contribute to promoting tumor growth in host cells, each of these proteins has distinct functions. As previously discussed, E1 and E2 are responsible for replication and regulation of viral transcription, whereas E6 and

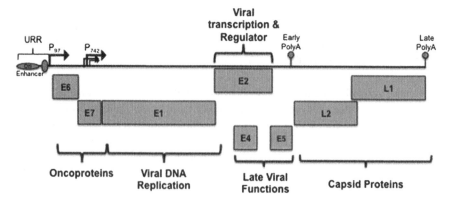

Fig. 2 Linear arrangement of HPV genome. The HPV genome is represented in linear form here for simplicity. The genomes are small, circular, double-stranded DNA genomes of 8 kilobases (*kb*) in size. There are on average 8 open reading frames (*ORFs*) (E1, E2, E4, E5, E6, E7, L1, and L2) expressed from a single polycistronic transcript transcribed from a single strand of DNA. The upstream regulatory region (*URR*) is located in the non-coding region and contains sequences responsible for regulating viral transcription and replication. Three general groups of HPV genes that are regulated during differentiation are the virus early promoter (P_{97}), the differentiation-dependent late promoter (P_{742}), and two polyadenylation signals (PolyA). P_{97} is located upstream of the *E6* ORF and directs expression of early genes in undifferentiated cells. P_{742} is located within the *E7* ORF, is activated upon differentiation of the host cells, and directs expression of late gene products. *E6* and *E7* are oncogenes involved in replication competence. *E1* and *E2* are genes involved in viral DNA replication and regulation of viral transcription. *E4* and *E5* are genes involved in late functions. *L1* and *L2* are the capsid proteins

E7 proteins are largely responsible for modulating cell cycle progression. In the HR-HPVs, E6 and E7 act as oncoproteins that are necessary for the development of genital carcinomas. Conversely, no such function has been demonstrated for LR-HPV proteins. While both E6 and E7 proteins are localized to the host nucleus, the E6 proteins are also detected in the cytoplasm of HPV-infected cells (Howley and Lowy 2007; Moody and Laimins 2010). Studies have shown that expression of E6 proteins is sufficient for immortalization of human mammary epithelial cells and transformation of NIH3T3 fibroblasts; however, expression of both E6 and E7 proteins is required for efficient immortalization of human keratinocytes (Howley and Lowy 2007).

E6 proteins are approximately 150 amino acids (18 kilodaltons) in size and contain two zinc-binding domains consisting of four Cys–X–X–Cys motifs (Howley and Lowy 2007). Studies have identified various E6-mediated activities that are mediated by interactions with over a dozen different proteins. One of the well-characterized interactions is the binding of E6 proteins to the tumor suppressor protein p53, affecting p53-dependent cell cycle regulation. The p53 protein is important in regulating the G_1/S and G_2/M cell cycle checkpoints following DNA damage (Slee et al. 2004; Oren 2003). Studies have shown that E6 proteins form complexes with an E3 cellular ubiquitin ligase, E6-associated protein

(E6AP), and p53, resulting in rapid proteasomal (26S) degradation of p53 (Scheffner et al. 1990). In a normal response to DNA damage or unscheduled induction of replication, p53 is activated via various modifications. The activation of this short-lived transcription factor results in modulation of the cell cycle and in, some cases, activation of apoptotic processes. Activated p53 forms a homotetramer that transcriptionally activates expression of cell cycle regulatory proteins, such as cyclin kinase inhibitor p21, which is responsible for inducing a G_1/S arrest (Ko and Prives 1996). The activation of p53 can also induce programmed cell death (apoptosis). Cell cycle arrest allows the cell to repair the damage to the DNA prior to entry to S phase. In the event that the damage is too extensive, the cell triggers apoptosis to prevent a cell from replicating damaged DNA. In addition, p53 is activated following viral infection. Given that HPV relies on host cell machinery and S phase entry to replicate its genome, the virus has devised a mechanism to disrupt normal p53 action. The E6-mediated proteasomal degradation of p53 results in the deregulation of the cell cycle, which allows the virus to persist and replicate its genome (Moody and Laimins 2010).

An additional HR E6 activity is its ability to interact with p300/CBP, a p53 co-activator (Patel et al. 1999; Zimmermann et al. 1999). The p300/CBP/E6 interaction prevents acetylation of p53, which down-regulates p53 activity, therefore blocking cell cycle arrest. The binding of p300/CBP occurs independently of E6-mediated degradation of p53 (Patel et al. 1999; Zimmermann et al. 1999). Interestingly, studies indicate that immortalization competency is not exclusively linked to p53-dependent mechanisms since E6 mutants incapable of degrading p53 are still able to immortalize cells; similarly, E6 mutants with normal degradation activity fail to immortalize cells (Kiyono et al. 1997).

One p53-independent function of E6 is the activation of telomerase by HR E6 proteins (Klingelhutz et al. 1996). Telomerase is an enzyme with four subunits that replicates telomeric DNA at the ends of chromosomes by adding hexamer repeats. Expression of its catalytic subunit, human telomerase reverse transcriptase (hTERT), plays an essential role in regulation of telomerase activity (Liu 1999). Over successive cell divisions, the telomeres become critically shortened and dysfunctional, leading to a limited cell proliferative lifespan because of the induction of senescence and irreversible cell growth arrest. In contrast, in cancers, hTERT expression is typically reactivated resulting in reconstitution of telomerase activity (Liu 1999). Studies have revealed that E6 increases expression of endogenous hTERT levels through transcriptional activation of the hTERT promoter through the action of NFX1-123, Myc and Sp-1 (Kyo et al. 2000; Howie et al. 2009; Gewin and Galloway 2001; Katzenellenbogen et al. 2009). While this activity is not the only function necessary for efficient immortalization of the host cell, it is a crucial function of E6.

Another interaction important for the ability of E6 proteins to immortalize cells is its association with several PDZ domain–containing proteins. PDZ domains are approximately 90 amino acids in size and are binding domains for a number of proteins including post-synaptic density protein (PSD-95), *Drosophila* disk large tumor suppressor (Dlg1), and zonula occludens-1 protein (zo-1). These domains are often found in proteins responsible for cell–cell adhesion as well as cell signaling and typically localized in areas of cell–cell contact. Studies have shown that PDZ proteins MUPP-1, hDLG, hScribble, and MAGI-1, 2, 3 bind to the extreme C-terminus of HR E6 proteins resulting in degradation of the PDZ protein (Lee et al. 1997, 2000). While PDZ protein interactions with E6 proteins may contribute to malignant progression, complete characterization of the mechanisms involved is still unclear.

HR E6 proteins have also been found to interact with various other cellular factors including paxillin, the putative calcium-binding protein E6-BP, and the interferon regulatory factor IRF-3 (Patel et al. 1999; Ronco et al. 1998). Several studies have identified many cellular binding partners for LR E6 proteins such as MCM7, Bak, zyxin, and GPS2 (Kuhne and Banks 1998; Kukimoto et al. 1998; Thomas and Banks 1999). It is clear from the characterization of numerous E6-binding partners and activities, that this viral protein is essential in the viral life cycle since knockout of E6 in genomes results in loss of ability to maintain episomes (Thomas et al. 1999).

The second oncoprotein, E7, is approximately 100 amino acids in size and is able to form dimers via its C-terminus. HR E7 proteins are comprised of three conserved regions: CR1 present at the N-terminus; CR2 containing an LXCXE motif that binds the retinoblastoma protein (Rb); and CR3, which contains two zinc finger-like motifs (Dyson et al. 1992). The CR1 and CR2 domains have sequence homology to the conserved regions CR1 and CR2 in adenovirus E1A. E7 is able to transform NIH3T3 fibroblasts by itself and with increased efficiency upon co-expression of E6. In contrast to E6, E7 oncoproteins are able to immortalize human keratinocytes when expressed alone, albeit, at a very low frequency (Howley and Lowy 2007; Munger et al. 1989; Riley et al. 2003).

A central activity of E7 proteins is their association with the Rb family of proteins (Dyson et al. 1989). Rb, p107, and p130 are the members of this family, and their expression occurs throughout the cell cycle, regulating cell cycle progression. In order to understand how E7 can circumvent Rb-regulated cell cycle progression, it is first important to appreciate how Rb regulates the cell cycle in normal circumstances. Unphosphorylated Rb proteins and the E2F/DP1 transcription factors form complexes to repress transcription of genes involved in S phase progression (DNA synthesis) or apoptosis. E2F transcription factors regulate the transcription of proteins required for normal cellular DNA synthesis. The transition from G_1 to S phase is triggered when cyclin kinase complexes phosphorylate the Rb proteins causing their release from the E2F complex relieving transcriptional repression of transcription of genes involved in DNA synthesis.

E7 alters the regulation of G_1/S by promoting the constitutive expression of E2F-regulated genes by binding Rb and sequestering it away from forming E2F/DP1 complexes. This relieves transcriptional repression and allows genes required for DNA synthesis to be transcribed (Edmonds and Vousden 1989; Weintraub et al. 1995). Additionally, E7 is able to target Rb for ubiquitin-mediated proteasomal degradation, which again allows for E2F-regulated genes to be constitutively transcribed, promoting DNA synthesis (Howley and Lowy 2007; Moody and Laimins 2010). Rb proteins are also responsible for controlling cell cycle exit during epithelial differentiation; hence, E7-mediated abrogation of Rb function maintains cell cycle activity. This activity is necessary for productive viral replication to occur in the differentiated epithelial cells (Thomas et al. 1999). The importance of the Rb-E7 is also essential for the virus' ability to maintain the genome as episomes in undifferentiated cells (Longworth and Laimins 2004b).

Another way in which E7 proteins can affect cell cycle progression is via its ability to associate with cyclins and cyclin-dependent kinases (cdk) inhibitors. For example, the association of E7 with cyclins A and E as well as cdk inhibitors p21 and p27 has been characterized in various studies (Davies et al. 1993; Funk et al. 1997; Jones et al. 1997; Ruesch and Laimins 1998; Tommasino et al. 1993; Zerfass-Thome et al. 1996). These cellular proteins affect the phosphorylation status of Rb proteins, and in doing so facilitate cell cycle progression. Specifically, E7 proteins can bind cyclins A, E, p21, and p27 resulting in an increase in cyclin A and E levels while causing a decrease in p21 and p27 levels (Jones et al. 1997; Ruesch and Laimins 1998; Tommasino et al. 1993). The net effect of these interactions is to drive progression of the cell cycle, which is necessary for facilitating the HPV life cycle in differentiating epithelia.

Alteration of Rb phosphorylation status is not the only means of affecting E2F-responsive promoters as they can also be repressed by action of histone deacetylases (HDACs). Targeting of class I HDACs is another mechanism in which E7 affects cell cycle regulation (Longworth and Laimins 2004a, b; Brehm et al. 1998). HDACs are ubiquitously expressed transcriptional co-repressors that remove acetyl groups from lysine-rich N-terminal tails of histone proteins in the nucleosome. Additionally, HDACs can directly deacetylated E2F-responsive factors resulting in the loss of their function. There are three classes of HDACs, which are classified according to sequence homology and localization of the factors in the cell. Class I HDACs are localized to the nucleus and include HDACs 1, 2, 3, and 8. Class I HDACs require binding to cofactor proteins that either modify their activity or localize them to site of action. HR E7 proteins indirectly associate with class I HDACs through direct binding to the auxiliary protein, Mi2β, via sequences in the zinc finger regions at the C-terminus (Brehm et al. 1998). This HDAC/E7 interaction specifically results in increased levels of E2F-responsive transcription in differentiating cells, which allows cells to maintain cell cycle activity (Longworth and Laimins 2004b). The binding of E7 to HDACs is also important in facilitating the viral life cycle as HPV 31 genomes with mutations abrogating the E7/HDAC interaction have slower growth, are defective in maintaining genomes as episomes, and have a limited life span (Longworth and Laimins 2004b).

Many cancers exhibit increased genomic instability, and similar effects are seen in HPV-induced cancers. HR E7 expression is responsible for inducing genomic instability in these cancers. In biopsies isolated from HPV-positive cancers, high levels of aneuploidy are observed suggesting that changes in chromosome numbers may promote the progression from low-grade lesions to malignancy. During normal cell division, centrosomes coordinate equal segregation of chromosomes to the daughter cells. Expression of E7 in human keratinocytes results in an increase in the number of cells harboring abnormal centrosome numbers, indicating E7 is a major inducer of chromosome missegregation (Duensing et al. 2000). Furthermore, E7 was found to induce chromosomal abnormalities in cells deficient of p130, Rb, and p107, indicating that this action is independent of E7's ability to bind and/or degrade Rb (Duensing and Munger 2003).

The third oncoprotein, E5, is a small, hydrophobic membrane protein that is localized primarily to the endoplasmic reticulum, but is also found in the Golgi and plasma membrane (Conrad et al. 1993; Disbrow et al. 2005). The HPV E5 proteins are approximately 84 amino acids in size and are primarily expressed late in the viral life cycle. The function of E5 in the viral life cycle is less well understood as compared to other viral proteins; however, studies have begun to illuminate E5's role in promoting cancer. In Bovine papillomavirus (BPV), E5 proteins exhibit efficient transforming ability of rodent fibroblasts when expressed alone (Petti et al. 1991). In contrast, HR-HPV E5 proteins exhibit very weak transforming ability when expressed alone. However, when HR-HPV E5 proteins are expressed in conjunction with E6 and E7 proteins, E5 proteins enhance transformation capacity of the cells (Stoppler et al. 1996; Valle and Banks 1995; Bouvard et al. 1994). The oncogenic potential of E5 proteins was most evidently demonstrated in estrogen-treated transgenic mice expressing E5 alone, which rapidly developed cervical cancer (Maufort et al. 2007). Although this is not the exclusive pathway by which HPVs promote tumorigenesis, these studies have shown that E5 proteins are important for the virus' survival, possibly serving as a putative target for cervical cancer therapies (Valle and Banks 1995; Bouvard et al. 1994; DiMaio and Mattoon 2001).

A number of studies have identified proteins that associate with E5. One such association is the binding and subsequent alteration of the activity of the epidermal growth factor receptor (EGFR) (Straight et al. 1995). In addition, E5 is able to interact with the 16 kDa subunit of the vacuolar proton-ATPase, which alters endosomal pH, endocytic trafficking, and may contribute to the alteration of EGFR turnover (Conrad et al. 1993; Disbrow et al. 2005; Straight et al. 1993). Studies have also illustrated the importance of E5 in activating late functions in the productive phase of the viral life cycle. This was shown using stable keratinocyte cells lines containing HPV 31 genomes harboring wild-type or translational termination mutant E5 sequences (Fehrmann et al. 2003). Recently, a split-ubiquitin yeast two-hybrid system yielded identification of novel E5-binding partners, including a B cell receptor protein (BAP31), which is involved in the regulation of membrane protein transport. This interaction has been shown to be important for maintaining proliferative capacity of HPV-infected cells following differentiation (Regan and Laimins 2008).

2   Summary

Human papillomaviruses are important human pathogens that are responsible for the induction of a variety of human cancers. Despite the introduction of vaccines against HPV, they only protect against initial infections and no therapeutics, other than surgery, are available to treat existing HPV lesions. The E6 and E7 viral proteins provide important functions in the viral life cycle and also provide major contributions to progression to malignancy. Among the E6 and E7, cellular targets are p53, Rb, p300, and telomerase as well as a variety of other factors. The membrane-associated E5 protein can also contribute to malignant progression though its mechanism of action is unclear. Understanding the modes of action of HPV oncoproteins can provide important targets for therapeutics to treat HPV-associated cancers.

References

Bodily J, Laimins LA (2011) Persistence of human papillomavirus infection: keys to malignant progression. Trends Microbiol 19(1):33–39. doi:10.1016/j.tim.2010.10.002

Bouvard V, Storey A, Pim D, Banks L (1994a) Characterization of the human papillomavirus E2 protein: evidence of trans-activation and trans-repression in cervical keratinocytes. EMBO J 13(22):5451–5459

Bouvard V, Matlashewski G, Gu ZM, Storey A, Banks L (1994b) The human papillomavirus type 16 E5 gene cooperates with the E7 gene to stimulate proliferation of primary cells and increases viral gene expression. Virology 203(1):73–80

Brehm A, Miska EA, McCance DJ, Reid JL, Bannister AJ, Kouzarides T (1998) Retinoblastoma protein recruits histone deacetylase to repress transcription. Nature 391(6667):597–601. doi:10.1038/35404

Conrad M, Bubb VJ, Schlegel R (1993) The human papillomavirus type 6 and 16 E5 proteins are membrane-associated proteins which associate with the 16-kilodalton pore-forming protein. J Virol 67(10):6170–6178

Davies R, Hicks R, Crook T, Morris J, Vousden K (1993) Human papillomavirus type 16 E7 associates with a histone H1 kinase and with p107 through sequences necessary for transformation. J Virol 67(5):2521–2528

DiMaio D, Mattoon D (2001) Mechanisms of cell transformation by papillomavirus E5 proteins. Oncogene 20(54):7866–7873. doi:10.1038/sj.onc.1204915

Disbrow GL, Hanover JA, Schlegel R (2005) Endoplasmic reticulum-localized human papillomavirus type 16 E5 protein alters endosomal pH but not trans-Golgi pH. J Virol 79(9):5839–5846. doi:10.1128/JVI.79.9.5839-5846.2005

Duensing S, Munger K (2003) Human papillomavirus type 16 E7 oncoprotein can induce abnormal centrosome duplication through a mechanism independent of inactivation of retinoblastoma protein family members. J Virol 77(22):12331–12335

Duensing S, Lee LY, Duensing A, Basile J, Piboonniyom S, Gonzalez S, Crum CP, Munger K (2000) The human papillomavirus type 16 E6 and E7 oncoproteins cooperate to induce mitotic defects and genomic instability by uncoupling centrosome duplication from the cell division cycle. Proce Natl Acad Sci USA 97(18):10002–10007. doi:10.1073/pnas.170093297

Dyson N, Howley PM, Munger K, Harlow E (1989) The human papilloma virus-16 E7 oncoprotein is able to bind to the retinoblastoma gene product. Science 243(4893):934–937 (New York)

Dyson N, Guida P, Munger K, Harlow E (1992) Homologous sequences in adenovirus E1A and human papillomavirus E7 proteins mediate interaction with the same set of cellular proteins. J Virol 66(12):6893–6902

Edmonds C, Vousden KH (1989) A point mutational analysis of human papillomavirus type 16 E7 protein. J Virol 63(6):2650–2656

Fehrmann F, Laimins LA (2003) Human papillomaviruses: targeting differentiating epithelial cells for malignant transformation. Oncogene 22(33):5201–5207. doi:10.1038/sj.onc.1206554

Fehrmann F, Klumpp DJ, Laimins LA (2003) Human papillomavirus type 31 E5 protein supports cell cycle progression and activates late viral functions upon epithelial differentiation. J Virol 77(5):2819–2831

Ferlay J, Shin HR, Bray F, Forman D, Mathers C, Parkin DM (2010) Estimates of worldwide burden of cancer in 2008: GLOBOCAN 2008. Int J Cancer 127(12):2893–2917. doi: 10.1002/ijc.25516

Funk JO, Waga S, Harry JB, Espling E, Stillman B, Galloway DA (1997) Inhibition of CDK activity and PCNA-dependent DNA replication by p21 is blocked by interaction with the HPV-16 E7 oncoprotein. Genes Dev 11(16):2090–2100

Gewin L, Galloway DA (2001) E box-dependent activation of telomerase by human papillomavirus type 16 E6 does not require induction of c-myc. J Virol 75(15):7198–7201. doi:10.1128/JVI.75.15.7198-7201.2001

Howie HL, Katzenellenbogen RA, Galloway DA (2009) Papillomavirus E6 proteins. Virology 384(2):324–334. doi:10.1016/j.virol.2008.11.017

Howley PM, Lowy DR (2007) Papillomaviruses. Wolters Kluwer Health/Lippincott Williams & Wilkins, Philadelphia

Jones DL, Alani RM, Munger K (1997) The human papillomavirus E7 oncoprotein can uncouple cellular differentiation and proliferation in human keratinocytes by abrogating p21Cip1-mediated inhibition of cdk2. Genes Dev 11(16):2101–2111

Katzenellenbogen RA, Vliet-Gregg P, Xu M, Galloway DA (2009) NFX1-123 increases hTERT expression and telomerase activity posttranscriptionally in human papillomavirus type 16 E6 keratinocytes. J Virol 83(13):6446–6456. doi:10.1128/JVI.02556-08

Kiyono T, Hiraiwa A, Fujita M, Hayashi Y, Akiyama T, Ishibashi M (1997) Binding of high-risk human papillomavirus E6 oncoproteins to the human homologue of the Drosophila discs large tumor suppressor protein. Proc Natl Acad Sci USA 94(21):11612–11616

Klingelhutz AJ, Foster SA, McDougall JK (1996) Telomerase activation by the E6 gene product of human papillomavirus type 16. Nature 380(6569):79–82. doi:10.1038/380079a0

Ko LJ, Prives C (1996) p53: puzzle and paradigm. Genes Dev 10(9):1054–1072

Kuhne C, Banks L (1998) E3-ubiquitin ligase/E6-AP links multicopy maintenance protein 7 to the ubiquitination pathway by a novel motif, the L2G box. J Biol Chem 273(51):34302–34309

Kukimoto I, Aihara S, Yoshiike K, Kanda T (1998) Human papillomavirus oncoprotein E6 binds to the C-terminal region of human minichromosome maintenance 7 protein. Biochem Biophys Res Commun 249(1):258–262. doi:10.1006/bbrc.1998.9066

Kyo S, Takakura M, Taira T, Kanaya T, Itoh H, Yutsudo M, Ariga H, Inoue M (2000) Sp1 cooperates with c-Myc to activate transcription of the human telomerase reverse transcriptase gene (hTERT). Nucleic Acid Res 28(3):669–677

Lee SS, Weiss RS, Javier RT (1997) Binding of human virus oncoproteins to hDlg/SAP97, a mammalian homolog of the Drosophila discs large tumor suppressor protein. Proc Natl Acad Sci USA 94(13):6670–6675

Lee SS, Glaunsinger B, Mantovani F, Banks L, Javier RT (2000) Multi-PDZ domain protein MUPP1 is a cellular target for both adenovirus E4-ORF1 and high-risk papillomavirus type 18 E6 oncoproteins. J Virol 74(20):9680–9693

Liu JP (1999) Studies of the molecular mechanisms in the regulation of telomerase activity. FASEB J 13(15):2091–2104

Longworth MS, Laimins LA (2004a) Pathogenesis of human papillomaviruses in differentiating epithelia. Microbiol Mol Biol Rev 68(2):362–372. doi:10.1128/MMBR.68.2.362-372.2004

Longworth MS, Laimins LA (2004b) The binding of histone deacetylases and the integrity of zinc finger-like motifs of the E7 protein are essential for the life cycle of human papillomavirus type 31. J Virol 78(7):3533–3541

Lorincz AT, Reid R, Jenson AB, Greenberg MD, Lancaster W, Kurman RJ (1992) Human papillomavirus infection of the cervix: relative risk associations of 15 common anogenital types. Obstet Gynecol 79(3):328–337

Markowitz LE, Dunne EF, Saraiya M, Lawson HW, Chesson H, Unger ER (2007) Quadrivalent human papillomavirus vaccine: recommendations of the advisory committee on immunization practices (ACIP). MMWR Recomm Rep 56(RR-2):1–24 (Morbidity and mortality weekly report Recommendations and reports/Centers for Disease Control)

Maufort JP, Williams SM, Pitot HC, Lambert PF (2007) Human papillomavirus 16 E5 oncogene contributes to two stages of skin carcinogenesis. Cancer Res 67(13):6106–6112. doi: 10.1158/0008-5472.CAN-07-0921

Moody CA, Laimins LA (2010) Human papillomavirus oncoproteins: pathways to transformation. Nat Rev 10(8):550–560. doi:10.1038/nrc2886

Munger K, Phelps WC, Bubb V, Howley PM, Schlegel R (1989) The E6 and E7 genes of the human papillomavirus type 16 together are necessary and sufficient for transformation of primary human keratinocytes. J Virol 63(10):4417–4421

Oren M (2003) Decision making by p53: life, death and cancer. Cell Death Differ 10(4):431–442. doi:10.1038/sj.cdd.4401183

Parkin DM, Bray F, Ferlay J, Pisani P (2000) Estimating the world cancer burden: Globocan. Int J Cancer 94(2):153–156

Patel D, Huang SM, Baglia LA, McCance DJ (1999) The E6 protein of human papillomavirus type 16 binds to and inhibits co-activation by CBP and p300. EMBO J 18(18):5061–5072. doi:10.1093/emboj/18.18.5061

Petti L, Nilson LA, DiMaio D (1991) Activation of the platelet-derived growth factor receptor by the bovine papillomavirus E5 transforming protein. EMBO J 10(4):845–855

Regan JA, Laimins LA (2008) Bap31 is a novel target of the human papillomavirus E5 protein. J Virol 82(20):10042–10051. doi:10.1128/JVI.01240-08

Riley RR, Duensing S, Brake T, Munger K, Lambert PF, Arbeit JM (2003) Dissection of human papillomavirus E6 and E7 function in transgenic mouse models of cervical carcinogenesis. Cancer Res 63(16):4862–4871

Ronco LV, Karpova AY, Vidal M, Howley PM (1998) Human papillomavirus 16 E6 oncoprotein binds to interferon regulatory factor-3 and inhibits its transcriptional activity. Genes Dev 12(13):2061–2072

Ruesch MN, Laimins LA (1998) Human papillomavirus oncoproteins alter differentiation-dependent cell cycle exit on suspension in semisolid medium. Virology 250(1):19–29. doi: 10.1006/viro.1998.9359

Scheffner M, Werness BA, Huibregtse JM, Levine AJ, Howley PM (1990) The E6 oncoprotein encoded by human papillomavirus types 16 and 18 promotes the degradation of p53. Cell 63(6):1129–1136. doi:0092-8674(90)90409-8

Slee EA, O'Connor DJ, Lu X (2004) To die or not to die: how does p53 decide? Oncogene 23(16):2809–2818. doi:10.1038/sj.onc.1207516

Stanley M (2008) Immunobiology of HPV and HPV vaccines. Gynecol Oncol 109(2):S15–S21. doi:10.1016/j.ygyno.2008.02.003

Stanley M (2010) HPV: immune response to infection and vaccination. Infect Agent Cancer 5:19. doi:10.1186/1750-9378-5-19

Stoppler MC, Straight SW, Tsao G, Schlegel R, McCance DJ (1996) The E5 gene of HPV-16 enhances keratinocyte immortalization by full-length DNA. Virology 223(1):251–254. doi: 10.1006/viro.1996.0475

Straight SW, Hinkle PM, Jewers RJ, McCance DJ (1993) The E5 oncoprotein of human papillomavirus type 16 transforms fibroblasts and effects the downregulation of the epidermal growth factor receptor in keratinocytes. J Virol 67(8):4521–4532

Straight SW, Herman B, McCance DJ (1995) The E5 oncoprotein of human papillomavirus type 16 inhibits the acidification of endosomes in human keratinocytes. J Virol 69(5):3185–3192

Thomas M, Banks L (1999) Human papillomavirus (HPV) E6 interactions with Bak are conserved amongst E6 proteins from high and low risk HPV types. J Gen Virol 80(Pt 6):1513–1517

Thomas JT, Hubert WG, Ruesch MN, Laimins LA (1999) Human papillomavirus type 31 oncoproteins E6 and E7 are required for the maintenance of episomes during the viral life cycle in normal human keratinocytes. Proc Natl Acad Sci USA 96(15):8449–8454

Tommasino M, Adamczewski JP, Carlotti F, Barth CF, Manetti R, Contorni M, Cavalieri F, Hunt T, Crawford L (1993) HPV16 E7 protein associates with the protein kinase p33CDK2 and cyclin A. Oncogene 8(1):195–202

Valle GF, Banks L (1995) The human papillomavirus (HPV)-6 and HPV-16 E5 proteins co-operate with HPV-16 E7 in the transformation of primary rodent cells. J Gen Virol 76(Pt 5):1239–1245

Weintraub SJ, Chow KN, Luo RX, Zhang SH, He S, Dean DC (1995) Mechanism of active transcriptional repression by the retinoblastoma protein. Nature 375(6534):812–815. doi: 10.1038/375812a0

Zerfass-Thome K, Zwerschke W, Mannhardt B, Tindle R, Botz JW, Jansen-Durr P (1996) Inactivation of the cdk inhibitor p27KIP1 by the human papillomavirus type 16 E7 oncoprotein. Oncogene 13(11):2323–2330

Zimmermann H, Degenkolbe R, Bernard HU, O'Connor MJ (1999) The human papillomavirus type 16 E6 oncoprotein can down-regulate p53 activity by targeting the transcriptional coactivator CBP/p300. J Virol 73(8):6209–6219

zur Hausen H (1996) Papillomavirus infections–a major cause of human cancers. Biochim Biophys Acta 1288(2):F55–F78

zur Hausen H (2009) Papillomaviruses in the causation of human cancers: a brief historical account. Virology 384(2):260–265. doi:10.1016/j.virol.2008.11.046

zur Hausen H, de Villiers EM (1994) Human papillomaviruses. Annu Rev Microbiol 48:427–447. doi:10.1146/annurev.mi.48.100194.002235

Control of HPV Infection and Related Cancer Through Vaccination

Nam Phuong Tran, Chien-Fu Hung, Richard Roden and T.-C. Wu

Abstract

Human papillomavirus (HPV), the most common sexually transmitted virus, and its associated diseases continue to cause significant morbidity and mortality in over 600 million infected individuals. Major progress has been made with preventative vaccines, and clinical data have emerged regarding the efficacy and cross-reactivity of the two FDA approved L1 virus like particle (VLP)-based vaccines. However, the cost of the approved vaccines currently limits their widespread use in developing countries which carry the greatest burden of HPV-associated diseases. Furthermore, the licensed preventive HPV vaccines only contain two high-risk types of HPV (HPV-16 and HPV-18) which can protect only up to 75 % of all cervical cancers. Thus, second generation preventative vaccine candidates hope to address the issues of cost and broaden

N. P. Tran · C.-F. Hung · R. Roden · T.-C. Wu
Department of Pathology, The Johns Hopkins School of Medicine, Cancer Research Building II, Room 310, 1550 Orleans Street, Baltimore, MD 21231, USA

C.-F. Hung · R. Roden · T.-C. Wu
Departments of Oncology, The Johns Hopkins School of Medicine, Cancer Research Building II, Room 307, 1550 Orleans Street, Baltimore, MD 21231, USA

R. Roden · T.-C. Wu
Departments of Obstetrics and Gynecology, The Johns Hopkins School of Medicine, Cancer Research Building II, Room 308, 1550 Orleans Street, Baltimore, MD MD 21231, USA

T.-C. Wu (✉)
Department Molecular Microbiology and Immunology, The Johns Hopkins Medical Insitutions, Cancer Research Building II, Room 309, 1550 Orleans Street, Baltimore, MD 21231, USA
e-mail: wutc@jhmi.edu

M. H. Chang and K.-T. Jeang (eds.), *Viruses and Human Cancer*,
Recent Results in Cancer Research 193, DOI: 10.1007/978-3-642-38965-8_9,
© Springer-Verlag Berlin Heidelberg 2014

protection through the use of more multivalent L1-VLPs, vaccine formulations, or alternative antigens such as L1 capsomers, L2 capsid proteins, and chimeric VLPs. Preventative vaccines are crucial to controlling the transmission of HPV, but there are already hundreds of millions of infected individuals who have HPV-associated lesions that are silently progressing toward malignancy. This raises the need for therapeutic HPV vaccines that can trigger T cell killing of established HPV lesions, including HPV-transformed tumor cells. In order to stimulate such antitumor immune responses, therapeutic vaccine candidates deliver HPV antigens in vivo by employing various bacterial, viral, protein, peptide, dendritic cell, and DNA-based vectors. This book chapter will review the commercially available preventive vaccines, present second generation candidates, and discuss the progress of developing therapeutic HPV vaccines.

Keywords

Human Papillomavirus (HPV) · Vaccines · Cancer · Immunotherapy · Virus like particle (VLP)

Contents

1   Introduction

Why is it important to have an effective method to control human papillomavirus (HPV) infections on a global scale? HPV currently infects an estimated 660 million people worldwide and is the most common viral infection of the human reproductive tract (Brooks 2010). The mucosal-specific HPVs cause a variety of diseases in the human body, ranging from benign warts to metastatic cervical cancer. The benign strains, notably HPV-6 and -11 cause anogenital condylomas and laryngeal papillomas. These lesions are not inherently carcinogenic in nature. The high-risk types of HPV strains (16, 18, 31, 33, 35, 45, etc.), however, are a necessary cause of 99.7 % of cervical cancer, 90 % of anogenital cancer, 40 % of penile cancers, and 42–60 % of oropharyngeal carcinomas [for review see (Brooks 2010; Kwak et al. 2010; Simard et al. 2012)].

Of the above HPV-associated diseases, cervical cancer causes the greatest number of deaths per year (Lowy and Schiller 2012). Cervical cancer is also the third most common cancer in women worldwide and kills over 400,000 women each year (Brooks 2010; Jemal et al., 2011). Women in developing countries bear 85 % of the burden of cervical cancer, yet they are also the least likely to have access to vaccination, screening, or treatment programs (Lowy and Schiller 2012). The prevalence of cervical cancer in developed countries has decreased anywhere between 27 and 77 % since the 1950s due to earlier detection and treatment; however, women in developed countries still have a greater than 80 % lifetime risk of infection with an oncogenic strain of HPV and treatment is associated with some morbidity [for review see (Echelman and Feldman 2012; Trimble and Frazer 2009)]. Additionally, the rising incidence of head and neck neoplasms attributable to HPV infections is alarming (Chaturvedi et al. 2011), especially considering the fact that there are no screening tests for oropharyngeal cancers equivalent to the Pap smear for cervical cancer (Trimble and Frazer 2009). Action needs to be taken to control such oncogenic viruses. The understanding of HPV as the etiological factor for HPV-associated malignancies has led to the notion of controlling these cancers through vaccination against HPV.

Our growing understanding of HPV molecular biology has driven the development of vaccines targeting HPV. HPVs are double-stranded DNA viruses that belong to the family of Papillomaviridae and specifically replicate in human hosts. HPV DNA is circular and comprises about 8000 base pairs which code for an assortment of early (E) and late (L) genes. The early genes (E1, E2, E4, E5, and E6) provide viral DNA replication and transcription control functions while the two late genes (L1 and L2) encode viral capsid proteins. The non-coding region (or URR) regulates expression and contains the viral origin of replication. Once inside of a host cell, early genes carry out processes essential to viral replication and transcription, but some of the early genes are directly involved in oncogenic transformation of host cells. The early proteins E1 and E2 are involved in viral DNA replication and viral RNA transcription, while E4 is involved in cytoskeleton reorganization. E5 is an oncogenic protein that has been found to enhance EGFR activation, limit recycling inside host cells, cause cell–cell fusion, and immortalize human keratinocytes [for review see (DiMaio and Mattoon 2001)]. E6 downregulates p53, a checkpoint protein in the cell cycle while E7 sequesters Rb protein—both processes instrumental in causing unregulated proliferation of infected cells and transformation into cervical cancer. In most cases of cervical cancer, the HPV genome integrates into the host chromosomal DNA and causes disruption of the viral E2 gene. E2 is a transcriptional regulator for the E6 and E7 genes. Thus, loss of E2 leads to the uncontrolled expression of E6 and E7 proteins, which in turn leads to disruption of the normal cell cycle regulation by interacting with p53 and Rb, respectively. This leads to an uncontrolled cell cycle, genomic instability, and the suppression of apoptosis, facilitating progression to HPV-associated cervical cancer [for review, see zur Hausen (2002)]. Among the late genes, L1 serves as the major structural protein of the viral capsid and is used in commercially available preventive vaccines. L2 is the minor viral capsid protein to be discussed later in the context of broadening the spectrum of future generation preventative vaccines [for review, see (Roden and Wu 2006)].

HPVs have tissue tropism in that they preferentially replicate in specific mucosal epithelia. The expression of early and late genes correlates with the differentiation of the keratinocyte: from newly infected basal cells expressing early proteins in stratum basale and keratinocytes in the middle layers to finally differentiated keratinocytes expressing both L1 and L2 viral capsid proteins used in viral shedding from superficial layers (Roden and Wu 2006). HPV infections are restricted to the epithelium. This makes it difficult for the human immune system to launch potent HPV-specific immune responses to eradicate an HPV infection since HPV lesions remain above the basement membrane and no HPV antigen is found in the circulatory or lymphatic systems. Furthermore, HPV does not elicit a potent inflammatory response, but rather seems to slip below the radar. In preventive vaccines, the goal is to deliver HPV L1 and/or L2 capsid antigens in order to stimulate immune production of neutralizing antibodies that can block HPV from infecting epithelial cells. Antibodies to HPV, though effective in preventing infection, are unable to kill established HPV-infected and/or transformed cells. The actions of specific cytotoxic T cells (CTLs) and T helper cells are needed in order to eliminate the infected and/or transformed cell [for review, see (Hung and Wu 2003)]. Therapeutic HPV vaccines rely upon activation of T-cell-mediated immunity by antigen-presenting cells (such as dendritic cells) that present HPV antigens through MHC class I and II molecules to prime HPV-antigen-specific T cells. These characteristics of the HPV life cycle are very important to keep in mind while devising strategies for preventive and therapeutic vaccines.

The choice of target antigen is also an important factor that needs to be considered. While L1 and L2 are suitable targets for the development of preventive vaccines, they are not ideal targets for therapeutic HPV vaccine development since they are not expressed in the basal cells infected with HPV. On the other hand, early viral proteins, particularly E6 and E7, are expressed early in viral infection and help drive malignant the progression. Therefore, therapeutic vaccines should aim to generate T-cell-mediated immune responses against the early proteins E6 and E7 [for review, see (Lin et al. 2010)]. Furthermore, E6 and E7 co-expression is essential for transformation, and they are not expressed in normal cells. Therefore, E6 and E7 represent ideal targets for the development of therapeutic HPV vaccines.

2   Current Commercially Available Preventive Vaccines

The successful development of two preventative HPV vaccines in the last decade, Cervarix™ from GlaxoSmithKline and Gardasil® from Merck, provides the opportunity to prevent the spread of HPV. Both vaccines are available commercially, each in three dose regimens spread out over six months. Additionally, both vaccines employ L1 virus like particles (VLPs): non-infectious papillomavirus particles without the viral genome [for review see (Campo and Roden 2010; Roden and Wu 2006)]. Cervarix™, a bivalent vaccine, contains HPV-16 and HPV-18 VLPs produced in insect cells (*Trichoplusia ni*) using a baculovirus expression vector system and incorporates Adjuvant System 04 (monophosphoryl lipid A and aluminum hydroxide salt). Gardasil® is a quadrivalent vaccine that contains HPV-6, 11, 16, and 18 VLPs produced in yeast cells (*Saccharomyces cerevisiae*) and an amorphous aluminum hydroxyphosphate salt adjuvant (Einstein et al. 2009). Thus, this quadrivalent vaccine protects patients from oncogenic HPV-16 and -18 as well as HPV-6 and -11, which cause common genital warts. In a comparison study of the 2 vaccines in which 1106 women, stratified by age, received either the bivalent (Cervarix™) or the quadrivalent (Gardasil®) vaccine, the bivalent vaccine induced 2.3–9.1-fold higher geometric mean titers (GMTs) of neutralizing antibodies for HPV-16 and HPV-18 across all age strata as well as higher numbers of circulating memory B cells than the quadrivalent vaccine (Einstein et al. 2009). In a follow-up study of these patients at months 12–24 after vaccination, Einstein et al. 2011 found that GMTs were 2.4–5.8-fold higher for HPV-16 and 7.7-9.4-fold higher for HPV-18 with the bivalent compared to the quadrivalent vaccine (Table 1).

Data regarding efficacy of each of the two vaccines have recently been published in 4 year follow-up studies of women from each vaccine's phase 3 clinical trials. In a 4-year trial with 17,622 women aged 16–26, the quadrivalent vaccine was found to be effective in preventing the development of lesions caused by HPV-6, 11, 16, and 18: 96 % for CIN1 (cervical intraepithelial neoplasia 1), 100 % for VIN1 and VAIN1 (vulva and vaginal intraepithelial neoplasia 1), and 99 % for condyloma (Dillner et al. 2010). However, protection against HPV strains not covered by the quadrivalent vaccine was less robust; efficacy against any lesion regardless of HPV type was 30, 75, 48, and 83 % for cervical, vulvar, vaginal intraepithelial neoplasia, and condyloma, respectively, suggesting that the quadrivalent vaccine has low cross-reactivity for other HPV strains. The bivalent vaccine also had its own four-year follow-up study. In a trial enrolling 18,644 women ages 15–25, the bivalent vaccine was found to be 100 % effective in preventing CIN3 and AIS caused by HPV strains 16 and 18 and 93.2 % effective in protecting against all CIN3 lesions regardless of HPV type. However, in a concurrent study analyzing the same data, the cross-protective efficacy of the bivalent vaccine in preventing CIN2 + lesions caused by non-HPV-16 and -18 strains was variable among HPV naïve patients who received all 3 doses of vaccine: 84.3 % effective for HPV-31, 59.4 % effective for HPV-33, and statistically insignificant for HPV-39, 45, 52, 58, 59, and 68 (Wheeler et al. 2012). Therefore, the cross-reactivity of the bivalent vaccine, like that of the quadrivalent vaccine, is also limited in nature. The observation that the bivalent vaccine is more effective in preventing AIS and CIN3 + lesions caused by any HPV infection but not as effective in preventing CIN2 + lesions caused by non-HPV-16 and -18 strains suggests that a significant proportion of higher grade lesions are caused by

Table 1 Comparison of therapeutic vaccines

Type	Advantage	Disadvantage	Developmental stage	References
Live vector-based	Highly immunogenic customizable	Vector-specific neutralizing antibodies or pre-existing vector-specific immunity, safety concerns, flu-like adverse effects	ADXS11-001 in phase II Trial	Maciag et al. (2009), NCI (2012b), Radulovic et al. (2009)
			MVA-E2 in phase II trial	Corona Gutierrez et al. (2004)
			TA-HPV completed phase I/II trial	Kaufmann et al. (2002)
			TG4001-R3484 completed phase II trial	Brun et al. (2011)
Peptide-based	Safe, easy to produce, stable	HLA restriction, Poorly immunogenic	HPV-16 E7 for HLA-A*0201 completed phase 1 Trial	Kenter et al. (2008), Muderspach et al. (2000)
			Overlapping peptides Clinical trials	Kenter et al. (2008, Kenter et al. (2009), Welters et al. (2010), Welters et al. (2008)
Protein-based	Safe, easy to produce, stable	Poorly immunogenic	HspE7 completed phase I and II trials	Derkay et al. (2005), Einstein et al. (2007), Goldstone et al. (2002), Van Doorslaer et al. (2009)
			TA-CIN completed phase II trials	Daayana et al. (2010)
DC-based	Highly immunogenic	Individualized and labor intensive, potentially oncogenic	HPV-16 and -18 E7 pulsed DC vaccine completed phase 1 trials	Wang et al. (2009)
DNA-based	Easy to produce, stable, prolonged expression of antigens, capacity for repeated administration	Poorly immunogenic, potentially oncogenic	ZYC101a completed phase II trial	Garcia et al. (2004), Matijevic et al. (2011), Sheets et al. (2003)
			VGX-300 in phase II trials	Inovio (2012)
			pNGVL4a-Sig/E7/HSP70 completed phase I trial	Trimble et al. (2009)

HPV-16 and -18 relative to other strains (Wheeler et al. 2012). Efficacy of the bivalent vaccine in preventing CIN2 + and CIN3 + lesions in both studies was found to be highest in the 15–17 age group and declined with increasing age, supporting the current recommendation of vaccinating at younger ages (Lehtinen et al. 2012; Wheeler et al. 2012).

Both commercially available vaccines are currently marketed as a 3-dose regimen. However, one recent randomized trial involving 960 subjects compared the bivalent vaccine at either 2 doses or the standard 3-dose regimen. The 3-dose regimen was considered superior if there was more than a twofold difference in the ratio of geometric mean antibody titers (GMTs). Results showed that 2 doses of the bivalent vaccine elicited antibody responses both at one month and 24 months after vaccination that were non-inferior to those of the same concentration vaccine given in the standard 3-dose regimen. This suggests that 2 doses of the current bivalent vaccine would yield equivalent protection to the standard 3-dose regimen up to 24 months post-vaccination (Romanowski et al. 2011); however, the longevity of protection beyond this has yet to be determined.

Although the primary purpose of commercial HPV vaccines is to prevent cervical cancer in women, the quadrivalent vaccine may be used to prevent condyloma and anogenital lesions, caused by HPV 6 and 11, in males. Males may also benefit from protection against HPV-16 and -18 that can cause cancers of the penis, anus, and oropharynx (Giuliano et al. 2011). In a randomized, placebo-controlled, double-blind trial of 4065 males age 16–26, the efficacy of the quadrivalent vaccine in preventing external genital lesions was tested. In the intention to treat group, comprised of subjects who may or may not have been seropositive for HPV before the trial and received at least one dose of either vaccine or placebo, 36 external genital lesions were seen in the HPV-vaccinated group compared to 89 lesions in the placebo group, yielding an overall efficacy of 60.2 % (95 % CI, 40.8–73.8). In the per-protocol population of males, who were seronegative before vaccination and completed all 3 doses of quadrivalent vaccine, the efficacy against HPV-6, 11, 16, and 18 lesions was 90.4 % (95 % CI, 69.2–98.1), suggesting that the quadrivalent vaccine is effective in preventing HPV-associated genital lesions in males (Giuliano et al. 2011).

3   Second Generation of Preventive Vaccine Development

The clinical results offered by available HPV vaccines are excellent for the types targeted, but issues remain. Eighty-five percent of the burden of cervical cancer is in developing countries (Jemal et al. 2011). Not only are most residents in these developing countries unable to afford the current commercial vaccines, many do not even live in areas which have the capability to store and distribute such vaccines (through cold chains). Another limiting factor of the available preventive vaccines is their limited cross-reactivity to many oncogenic HPV strains. Patients vaccinated with available vaccines will be protected from HPV-16 and -18, which

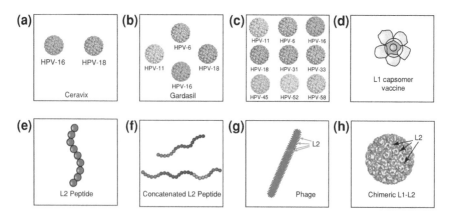

Fig. 1 Schematic diagram to depict the next generation of preventive HPV vaccines. **a** Cervarix composed of HPV-16 and HPV-18 VLPs. **b** Gardasil composed of HPV-6, HPV-11, HPV-16, and HPV-18 VLPs. **c** Multivalent VLP vaccines composed of HPV-6, HPV-11, HPV-16, HPV-18, HPV-31, HPV-33, HPV-45, HPV-52, and HPV-58 VLPs. **d** L1 capsomer vaccine. **e** L2 peptide vaccine. **f** Concatenated L2 peptide vaccine. **g** Phage vaccine with L2 on the surface of the phage. **h** Chimeric L1-L2 VLP vaccine with L2 on the surface

cause more than 70 % of cervical cancers, but there are a dozen other oncogenic strains against which patients have limited or no defense. Broader coverage against oncogenic HPV strains is a desirable attribute in future generations of HPV vaccines. Therefore, efforts to improve upon the already available HPV vaccines have much merit.

There have been several approaches for the next generation of preventative HPV vaccines, which include as follows: (1) more multivalent VLP-based vaccines, (2) L1-based capsomers, (3) L2-based vaccines, and (4) chimeric L1-L2 vaccines. Figure 1 summarizes the various strategies in the current as well as the next generation of preventive HPV vaccines. Multivalent VLP vaccines build upon the original idea used to make commercially available bivalent and quadrivalent vaccines (Cervarix™ and Gardasil®). If VLPs of high-risk strains of HPV-16 and HPV-18 are protective, then why not add more VLPs in order to broaden the coverage? Currently, there are ongoing trials to compare nonavalent VLP vaccines to the quadrivalent vaccine (Merck 2011). The addition of more VLPs will likely increase the costs of production; however, there are several efforts to cost-effectively produce HPV VLPs inside of bacteria.

Another potential avenue for new preventive vaccines involves the creation of L1 capsomer vaccines, a potential cost effective alternative to VLP vaccines. Capsomers are basic structural components of viral capsids, whereas HPV L1 VLP is composed of 360 copies of the L1 protein, only five L1 monomers are needed to assemble a pentavalent capsomer. Additionally, these capsomer proteins can be made in bacteria and thus are cheaper to produce than VLPs made in insect or yeast cells. L1 capsomer proteins made in *E. coli,* recombinant measles virus, or recombinant *Salmonella enterica* serovar Typhimurium have successfully induced protective antibodies in preclinical models and these genetically modified live vaccines might be another approach to lower costs and perhaps reduce the number of immunizations (Chen et al. 2000; Fraillery et al. 2007; Li et al. 1997; Rose et al. 1998).

Another approach for second generation preventive vaccines involves the use of L2 instead of L1 VLPs. L2 is highly conserved among different HPV types and the L2 from one strain of HPV can possibly induce broader protection through cross-neutralizing antibodies, even across species. It has been shown that immunization of animals with the amino-terminal peptide of L2 produced in *E. coli* elicited neutralizing antibodies that protect against challenge with cognate papillomavirus types in vivo (Embers et al. 2002; Gaukroger et al. 1996) cross-neutralize heterologous types in vitro (Kawana et al. 1999; Pastrana et al. 2005; Roden et al. 2000), and confer cross-protection in vivo (Gambhira et al. 2007).

L2 has also been used in other forms. In a preclinical experiment, concatenated multitype L2 fusion proteins were used in different combinations (L2 residues 11–200 from 3 HPV types (6, 16, and 18), L2 residues 11–88 from 5 HPV types (1, 5, 6, 16, and 18), or L2 residues 17–36 of five cutaneous, two mucosal low-risk, and 15 oncogenic HPV types) (Jagu et al. 2009). Vaccination in both mice and rabbits with the concatenated multitype L2 with different adjuvants elicited higher neutralizing antibody titers than in animals vaccinated with only HPV-16 amino-terminal L2 polypeptides, HPV-16 L1 VLP, Gardasil®, or the negative control (Jagu et al. 2009). Mice vaccinated with these concatenated multitype L2 fusion proteins also had immunity against challenge with HPV-16 pseudovirions 4 months later. Furthermore, the HPV-16 L2 peptides were able to generate antisera that neutralized HPV-18, 31, 45, and 58, thus confirming the cross-reactivity of L2 (Jagu et al. 2009). Generally, L2 is less immunogenic than L1, and this limitation prompted Tumban et al. 2011 to create a vaccine consisting of PP7 bacteriophage VLPs that displayed the neutralizing epitope from L2 proteins of 8 different HPV types on their surfaces (Tumban et al.). Mice vaccinated twice with HPV-16 L2 peptide displaying PP7 and HPV-18 L2 peptide displaying PP7 produced GMTs ranging from 10^4 to 10^5 of broadly reactive anti-L2 IgGs. Additionally, these anti-L2 antibodies protected against vaginal challenge with HPV-16 or -18 pseudovirions and cross-reacted with synthetically engineered HPV L2 peptides (Tumban et al. 2011). Although they have the potential to confer greater cross-reactivity, the immunogenicity of L2 proteins is low and more potent adjuvants and display technologies are currently being explored.

The advantages of highly immunogenic L1 vaccines and the broad cross-protection mediated by L2 can potentially be combined in the form of chimeric L1-L2 VLPs. Since L2 is a less abundant protein that is found predominantly in the interior of the viral particle, replacing some regions of the VLP surface normally occupied by L1 immunodominant epitopes with a neutralizing epitope of L2 may generate more immunogenic, cross-protective immune responses against multiple HPV genotypes. The surface expression of the neutralizing epitope of L2 on L1 VLP in chimeric models is essential since VLPs with L1 and L2 s in their normal positions do not readily generate L2 antibodies. Immunization of rabbits with chimeric L1-L2 VLPs with L2 peptides on their surface was shown to induce not only antibodies neutralizing HPV-16 but also antibodies cross-neutralizing HPV-18, 31, 52, 58 pseudovirions (Kondo et al. 2008). Additionally, vaccination with a chimeric HPV-16 L1-L2 VLP vaccine in both rabbits and mice also exhibited cross-neutralization to high-risk and low-risk HPV types evolutionarily divergent from HPV-16 (Schellenbacher et al. 2009). The inclusion of the immunodominant neutralizing epitopes of L2 into the L1 VLPs represents a promising direction for the next generation of preventative HPV vaccines to induce broad-spectrum neutralizing antibodies against HPV.

4 Strategies of Therapeutic Vaccines in Clinical Development

4.1 Concept and Goals of Therapeutic Vaccines

HPV preventive vaccines, though effective in blocking HPV infections, cannot eliminate established HPV infections or HPV-associated lesions and thus, have no therapeutic activity (Lin et al. 2010). Given the fact that almost all cervical cancers are caused by high-risk HPVs, therapeutic vaccines targeting HPVs offer great potential for highly specific eradication of HPV-infected cells and HPV-associated tumors through immunotherapy. As discussed in the preceding chapter, the progression of latent HPV infections into invasive cancer takes many years and fortunately, most women clear their infections. However, if women with high-grade CIN lesions suffer few symptoms and are not screened, they may be unaware that cancer is progressing inside of their cervix. In a prolonged chronic infection, there is a considerable window for secondary preventive treatment for infections caught by cytologic screening/HPV DNA testing. Effective therapeutic HPV vaccines that can actively attack HPV-infected cells during this long phase of tumor progression are greatly needed to prevent the development of and to treat advanced cancer. If therapeutic vaccines can eradicate transformed cells early, the morbidity and mortality associated with HPV-associated malignancies and invasive treatments may dramatically diminish.

Many different platforms have been used in the development of therapeutic HPV vaccines, each in various phases of testing and clinical trials. In order to compare the potential of various candidates, one has to keep in mind the characteristics of an ideal therapeutic vaccine. These qualities include: (1) safety, (2) the ability to trigger effective HPV-antigen-specific cytotoxic T-cell-mediated responses that target the lesion, (3) specificity for tumor cells, (4) long duration of efficacy, (5) cheaper mass production and storage, and (6) fewer number of required doses. The ability of a therapeutic vaccine to trigger specific T-cell-mediated killing is essential to a vaccine's efficacy in targeting transformed cells vs. healthy cells. Reduced cost of production and storage is a highly desirable trait that will increase access to cancer treatments for patients worldwide. Currently, the compliance rate of patients who start *and* complete all three shots of either available bivalent or quadrivalent vaccine is roughly 25 % [for review, see Trimble and Frazer (2009)]. Thus, simplicity of delivery will greatly increase the compliance rates of a therapeutic vaccine.

As aforementioned in Sect. 1, E6 and E7 are ideal targets for therapeutic vaccines. There has been much research and development in the field of therapeutic HPV vaccines, including live-vector-based, peptide-based, protein-based, dendritic cell-based, DNA-based, and combination vaccines targeting E6 and/or E7 antigens.

4.2   Live Vector-Based Therapeutic HPV Vaccines

The concept of live vector vaccines encompasses the use of bacterial and viral vectors, which effectively infect host cells and provide a source of E6 and/or E7 antigens for antigen-presenting cells. Several bacterial vectors have been explored for therapeutic HPV vaccines, and among these, *L. monocytogenes*, a gram-positive intracellular bacterium, has generated significant interest since it is able to invade macrophages and evade phagocytosis within the phagosome using the pore-forming toxin, listeriolysin O (LLO). Antigens from *L. monocytogenes* infection can be processed in both MHC class I and II pathways. *L. monocytogenes*-based HPV E7 vaccines have been shown to stimulate a greater number of E7-specific CD8+ T cells and resolve solid tumors in both transgenic and wild type mice (Gunn et al. 2001; Hussain and Paterson 2004; Sewell et al. 2004; Souders et al. 2007). There has been translation of this research into clinical trials. ADXS11-001, formerly known as Lovaxin C or Lm-LLO-E7, is a live, attenuated *L. monocytogenes* bacterial vector secreting HPV-16 E7 fused to LLO. In phase I trials, ADXS11-001 was found to be safe and well-tolerated with a dose limiting toxicity linked to flu-like symptoms (Maciag et al. 2009; Radulovic et al. 2009). Currently, ADXS11-001 is being studied in two clinical trials with active enrollment. The primary goal of one randomized, single blind, placebo-controlled phase 2 study is to determine whether 3 doses of ADXS11-001 can safely reverse disease in subjects with CIN2/3 in whom surgery is indicated (NCI 2012d). Another ongoing phase 2 trial involves testing the safety and efficacy of ADXS11-001 in treating and increasing the one-year survival rate of subjects with persistent or recurrent carcinoma of the cervix (NCI 2012b).

In addition to bacteria, live viral vectors have been investigated due to their high immunogenicity. There have been several viral vectors used to deliver HPV E7 antigens, but the *Vaccina* virus, an enveloped, double-stranded DNA virus within the Poxviridae family, represents a particularly promising viral vector of interest because of its large genome and high infectivity. A recombinant *Vaccinia* virus expressing HPV-16 and HPV-18 E6/E7 antigens (TA-HPV) has been evaluated in phase I/II clinical trials in patients with early-stage cervical cancer (Kaufmann et al. 2002), late-stage cervical cancer (Borysiewicz et al. 1996), vulvar intraepithelial neoplasia (Davidson et al. 2003), and vaginal intraepithelial neoplasia (Baldwin et al. 2003). TA-HPV was found to be safe and potent in stimulating vaccinia-specific antibody and HPV-antigen-specific CTL responses (Adams et al. 2001; Borysiewicz et al. 1996; Kaufmann et al. 2002).

Another vaccinia-based therapeutic HPV vaccine is MVA-E2, an attenuated, recombinant vaccinia virus encoding bovine papillomavirus type 1 (BPV-1) E2 (Corona Gutierrez et al. 2004). In a phase I/II clinical trial, four intrauterine doses of MVA E2 led to significant effects in patients with CIN lesions. It was reported that 34 of 36 subjects in the trial showed complete CIN lesion regression, while the other 2 subjects had a reduction in the lesion grade from CIN3 to CIN1 (Corona Gutierrez et al. 2004). Several possible explanations may account for the observed therapeutic effects. First, in the early stage of infection, for example CIN lesions when the HPV genome in infected cells has not yet been integrated into the host genome, infected cells may still express the E2 antigenic peptide on their sur-faces—enabling E2-specific CTLs, generated through treatment, to hone in and attack. However, it is not clear whether the BPV-1 E2-specific CTLs generated through treatment with MVA-E2 are capable of recognizing HPV E2 targets on the infected cells of the CIN lesions. Second, the E2 from MVA-E2 may act as a transcriptional repressor that inhibits the expression of E6 and E7. Lastly, local administration of a highly immunogenic live vector, such as MVA-E2, may stimulate stronger immune responses in the microenvironment of HPV-associated lesions. While the initial observation is encouraging, further validation is required to establish this approach for the control of CIN lesions.

Another modified vaccinia-based vaccine uses the Ankara vector to express HPV-16 E6 and E7 antigens and adjuvant IL-2 (TG4001/R3484). In phase II clinical trials, TG4001/R3484 was found to be both safe and encouraging in generating clinical responses in women with HPV-16-positive CIN 2/3. Ten of 21 women in the trial no longer had detectable levels of CIN 2/3 six months after vaccination. At the twelve-month follow-up, no relapse or HPV-16 persistence was observed in these women (Brun et al. 2011). Although live vector vaccines have shown promising results, they inherently pose a potential safety risk, particularly to immunocompromised individuals. Live vectors may also face limited capacity for repeated administration due to induction of vector-specific neutralizing anti-bodies and/or pre-existing vector-specific immunity.

4.3   Peptide-Based Therapeutic HPV Vaccines

Short peptides of HPV antigens along with adjuvant can be delivered to dendritic cells, which then present those antigens onto MHC class I molecules and thus activate antigen-specific T-cell immunity. Production of these short-peptide vac-cines involves the prior identification of specific CTL and CD4+ T helper epitopes of HPV antigens. Peptide-based vaccines are advantageous in that they are stable, easy to produce, and have a high safety profile. Current research focuses on addressing the main limitations of these vaccines—namely low immunogenicity and MHC restriction.

The immunogenicity of HPV peptide-based vaccines can be enhanced with adjuvants including immunoglobulin G fragment (Qin et al. 2005), streptavidin fused to the extracellular domain of murine 4-1BBL (Sharma et al. 2009), dendritic cell stimulatory cytokine bryostatin (Yan et al. 2010) and toll like receptor (TLR) agonists (Daftarian et al. 2006; Wu et al. 2010; Zhang et al. 2010; Zwaveling et al. 2002). For example, a prime boost regimen of HPV-16 E7 (a.a. 43–77), which contains both CTL and Th epitopes, adjuvanted with TLR9 agonist CpG oligodeoxynucleotide, was shown to enhance priming of E7-specific CD4+ and CD8+ T cells and produce potent antitumor effects in vaccinated mice (Zwaveling et al. 2002). MHC restriction is another problem that has to be overcome. Since HLA-A*0201 is the most common human MHC class I molecule, peptide-based therapeutic HPV vaccine studies have focused on HPV-16 E7 HLA-A*0201 CTL epitopes. A phase I clinical trial found that a vaccine with HPV-16 E7 peptide specific for HLA-A*0201 along with adjuvant was able to stimulate an immune response in 10 of 16 HLA-A2-positive patients with cervical or vulva intraepithelial neoplasia (CIN/VIN) stage 2/3 and complete regression of CIN lesions in 3 of 18 patients (Muderspach et al. 2000). However, these peptide vaccines are only effective in patients with HLA-A*0201.

Long overlapping peptides circumvent the problem of MHC restriction by including a range of antigenic epitopes of HPV E6 and E7 proteins for CTLs and additionally provides CD4 Th epitopes. A vaccine comprised of 13 overlapping peptides representing HPV-16 E6 and E7 antigens with adjuvant was able to elicit a broad T cell response in end-stage cervical cancer patients (Kenter et al. 2008). A vaccine with a broad array of epitopes generated increased HPV-16-specific CD4+ and CD8+ T cell responses in early-stage cervical cancer patients relative to unvaccinated patients (Welters et al. 2008). A nonplacebo-controlled phase II clinical trial of this same vaccine demonstrated great efficacy in subjects with HPV-16-positive-high-grade vulva intraepithelial neoplasia (VIN): Half of the patients with histologically confirmed HPV-16-positive VIN3 displayed a complete regression of their lesion (Kenter et al. 2008). One possible explanation for outcomes of lesion regression could be attributed to the ratio of the number of HPV-16-specific primed effector T cells to the number of HPV-16-specific CD4+CD25+Foxp3+T reg cells. Foxp3+ T cells have been associated previously with impaired immunity in malignancies (Welters et al. 2010).

4.4   Protein-Based Vaccines

Protein-based vaccines have also been developed as potential therapeutic vaccines for cervical cancer. They are safer than live vector-based vaccines and have an advantage over peptide vaccines in that they include epitopes that bind to all haplotypes of MHC class I and class II molecules. However, disadvantages of protein-based vaccines include poor immunogenicity and generation of predominantly antibody rather than CTL responses. Numerous strategies to increase the potency of protein-based vaccines have been studied, including the use of a diverse array of adjuvants and fusion proteins. These adjuvants include liposome-polycation-DNA (LPD) adjuvant (Cui and Huang 2005), saponin-based ISCOMATRIX (Frazer et al. 2004), and toll like receptor agonists (Kang et al. 2010).

Among the several protein-based therapeutic HPV candidates explored in clinical trials, HspE7, a chimeric protein of bacille Calmette-Guerin heat shock protein (Hsp65) linked to HPV-16 E7, has generated considerable interest. HspE7 has been found to be well tolerated in both phase I and phase II clinical trials of various types of lesions caused by HPV. In a phase II trial in which 27 subjects with high-grade anal squamous intraepithelial lesions were vaccinated with HspE7, 21 % of subjects had complete, and 71 % had nearly complete lesion regression (Goldstone et al. 2002). In a phase II trial studying CIN3 lesion regression, 13 of 58 subjects had complete pathologic regression while 32 of 58 subjects had partial regression following subcutaneous administration of HspE7 vaccination (Einstein et al. 2007; Roman et al. 2007; Van Doorslaer et al. 2009). However, there was no significant difference in lesion regression in subjects infected with HPV-16 vs. non-HPV-16 infected subjects, so it is not clear as to the exact cause of lesion regression (Einstein et al. 2007). In a different phase II trial involving 27 subjects with recurrent respiratory papillomatosis (the majority of which is caused by HPV-6 and -11), HspE7 was able to reduce the frequency of required surgeries by increasing the time between surgeries by 93 % (Derkay et al. 2005). These results could be due to the non-specific immunogenicity of Hsp or the ability of the E7 protein from HPV-16 to provide cross protection against that of HPV-6 and -11.

Another protein-based vaccine that has shown efficacy against HPV infections is TA-CIN, a fusion protein-based vaccine consisting of HPV-16 L2-E6-E7. A recent phase II clinical trial assessed intramuscular administration of TA-CIN combined with topical application of imiquimod in patients with high-grade VIN. This combination was well tolerated without adverse effects, and "responders" to the therapy demonstrated high levels of CD4+ and CD8+ T cells locally as well as within HPV-associated lesions. Imiquimod was shown to increase T cell infiltration, leading to the complete regression of VIN lesions in 63 % of patients one year post-treatment. Additionally, 36 % of patients with VIN lesions showed complete HPV clearance, and 79 % of women remained symptom free (Daayana et al. 2010). Phase III clinical trials will be needed to assess the efficacy of this combinatorial approach.

4.5 Dendritic Cell-Based Vaccines

Dendritic cells (DCs) are professional antigen-presenting cells able to induce the adaptive immune response by processing antigens to prime antigen-specific T cells. DCs can be pulsed with peptides, proteins, or DNA encoding antigens ex vivo. The prepared DCs are then reintroduced into the body in order to elicit a cell-mediated immune response. Understanding DC maturation and antigen presentation has proven fruitful in exploring DCs as a means of immunotherapy [for a review, see (Santin et al. 2005)].

Strategies to increase the immunogenicity and efficacy of DC-based vaccines have emerged in recent years, including a range of adjuvants such as cholera toxin (Nurkkala et al. 2010) and toll like receptor (TLR) agonists (Chen et al. 2010). A phase I clinical trial of a DC-based HPV vaccine using HPV-16 E7 and/or HPV-18 E7 proteins assessed the safety and immunogenicity of the therapeutic vaccine in patients diagnosed with early-stage cervical cancer. Isolated dendritic cells were primed with HPV proteins E6 and E7 and then re-administered to patients to elicit specific T-cell responses. Subjects generated HPV-16 E7-specific CD4+ T-cell immunity to three immunodominant E7-specific regions: amino acid sequences 46–70, 47–70, and 76–98. However, one patient generated a T-cell response to a novel E7 antigen of amino acids 58–68 (Wang et al. 2009). This newly identified CD4+ Th cell epitope may be used in future DC-based HPV therapeutic vaccine strategies as well as in peptide-based HPV vaccines.

4.6   DNA-Based Vaccines

Among the different forms of therapeutic HPV vaccines, DNA vaccines have emerged as an attractive approach for their safety, stability, and simplicity. DNA vaccination involves injecting plasmid DNA encoding antigens of interest into host cells. The expression of the encoded antigens can lead to cell-mediated and/or humoral immune responses against the encoded antigen. DNA vaccines have less acute safety risks than live vectors such as bacteria or viruses but integration remains a potential issue. Naked DNA is relatively easy to manufacture and can sustain expression of antigens inside target cells for a longer period of time. Additionally, DNA vaccines do not elicit neutralizing antibodies like live-vector-based vaccines, therefore yielding the capacity for repeated administration. DNA vaccines, however, have shown limited immunogenicity and require strategies to improve delivery to target dendritic cells (DCs), the potent activators of antigen-specific immune responses. In general, these strategies can be classified as: (1) increasing the number of antigen-expressing/antigen-loaded DCs, (2) improving HPV antigen expression, processing, and presentation in DCs, and (3) enhancing DC and T cell interaction [for review, see (Lin et al., 2010)].

Strategies to increase the number of antigen-expressing/antigen-loaded DCs include enhanced delivery methods of DNA vaccines to DCs via gene gun, microencapsulation, and electroporation. Of particular interest is ZYC101, a plasmid encoding a HPV-16 E7 HLA-A2 restricted peptide encapsulated in microparticles composed of poly-lactide co-glycolide (PLG). A phase I trial showed that ZYC101 produced complete histological regression in 5 of 15 patients, and significant HPV-specific T cell responses with no serious adverse effects were found in 11 of 15 patients (Sheets et al. 2003). A more recent version, ZYC101a (amolimogene bepiplasmid encoding antigenic peptides derived from HPV-16 and -18 E6 and E7), was used in a phase II clinical trial involving 127 subjects with high-grade CIN. The vaccine was well tolerated and promoted CIN 2/3 resolution in patients under 25 years of age compared to the placebo group (70 vs. 23 %) (Garcia et al. 2004). Recently, a phase II trial of ZYC101a involving 21 subjects with CIN 2/3 showed elevated CD8+ T-cell responses to HPV-16 and -18 in 11 subjects; however, there was no significant difference between the placebo and vaccinated groups in clearance of CIN lesions (Matijevic et al. 2011).

Besides microencapsulation of DNA vaccines, intramuscular injection with subsequent electroporation has also been assessed as a delivery method (Inovio 2012). VGX-3100, a DNA vaccine incorporating plasmids targeting HPV-16 and -18 E6 and E7 proteins is delivered via intramuscular injection, and the vaccinated area is then electroporated using an electroporation device. In a phase I trial of subjects with high-grade CIN lesions, VGX-3100 was able to elicit a T cell response in 13 of 18 subjects. Individuals with histologically confirmed HPV-16 and -18 associated high-grade CIN are currently being recruited for a double-blinded, randomized, placebo-controlled phase II clinical trial (Inovio 2012).

In hopes of enhancing antigen delivery to DCs, different methods of DNA vaccine delivery are being tested in clinical trials. Currently, a head-to-head comparison of the immunogenicity of three routes of administration of a DNA vaccine encoding calreticulin linked to a non-oncogenic form of HPV-16 E7 [pNGVL4a-CRT/E7 (Detox)] is recruiting women with HPV-16 + CIN2/3. Subjects will be divided into groups and receive various doses of the same vaccine either intradermally via gene gun, intramuscularly, or intralesionally in the cervix in order to compare their ability to generate HPV-16 E7 antigen-specific immune responses and therapeutic efficacies (NCI 2012c).

Another strategy to enhance DNA vaccine efficacy is to employ HPV antigen linked to a molecule capable of targeting professional antigen-presenting cells to improve cross-priming of the linked antigen. One of such molecules is heat shock protein 70 (HSP 70). A phase I clinical trial studied the effects of intramuscular injection with pNGVL4a-Sig/E7(detox)/HSP70 (a vaccine consisting of DNA encoding a signal sequence localized to the endoplasmic reticulum (Sig), an attenuated form of HPV-16 E7, and HSP70) in fifteen subjects with high-grade CIN (Trimble et al. 2009). Results demonstrated evidence of HPV-16 E7 specific T cell responses post-vaccination in 4 of 15 subjects and in 5 of 9 subjects six months later in those who received the highest dose. Despite weak immunologic responses, complete histologic regression occurred in 33 % of patients vaccinated with the highest dose of pNGVL4a-Sig/E7(detox)/HSP70 (3 mg) (Trimble et al. 2009). The same vaccine has also been explored in patients with HPV-16+ head and neck cancer, 20 % of which are related to HPV infection (Gilison and Wu personal communication). These early phase clinical trials with pNGVL4a-Sig/E7(detox)/HSP70 demonstrate great safety and few side effects in both CIN 2/3 and head and neck cancers caused by HPV.

Therapeutic HPV DNA vaccines have also been used in conjunction with adjuvants to improve DNA vaccine potency (Chuang et al. 2010). One such adjuvant is a TLR7 agonist, imiquimod, which has been shown to promote activation of antigen presenting cells and lead to the production of cytokines (IFN-alpha, IL-6, and TNF-alpha) that facilitate adaptive immune cell differentiation (Bilu and Sauder 2003). Currently, an ongoing phase I clinical trial combines the use of topical imiquimod along with pNGVL4a-Sig/E7(detox)/HSP70 DNA vaccine priming and TA-HPV vaccine boosting in patients with HPV-16+ CIN3 lesions (NCI 2012a). This use of multiple vaccines and adjuvant tested together is an attempt to generate increased HPV antigen-specific immune responses and better therapeutic effects.

4.7   Combination Strategies

Therapeutic HPV vaccines can be used in combination with other therapies to improve vaccine potency. Combinatorial approaches of therapeutic HPV vaccines along with chemotherapy, radiation therapy, and/or surgery may potently enhance treatment for cancer cells, particularly for minimal residual disease. For example, treatment of mice with DNA encoding calreticulin (CRT) fused to HPV-16 E7 (CRT/E7) in combination with radiation therapy showed a significant increase in the number of E7-specific CD8+ T-cell responses and antitumor effects against E7-expressing tumors as compared to treatment with therapeutic HPV DNA vaccine or radiation alone (Tseng et al. 2009). Another combination therapy involves the use of a recombinant, HPV E6 and E7-expressing adenovirus vaccine (Ad-p14) along with systemic administration of one of the following immuno-modulating reagents: imiquimod, anti-CD4, alpha-interferon, or anti-GITR. Significant antitumor results were found in the E6/E7-expressing TC-1 mouse model whereby combining Ad-p14 with anti-GITR resulted in complete and permanent eradication of all TC-1 tumors (Hoffmann et al. 2010). Chemotherapy in conjunction with therapeutic HPV DNA-based vaccines is another promising approach. One preclinical study in tumor-bearing mice combined the use of a common chemotherapeutic drug, cisplatin, along with a therapeutic HPV DNA vaccine (CRT/E7). The use of cisplatin led to cell-mediated lysis of E7-expressing tumor cells, thus increasing the available E7 antigens for T-cell priming. Results also showed an increased number of E7-specific CD8+ T-cell precursors that were able to proliferate and migrate into tumor locations (Tseng et al. 2008). Recently, the chemotherapeutic agent apigenin, a flavonoid with antioxidant, anti-inflammatory and anti-cyclooxygenase activity, was used concurrently with a HPV DNA vaccine encoding HPV-16 E7 linked to heat shock protein 70 (E7/HSP70) (Chuang et al. 2009). Vaccination of E7/HSP70 DNA in conjunction with apigenin chemotherapy in TC-1 tumor-bearing mice demonstrated the greatest increase in the number of E7-specific effector and memory CD8+ T cells compared to controls. Apigenin treatment also increased tumor cell apoptosis in a dose-dependent manner. Vaccination and chemotherapy most likely caused tumor susceptibility to E7-specific cytotoxic immune responses, which led to a reduction in tumor size and an increase in survival rates (Chuang et al. 2009). The combination of chemotherapy and/or radiation therapy with therapeutic HPV vaccines has produced encouraging anti-tumor results in preclinical models and therefore warrants future testing in patients with HPV-associated malignancies.

5 Conclusion

Although HPV infections are incredibly common throughout the world, the development of accessible, efficacious and highly specific methods of controlling infection and malignant transformation due to HPV infections remains challenging. The current commercially available preventive vaccines represent a significant triumph for the HPV research field. However, there is still much room for improvement in the control of cervical cancer in terms of: (1) making broader spectrum, low cost, one dose, heat stable and widely accessible preventative vaccines, and (2) employing our knowledge of the immunology of HPV infections to create effective therapeutic HPV vaccines. A variety of innovative strategies have been used to develop therapeutic HPV vaccines, resulting in several phase I/II clinical trials. It is important to further advance these trials in order to identify suitable therapeutic HPV vaccine candidates for the control of established HPV infections and HPV-associated lesions. The control of advanced HPV-associated malignancies will most likely require combinatorial approaches using therapeutic HPV vaccines in conjunction with conventional therapies such as surgery, chemotherapy, and radiation therapy. The advances in both preventive and therapeutic HPV vaccines will undoubtedly offer relief to the hundreds of millions of individuals exposed to HPV each year.

Acknowledgments We thank Dr. Shiwen Peng for helpful discussion. This review is not intended to be an encyclopedic one, and the authors apologize to those not cited. This work was funded by the National Institutes of Health Cervical Cancer SPORE and Head and Neck Cancer SPORE (P50 CA098252 and P50 CA96784-06).

References

Adams M, Borysiewicz L, Fiander A, Man S, Jasani B, Navabi H, Lipetz C, Evans AS, Mason M (2001) Clinical studies of human papilloma vaccines in pre-invasive and invasive cancer. Vaccine 19:2549–2556
Baldwin PJ, van der Burg SH, Boswell CM, Offringa R, Hickling JK, Dobson J, Roberts JS, Latimer JA, Moseley RP, Coleman N et al (2003) Vaccinia-expressed human papillomavirus 16 and 18 e6 and e7 as a therapeutic vaccination for vulval and vaginal intraepithelial neoplasia. Clin Cancer Res 9:5205–5213
Bilu D, Sauder DN (2003) Imiquimod: modes of action. Br J Dermatol 149(Suppl 66):5–8
Borysiewicz LK, Fiander A, Nimako M, Man S, Wilkinson GW, Westmoreland D, Evans AS, Adams M, Stacey SN, Boursnell ME et al (1996) A recombinant vaccinia virus encoding

human papillomavirus types 16 and 18, E6 and E7 proteins as immunotherapy for cervical cancer. Lancet 347:1523–1527

Brooks G, Carroll KC, Butel JS, Morse SA, Mietzneron TA (eds) (2010) Jawetz, Melnick, & Adelberg's Medical Microbiology, 25th edn. McGraw-Hill Medical, New York

Brun JL, Dalstein V, Leveque J, Mathevet P, Raulic P, Baldauf JJ, Scholl S, Huynh B, Douvier S, Riethmuller D et al (2011) Regression of high-grade cervical intraepithelial neoplasia with TG4001 targeted immunotherapy. Am J Obstet Gynecol 204:161–168

Campo MS, Roden RB (2010) Papillomavirus prophylactic vaccines: established successes, new approaches. J Virol 84:1214–1220

Chaturvedi AK, Engels EA, Pfeiffer RM, Hernandez BY, Xiao W, Kim E, Jiang B, Goodman MT, Sibug-Saber M, Cozen W et al (2011) Human papillomavirus and rising oropharyngeal cancer incidence in the United States. J Clin Oncol 29:4294–4301

Chen XS, Garcea RL, Goldberg I, Casini G, Harrison SC (2000) Structure of small virus-like particles assembled from the L1 protein of human papillomavirus 16. Mol Cell 5:557–567

Chen XZ, Mao XH, Zhu KJ, Jin N, Ye J, Cen JP, Zhou Q, Cheng H (2010) Toll like receptor agonists augment HPV 11 E7-specific T cell responses by modulating monocyte-derived dendritic cells. Arch Dermatol Res 302:57–65

Chuang CM, Monie A, Wu A, Hung CF (2009) Combination of apigenin treatment with therapeutic HPV DNA vaccination generates enhanced therapeutic antitumor effects. J Biomed Sci 16:49

Chuang CM, Monie A, Hung CF, Wu TC (2010) Treatment with imiquimod enhances antitumor immunity induced by therapeutic HPV DNA vaccination. J Biomed Sci 17:32

Corona Gutierrez CM, Tinoco A, Navarro T, Contreras ML, Cortes RR, Calzado P, Reyes L, Posternak R, Morosoli G, Verde ML et al (2004) Therapeutic vaccination with MVA E2 can eliminate precancerous lesions (CIN 1, CIN 2, and CIN 3) associated with infection by oncogenic human papillomavirus. Hum Gene Ther 15:421–431

Cui Z, Huang L (2005) Liposome-polycation-DNA (LPD) particle as a carrier and adjuvant for protein-based vaccines: therapeutic effect against cervical cancer. Cancer Immunol Immunother 54:1180–1190

Daayana S, Elkord E, Winters U, Pawlita M, Roden R, Stern PL, Kitchener HC (2010) Phase II trial of imiquimod and HPV therapeutic vaccination in patients with vulval intraepithelial neoplasia. Br J Cancer 102:1129–1136

Daftarian P, Mansour M, Benoit AC, Pohajdak B, Hoskin DW, Brown RG, Kast WM (2006) Eradication of established HPV 16-expressing tumors by a single administration of a vaccine composed of a liposome-encapsulated CTL-T helper fusion peptide in a water-in-oil emulsion. Vaccine 24:5235–5244

Davidson EJ, Boswell CM, Sehr P, Pawlita M, Tomlinson AE, McVey RJ, Dobson J, Roberts JS, Hickling J, Kitchener HC et al (2003) Immunological and clinical responses in women with vulval intraepithelial neoplasia vaccinated with a vaccinia virus encoding human papillomavirus 16/18 oncoproteins. Cancer Res 63:6032–6041

Derkay CS, Smith RJ, McClay J, van Burik JA, Wiatrak BJ, Arnold J, Berger B, Neefe JR (2005) HspE7 treatment of pediatric recurrent respiratory papillomatosis: final results of an open-label trial. Ann Otol Rhinol Laryngol 114:730–737

Dillner J, Kjaer SK, Wheeler CM, Sigurdsson K, Iversen OE, Hernandez-Avila M, Perez G, Brown DR, Koutsky LA, Tay EH et al (2010) Four year efficacy of prophylactic human papillomavirus quadrivalent vaccine against low grade cervical, vulvar, and vaginal intraepithelial neoplasia and anogenital warts: randomised controlled trial. BMJ 341:c3493

DiMaio D, Mattoon D (2001) Mechanisms of cell transformation by papillomavirus E5 proteins. Oncogene 20:7866–7873

Echelman D, Feldman S (2012) Management of cervical precancers: a global perspective. Hematol Oncol Clin North Am 26:31–44

Einstein MH, Kadish AS, Burk RD, Kim MY, Wadler S, Streicher H, Goldberg GL, Runowicz CD (2007) Heat shock fusion protein-based immunotherapy for treatment of cervical intraepithelial neoplasia III. Gynecol Oncol 106:453–460

Einstein MH, Baron M, Levin MJ, Chatterjee A, Edwards RP, Zepp F, Carletti I, Dessy FJ, Trofa AF, Schuind A et al (2009) Comparison of the immunogenicity and safety of Cervarix and Gardasil human papillomavirus (HPV) cervical cancer vaccines in healthy women aged 18–45 years. Hum Vaccin 5:705–719

Einstein MH, Baron M, Levin MJ, Chatterjee A, Fox B, Scholar S, Rosen J, Chakhtoura N, Meric D, Dessy FJ et al (2011) Comparative immunogenicity and safety of human papillomavirus (HPV)-16/18 vaccine and HPV-6/11/16/18 vaccine: Follow-up from Months 12–24 in a Phase III randomized study of healthy women aged 18–45 years. Hum Vaccin 7:1343–1358

Embers ME, Budgeon LR, Pickel M, Christensen ND (2002) Protective immunity to rabbit oral and cutaneous papillomaviruses by immunization with short peptides of L2, the minor capsid protein. J Virol 76:9798–9805

Fraillery D, Baud D, Pang SY, Schiller J, Bobst M, Zosso N, Ponci F, Nardelli-Haefliger D (2007) Salmonella enterica serovar Typhi Ty21a expressing human papillomavirus type 16 L1 as a potential live vaccine against cervical cancer and typhoid fever. Clin Vaccine Immunol 14:1285–1295

Frazer IH, Quinn M, Nicklin JL, Tan J, Perrin LC, Ng P, O'Connor VM, White O, Wendt N, Martin J et al (2004) Phase 1 study of HPV16-specific immunotherapy with E6E7 fusion protein and ISCOMATRIX adjuvant in women with cervical intraepithelial neoplasia. Vaccine 23:172–181

Gambhira R, Karanam B, Jagu S, Roberts JN, Buck CB, Bossis I, Alphs H, Culp T, Christensen ND, Roden RB (2007) A protective and broadly cross-neutralizing epitope of human papillomavirus L2. J Virol 81:13927–13931

Garcia F, Petry KU, Muderspach L, Gold MA, Braly P, Crum CP, Magill M, Silverman M, Urban RG, Hedley ML et al (2004) ZYC101a for treatment of high-grade cervical intraepithelial neoplasia: a randomized controlled trial. Obstet Gynecol 103:317–326

Gaukroger JM, Chandrachud LM, O'Neil BW, Grindlay GJ, Knowles G, Campo MS (1996) Vaccination of cattle with bovine papillomavirus type 4 L2 elicits the production of virus-neutralizing antibodies. J Gen Virol 77:1577–1583

Giuliano AR, Palefsky JM, Goldstone S, Moreira ED Jr, Penny ME, Aranda C, Vardas E, Moi H, Jessen H, Hillman R et al (2011) Efficacy of quadrivalent HPV vaccine against HPV Infection and disease in males. N Engl J Med 364:401–411

Goldstone SE, Palefsky JM, Winnett MT, Neefe JR (2002) Activity of HspE7, a novel immunotherapy, in patients with anogenital warts. Dis Colon Rectum 45:502–507

Gunn GR, Zubair A, Peters C, Pan ZK, Wu TC, Paterson Y (2001) Two Listeria monocytogenes vaccine vectors that express different molecular forms of human papilloma virus-16 (HPV-16) E7 induce qualitatively different T cell immunity that correlates with their ability to induce regression of established tumors immortalized by HPV-16. J Immunol 167:6471–6479

Hoffmann C, Stanke J, Kaufmann AM, Loddenkemper C, Schneider A, Cichon G (2010) Combining T-cell vaccination and application of agonistic anti-GITR mAb (DTA-1) induces complete eradication of HPV oncogene expressing tumors in mice. J Immunother 33:136–145

Hung CF, Wu TC (2003) Improving DNA vaccine potency via modification of professional antigen presenting cells. Curr Opin Mol Ther 5:20–24

Hussain SF, Paterson Y (2004) CD4+CD25+ regulatory T cells that secrete TGFbeta and IL-10 are preferentially induced by a vaccine vector. J Immunother 27:339–346

Inovio (2012) A study of VGX-3100 DNA vaccine with electroporation in patients with cervical intraepithelial neoplasia grade 2/3 or 3 (HPV-003). http://clinicaltrials.gov/ct2/show/NCT01304524?term=NCT01304524&rank=1

Jagu S, Karanam B, Gambhira R, Chivukula SV, Chaganti RJ, Lowy DR, Schiller JT, Roden RB (2009) Concatenated multitype L2 fusion proteins as candidate prophylactic pan-human papillomavirus vaccines. J Natl Cancer Inst 101:782–792

Jemal A, Bray F, Center MM, Ferlay J, Ward E, Forman D (2011) Global cancer statistics. CA Cancer J Clin 61:69–90

Kang TH, Monie A, Wu LS, Pang X, Hung CF, Wu TC (2010) Enhancement of protein vaccine potency by in vivo electroporation mediated intramuscular injection. Vaccine 29:1082–1089

Kaufmann AM, Stern PL, Rankin EM, Sommer H, Nuessler V, Schneider A, Adams M, Onon TS, Bauknecht T, Wagner U et al (2002) Safety and immunogenicity of TA-HPV, a recombinant vaccinia virus expressing modified human papillomavirus (HPV)-16 and HPV-18 E6 and E7 genes, in women with progressive cervical cancer. Clin Cancer Res 8:3676–3685

Kawana K, Yoshikawa H, Taketani Y, Yoshiike K, Kanda T (1999) Common neutralization epitope in minor capsid protein L2 of human papillomavirus types 16 and 6. J Virol 73:6188–6190

Kenter GG, Welters MJ, Valentijn AR, Lowik MJ, Berends-van der Meer DM, Vloon AP, Drijfhout JW, Wafelman AR, Oostendorp J, Fleuren GJ et al (2008) Phase I immunotherapeutic trial with long peptides spanning the E6 and E7 sequences of high-risk human papillomavirus 16 in end-stage cervical cancer patients shows low toxicity and robust immunogenicity. Clin Cancer Res 14:169–177

Kenter GG, Welters MJ, Valentijn AR, Lowik MJ, Berends-van der Meer DM, Vloon AP, Essahsah F, Fathers LM, Offringa R, Drijfhout JW et al (2009) Vaccination against HPV-16 oncoproteins for vulvar intraepithelial neoplasia. N Engl J Med 361:1838–1847

Kondo K, Ochi H, Matsumoto T, Yoshikawa H, Kanda T (2008) Modification of human papillomavirus-like particle vaccine by insertion of the cross-reactive L2-epitopes. J Med Virol 80:841–846

Kwak K, Yemelyanova A, Roden RB (2010) Prevention of cancer by prophylactic human papillomavirus vaccines. Curr Opin Immunol 23:244–251

Lehtinen M, Paavonen J, Wheeler CM, Jaisamrarn U, Garland SM, Castellsague X, Skinner SR, Apter D, Naud P, Salmeron J et al (2012) Overall efficacy of HPV-16/18 AS04-adjuvanted vaccine against grade 3 or greater cervical intraepithelial neoplasia: 4-year end-of-study analysis of the randomised, double-blind PATRICIA trial. Lancet Oncol 13:89–99

Li M, Cripe TP, Estes PA, Lyon MK, Rose RC, Garcea RL (1997) Expression of the human papillomavirus type 11 L1 capsid protein in Escherichia coli: characterization of protein domains involved in DNA binding and capsid assembly. J Virol 71:2988–2995

Lin K, Roosinovich E, Ma B, Hung CF, Wu TC (2010) Therapeutic HPV DNA vaccines. Immunol Res 47:86–112

Lowy DR, Schiller JT (2012) Reducing HPV-associated cancer globally. Cancer Prev Res (Phila) 5:18–23

Maciag PC, Radulovic S, Rothman J (2009) The first clinical use of a live-attenuated Listeria monocytogenes vaccine: a Phase I safety study of Lm-LLO-E7 in patients with advanced carcinoma of the cervix. Vaccine 27:3975–3983

Matijevic M, Hedley ML, Urban RG, Chicz RM, Lajoie C, Luby TM (2011) Immunization with a poly (lactide co-glycolide) encapsulated plasmid DNA expressing antigenic regions of HPV 16 and 18 results in an increase in the precursor frequency of T cells that respond to epitopes from HPV 16, 18, 6 and 11. Cell Immunol 270:62–69

Merck (2011) Phase III clinical trial: broad spectrum HPV (Human Papillomavirus) vaccine study in 16-to 26-year-old women (V503-001 AM2)

Muderspach L, Wilczynski S, Roman L, Bade L, Felix J, Small LA, Kast WM, Fascio G, Marty V, Weber J (2000) A phase I trial of a human papillomavirus (HPV) peptide vaccine for women with high-grade cervical and vulvar intraepithelial neoplasia who are HPV 16 positive. Clin Cancer Res 6:3406–3416

NCI (2012a) A Phase I efficacy and safety study of HPV16-specific therapeutic DNA-vaccinia vaccination in combination with topical imiquimod, in patients with HPV16+ high grade cervical dysplasia (CIN3). http://www.cancer.gov/clinicaltrials/search/view?cdrid=617261&version=HealthProfessional&protocolsearchid=10105493

NCI (2012b) Phase II study of live-attenuated listeria monocytogenes cancer vaccine ADXS11-001 in patients with persistent or recurrent squamous cell or non-squamous cell carcinoma of the cervix. http://cancer.gov/clinicaltrials/search/view?cdrid=691288&version=healthprofessional

NCI (2012c) A pilot study of pnGVL4a-CRT/E7 (Detox) for the treatment of patients with HPV16+ cervical intraepithelial neoplasia 2/3 (CIN2/3). http://clinicaltrials.gov/ct2/show/NCT00988559?term=trimble&rank=1

NCI (2012d) A randomized, single blind, placebo controlled phase 2 study to assess the safety of ADXS11-001 for the treatment of cervical intraepithelial neoplasia grade 2/3. http://clinicaltrials.gov/ct2/show/NCT01116245

Nurkkala M, Wassen L, Nordstrom I, Gustavsson I, Slavica L, Josefsson A, Eriksson K (2010) Conjugation of HPV16 E7 to cholera toxin enhances the HPV-specific T-cell recall responses to pulsed dendritic cells in vitro in women with cervical dysplasia. Vaccine 28:5828–5836

Pastrana DV, Gambhira R, Buck CB, Pang YY, Thompson CD, Culp TD, Christensen ND, Lowy DR, Schiller JT, Roden RB (2005) Cross-neutralization of cutaneous and mucosal Papillomavirus types with anti-sera to the amino terminus of L2. Virology 337:365–372

Qin Y, Wang XH, Cui HL, Cheung YK, Hu MH, Zhu SG, Xie Y (2005) Human papillomavirus type 16 E7 peptide(38–61) linked with an immunoglobulin G fragment provides protective immunity in mice. Gynecol Oncol 96:475–483

Radulovic S, Brankovic-Magic M, Malisic E, Jankovic R, Dobricic J, Plesinac-Karapandzic V, Maciag PC, Rothman J (2009) Therapeutic cancer vaccines in cervical cancer: phase I study of Lovaxin-C. J Buon 14(Suppl 1):S165–S168

Roden R, Wu TC (2006) How will HPV vaccines affect cervical cancer? Nat Rev Cancer 6:753–763

Roden RB, Yutzy WHt, Fallon R, Inglis S, Lowy DR, Schiller JT (2000) Minor capsid protein of human genital papillomaviruses contains subdominant, cross-neutralizing epitopes. Virology 270:254–257

Roman LD, Wilczynski S, Muderspach LI, Burnett AF, O'Meara A, Brinkman JA, Kast WM, Facio G, Felix JC, Aldana M et al (2007) A phase II study of Hsp-7 (SGN-00101) in women with high-grade cervical intraepithelial neoplasia. Gynecol Oncol 106:558–566

Romanowski B, Schwarz TF, Ferguson LM, Peters K, Dionne M, Schulze K, Ramjattan B, Hillemanns P, Catteau G, Dobbelaere K et al (2011) Immunogenicity and safety of the HPV-16/18 AS04-adjuvanted vaccine administered as a 2-dose schedule compared to the licensed 3-dose schedule: Results from a randomized study. Hum Vaccin 7:1374–1386

Rose RC, White WI, Li M, Suzich JA, Lane C, Garcea RL (1998) Human papillomavirus type 11 recombinant L1 capsomeres induce virus-neutralizing antibodies. J Virol 72:6151–6154

Santin AD, Bellone S, Roman JJ, Burnett A, Cannon MJ, Pecorelli S (2005) Therapeutic vaccines for cervical cancer: dendritic cell-based immunotherapy. Curr Pharm Des 11:3485–3500

Schellenbacher C, Roden R, Kirnbauer R (2009) Chimeric L1-L2 virus-like particles as potential broad-spectrum human papillomavirus vaccines. J Virol 83:10085–10095

Sewell DA, Shahabi V, Gunn GR 3rd, Pan ZK, Dominiecki ME, Paterson Y (2004) Recombinant Listeria vaccines containing PEST sequences are potent immune adjuvants for the tumor-associated antigen human papillomavirus-16 E7. Cancer Res 64:8821–8825

Sharma RK, Elpek KG, Yolcu ES, Schabowsky RH, Zhao H, Bandura-Morgan L, Shirwan H (2009) Costimulation as a platform for the development of vaccines: a peptide-based vaccine containing a novel form of 4-1BB ligand eradicates established tumors. Cancer Res 69:4319–4326

Sheets EE, Urban RG, Crum CP, Hedley ML, Politch JA, Gold MA, Muderspach LI, Cole GA, Crowley-Nowick PA (2003) Immunotherapy of human cervical high-grade cervical intraepithelial neoplasia with microparticle-delivered human papillomavirus 16 E7 plasmid DNA. Am J Obstet Gynecol 188:916–926

Simard EP, Ward EM, Siegel R, Jemal A (2012) Cancers with increasing incidence trends in the United States: 1999 through 2008. CA Cancer J Clin. doi:10.3322/caac.20141

Souders NC, Sewell DA, Pan ZK, Hussain SF, Rodriguez A, Wallecha A, Paterson Y (2007) Listeria-based vaccines can overcome tolerance by expanding low avidity CD8+ T cells capable of eradicating a solid tumor in a transgenic mouse model of cancer. Cancer Immun 7:2–14

Trimble CL, Frazer IH (2009) Development of therapeutic HPV vaccines. Lancet Oncol 10:975–980

Trimble CL, Peng S, Kos F, Gravitt P, Viscidi R, Sugar E, Pardoll D, Wu TC (2009) A phase I trial of a human papillomavirus DNA vaccine for HPV16 + cervical intraepithelial neoplasia 2/3. Clin Cancer Res 15:361–367

Tseng CW, Hung CF, Alvarez RD, Trimble C, Huh WK, Kim D, Chuang CM, Lin CT, Tsai YC, He L et al (2008) Pretreatment with cisplatin enhances E7-specific CD8 + T-Cell-mediated antitumor immunity induced by DNA vaccination. Clin Cancer Res 14:3185–3192

Tseng CW, Trimble C, Zeng Q, Monie A, Alvarez RD, Huh WK, Hoory T, Wang MC, Hung CF, Wu TC (2009) Low-dose radiation enhances therapeutic HPV DNA vaccination in tumor-bearing hosts. Cancer Immunol Immunother 58:737–748

Tumban E, Peabody J, Peabody DS, Chackerian B (2011) A pan-HPV vaccine based on bacteriophage PP7 VLPs displaying broadly cross-neutralizing epitopes from the HPV minor capsid protein, L2. PLoS One 6:e23310

Van Doorslaer K, Reimers LL, Studentsov YY, Einstein MH, Burk RD (2009) Serological response to an HPV16 E7 based therapeutic vaccine in women with high-grade cervical dysplasia. Gynecol Oncol 116(2):208–212

Wang X, Santin AD, Bellone S, Gupta S, Nakagawa M (2009) A novel CD4 T-cell epitope described from one of the cervical cancer patients vaccinated with HPV 16 or 18 E7-pulsed dendritic cells. Cancer Immunol Immunother 58:301–308

Welters MJ, Kenter GG, Piersma SJ, Vloon AP, Lowik MJ, Berends-van der Meer DM, Drijfhout JW, Valentijn AR, Wafelman AR, Oostendorp J et al (2008) Induction of tumor-specific CD4+ and CD8+ T-cell immunity in cervical cancer patients by a human papillomavirus type 16 E6 and E7 long peptides vaccine. Clin Cancer Res 14:178–187

Welters MJ, Kenter GG, de Vos van Steenwijk PJ, Lowik MJ, Berends-van der Meer DM, Essahsah F, Stynenbosch LF, Vloon AP, Ramwadhdoebe TH, Piersma SJ et al (2010) Success or failure of vaccination for HPV16-positive vulvar lesions correlates with kinetics and phenotype of induced T-cell responses. Proc Natl Acad Sci U S A 107:11895–11899

Wheeler CM, Castellsague X, Garland SM, Szarewski A, Paavonen J, Naud P, Salmeron J, Chow SN, Apter D, Kitchener H et al (2012) Cross-protective efficacy of HPV-16/18 AS04-adjuvanted vaccine against cervical infection and precancer caused by non-vaccine oncogenic HPV types: 4-year end-of-study analysis of the randomised, double-blind PATRICIA trial. Lancet Oncol 13:100–110

Wu CY, Monie A, Pang X, Hung CF, Wu TC (2010) Improving therapeutic HPV peptide-based vaccine potency by enhancing CD4+ T help and dendritic cell activation. J Biomed Sci 17:88

Yan W, Chen WC, Liu Z, Huang L (2010) Bryostatin-I: A dendritic cell stimulator for chemokines induction and a promising adjuvant for a peptide based cancer vaccine. Cytokine 52:238–244

Zhang YQ, Tsai YC, Monie A, Hung CF, Wu TC (2010) Carrageenan as an adjuvant to enhance peptide-based vaccine potency. Vaccine 28:5212–5219

zur Hausen H (2002) Papillomaviruses and cancer: from basic studies to clinical application. Nat Rev Cancer 2:342–350

Zwaveling S, Ferreira Mota SC, Nouta J, Johnson M, Lipford GB, Offringa R, van der Burg SH, Melief CJ (2002) Established human papillomavirus type 16-expressing tumors are effectively eradicated following vaccination with long peptides. J Immunol 169:350–358

Prevention and Treatment for Epstein–Barr Virus Infection and Related Cancers

Françoise Smets and Etienne M. Sokal

Abstract

Epstein–Barr virus (EBV) was the first herpes virus described as being oncogenic in humans. EBV infection is implicated in post-transplant lymphoproliferative diseases (PTLD) and several other cancers in non-immunocompromised patients, with more than 200,000 new cases per year. While prevention of PTLD is improving, mainly based on EBV monitoring and preemptive tapering of immunosuppression, early diagnosis remains the best current option for the other malignancies. Significant progress has been achieved in treatment, with decreased mortality and morbidity, but some challenges are still to face, especially for the more aggressive diseases. Possible prevention by EBV vaccination would be a more global approach of this public health problem, but further active research is needed before this goal could be reached.

Keywords

Epstein–Barr virus (EBV) · Post-transplant lymphoproliferative diseases (PTLD) · Viral load · Specific immunity · Cell therapy · Cancer

Contents

F. Smets (✉)
Université Catholique de Louvain, Cliniques Universitaires St-Luc,
Avenue Hippocrate 10/1301, 1200 Brussels, Belgium
e-mail: francoise.smets@uclouvain.be

E. M. Sokal
Université Catholique de Louvain, Cliniques Universitaires St-Luc, Brussels, Belgium
e-mail: etienne.sokal@uclouvain.be

M. H. Chang and K.-T. Jeang (eds.), *Viruses and Human Cancer*,
Recent Results in Cancer Research 193, DOI: 10.1007/978-3-642-38965-8_10,
© Springer-Verlag Berlin Heidelberg 2014

1   Introduction

Epstein–Barr virus (EBV) was the first virus of the gamma herpes family shown to be oncogenic in humans (Schiller and Lowy 2010). EBV infection is implicated in more than 200,000 new cases of cancer per year. Tumors can develop because of deficient immunity as in post-transplant lymphoproliferative diseases (PTLD). In non-immunocompromised patients, the high oncogenic potential of EBV gives rise to aggressive cancers such as gastric carcinoma, nasopharyngeal carcinoma, Burkitt's and Hodgkin's lymphoma, or T/NK cell lymphoproliferation.

EBV is ubiquitous and infects more than 90 % of the population before adulthood (Cohen 1999; Steven 1997). Primary infection most often occurs in early childhood, often asymptomatic or with mild non-specific symptoms. In adolescents and young adults, the EBV infection is related to infectious mono-nucleosis (IM) in which symptoms are classically described as the result of the immune response to the virus. The first step of natural EBV infection is a lytic phase, mainly located in the epithelial cells of the oropharynx, leading to new virus production and death of the infected cells. After the primary infection, the EBV lytic cycle is usually inactive in healthy carriers but may be reactivated in some conditions, such as immunosuppression. Concomitant with the productive phase, lifelong infection is promoted by latent infection in B lymphocytes, leading to EBV-infected B lymphoblasts in which EBV genome is integrated as an episome and multiplies with the host cell. Early expansion of the infected blasts may be important for viral survival. Latent EBV has a complex survival strategy involving (1) the production of viral proteins that can transform human cells or inhibit antigen processing, (2) the secretion of a cytokine and soluble receptor that act against the cytotoxic cellular immunity, and (3) the down-regulation of its gene expression (Cohen 1999; Liebowitz 1998; Rickinson 1998). In this latent pathway, EBV is a potentially transforming virus and its oncogenicity is closely related to its ability to escape immune surveillance (Okano 2012; Schiller and Lowy 2010). However, the fact that a minority of immunocompetent EBV-infected individuals develop tumor underlies that infection alone is not sufficient and that other factors are involved (genetic and epigenetic, environmental…). Finally, some EBV-infected B lymphoblasts differentiate in memory B cells that present no more viral specific antigens and therefore are invisible to the cellular immunity, representing the final EBV reservoir (Steven 1997).

2 Post-Transplant Lymphoproliferative Diseases

2.1 Introduction

Post-transplant lymphoproliferative diseases (PTLD) occur after either solid organ or bone marrow/stem cell transplantation (Gulley and Tang 2010; Landgren et al. 2009; Reddy et al. 2011; Smets and Sokal 2002a; Smets et al. 2000). PTLD are classified into 4 categories according to WHO classification from 2008 (Swerdlow et al. 2008): early lesions (plasmacytic hyperplasia, IM), polymorphic PTLD, monomorphic PTLD (B- or T-cell lymphomas), and classic Hodgkin's lymphoma-like PTLD. The early lesions and polymorphic PTLD are the commonest presentations, usually arise within 1 year of transplantation and are mostly associated with EBV, while the more monomorphic one will present later and can be EBV negative in up to 70 % of cases. Incidence varies according to type of transplant and age of recipient (Table 1), reflecting underlying pathogenic mechanisms and following risk factors. In early PTLD, lymphoproliferation of EBV-infected cells is enabled by inadequate immune surveillance. Indeed, some EBV-naïve patients are unable to mount specific immunity after solid organ transplantation because of immunosuppression (IS). After bone marrow/stem cell transplant, PTLD appear while awaiting T-cell immunity restoration. Later on, PTLD can occur despite sufficient immunity thanks to several escape mechanisms of the virus (antigens under-expression, genetic mutations, or polymorphisms…). Diagnosis should rely on detailed histology of the tumor (including lymphocytes typing, search for monoclonality, and EBV staining), evaluation of performance status, and staging of the disease according to clinical presentation (radiology, bone marrow, or cerebrospinal fluid examination…). The role of EBV monitoring is detailed below.

Table 1 Incidence of PTLD

	Incidence (%)	
Type of organ	Children	Adults or all
Liver	4–15	1.6–5
Kidney	1–10	0.5–1
Heart–lung	6–20	1.9–10
Intestine	12	–
Bone marrow or stem cells	–	0.5–8

Gulley and Tang (2010), Reddy et al. (2011), and Landgren et al. (2009)

Table 2 Risk factors for PTLD

	Type of transplantation	
Early PTLD	Solid organ[a]	Bone marrow/stem cells[b]
Pre-transplant EBV status	EBV-naïve patients	Indifferent
Age at transplantation	Young children or advanced age	> 50 years
Immunosuppression (IS)	Anti-T-cell antibodies (ATG, OKT3), high degree of IS	ATG, T-cell depletion of graft
Other	Past graft rejection(s)	Unrelated or HLA-mismatched grafts, graft-versus-host disease (GVHD), 2nd transplantation
	Type of HLA (HLA-A3)	
	Cytokine polymorphisms (IFN-γ, TGF-β, IL-10)	T- and B-cell depletion of graft = protective
	Protection by rapamycin?	
	CMV mismatch?	
	Type of transplantation	
Late PTLD	Solid organ	Bone marrow/stem cells
Pre-transplant EBV status	EBV-naïve patients	Indifferent
Age at transplantation	Indifferent	Indifferent
Immunosuppression (IS)	Duration of IS	Cyclosporine or combined cycloazathioprine
Other	EBV genomic polymorphism	Chronic GVHD
	Past history of PTLD	
	Chronic high EBV load	

[a] Kamdar et al. (2011), Smets and Sokal (2002a, b), Cockfield (2001), Gulley and Tang (2010), Weiner et al. (2012), D'Antiga et al. (2007), Bakker et al. (2007), and Bingler et al. (2008)
[b] Lynch et al. (2003) and Landgren et al. (2009)

2.2   Risk Factors and Prevention

Based on pathophysiology and known risk factors (Table 2), several strategies of prevention are described and aim to decrease incidence of PTLD. Current results, however, mainly allow earlier diagnosis, leading to decreased mortality and morbidity. Clear impact on incidence remains to be demonstrated by well-designed prospective studies, when ethically acceptable.

One corner stone of prevention relies on EBV load monitoring by quantitative polymerase chain reaction (PCR), as high viral load is clearly associated with

PTLD. Numerous studies report the characteristics of EBV load after transplantation, measured in peripheral blood mononuclear cells (PBMC), plasma or whole blood (reviewed in Gulley and Tang 2010). Viral load is usually higher in patients under immunosuppression as compared to healthy controls, while the highest viral load is described after transplantation in subjects with acute primary infection, disease reactivation, and PTLD. In pediatric solid organ recipients, viral load >5,000/µg PBMC DNA was a diagnostic criterion for PTLD with a positive predictive value (PPV) of 89 % and negative predictive value (NPV) of 100 %, while both PPV and NPV were 100 % for an EBV load >1,000/100µl plasma (Wagner et al. 2001). High viral load was correlated with PTLD and normalized after treatment. Similar differences were described in other types of transplantation and led to preemptive assays. In Houston (Lee et al. 2005), after 2001, IS was decreased in liver-transplanted children after two EBV loads >4,000/µg PBMC DNA (previously described as giving PPV for PTLD of 56 % and NPV of 100 %). Before and after 2001, PTLD incidence was, respectively, 16 % (within 30 patients) and 2 % (out of 43 patients, 29 EBV-naïve patients before transplantation, $p < 0.05$). Preemptive IS alleviation (tacrolimus tapering, corticoids arrest) induced lowering of viral load in all and led to rejection in 1/11 patient. Another retrospective comparison of children after liver transplantation with or without EBV load monitoring (2001–2005 and 1993–1997), with preemptive IS tapering in case of high viral load, showed no difference in PTLD incidence (7.5 and 10.9 %, respectively) (Kerkar et al. 2010). PTLD occurring in the 2001–2005 era were, however, less advanced (polymorphic versus monomorphic, $p = 0.03$), with no mortality as compared to 3/10 deaths before. One problem is that each laboratory has to define its own cutoff for "high viral load" and risk assessment. The major drawback for using EBV load, as a predictive marker of PTLD, is the intermediate specificity and PPV that could lead to needless IS tapering and risk of graft rejection. The same group showed, always in liver-transplanted children, that reducing IS preemptively for EBV active infection led to graft rejection in 27.9 % of cases (with 1 graft loss), while IS tapering or cessation in case of PTLD was complicated by rejection in 83.3 % of patients, without graft loss (Weiner et al. 2012). Both the risk of PTLD and the risk of rejection after IS reduction were higher in case of past history of rejection. Finally, chronic high EBV load was also described as risk factor for late PTLD. EBV load $>500/10^5$ PBMC was found in 12/18 EBV-seronegative children who underwent primary infection after liver transplantation, as compared to 0/13 seropositive children before transplant (D'Antiga et al. 2007). Sustained viral detection, meaning positive for more than 6 months, was only seen in those EBV-naïve patients (14/18), and three of them developed late PTLD (median time 47 months, range 15–121). In 71 cardiac-transplanted children, chronic high viral load was defined as EBV load >16,000/mL whole blood in >50 % of samples over at least 6 months (Bingler et al. 2008) and was found in 20 of them (8 with prior PTLD, 7 with prior symptomatic EBV). Late PTLD occurred in 9 of them (45 %) compared to 2 in the 51 controls (4 %, $p < 0.001$). Multivariate analysis showed independently increased odds ratio (ORs) to develop late PTLD for chronic high viral load (OR = 12.4, 95 % CI

2.1–74.4) and prior history of PTLD (OR = 10.7, 95 % CI 1.9–60.6). Therefore, experts recommend serial monitoring of EBV load in EBV-naïve patients undergoing primary EBV infection after transplantation, to assess the risk to develop early or late PTLD (Gulley and Tang 2010; KDIGO 2009).

Higher viral load is also described in patients developing PTLD after bone marrow/stem cell transplantation, and same recommendations are applied. Additional factors are important to determine high-risk patient as EBV status before transplantation does not influence the occurrence of PTLD in this type of graft. In the large series from Landgren et al. (26.901 patients after allogeneic hematopoietic cell transplantation, 127 PTLD, 105 in the first-year post-transplant), the cumulative incidence of PTLD was 0.2 % in patients with no major risk factors (T-cell-depleted graft, HLA-mismatched graft, ATG, age >50 years), 1.1 % if 1 risk factor, 3.6 % if 2, and 8.1 % if 3 or more (Landgren et al. 2009). They underlined the theoretical interest to prospectively monitor EBV infection in this sub-group of high-risk patients, aiming to early treatment intervention. Additionally, preemptive strategy by low-dose rituximab in the conditioning regimen was proposed (Savani et al. 2009; Bacigalupo et al. 2010). This was based on one study showing no PTLD in 27 adult recipients of intestinal and multivisceral transplantation whom induction IS included one single dose of rituximab 150 mg/m^2 (Vianna et al. 2008). Moreover, 14 patients having developed EBV reactivation after allogeneic stem cell transplantation received preemptive rituximab therapy (standard doses) and did not develop PTLD (Kindwall-Keller et al. 2009). Finally, PTLD did not occur in 38 patients after rituximab-containing conditioning regimen or prior rituximab, even in 8 of them with 3 high risk factors (Savani et al. 2009). These preliminary data were confirmed by a retrospective study comparing 55 patients preemptively treated with 200 mg/m^2 rituximab at day+5 of conditioning to 68 who did not receive prophylaxis (Dominietto et al. 2012). Both groups received ATG. EBV viremia was found in 56 % of rituximab-treated patients as compared to 85 % of the controls ($p = 0.0004$). The maximal median viral load was also lower (91 vs 1321/10^5 cells, $p = 0.003$), and the risk to have EBV >1,000/10^5 cells, which is a known risk factor for PTLD, was found in 14 versus 49 % of patients ($p = 0.0001$). No PTLD were observed in the treated group as compared to 2 in the controls. No side effects were reported although leukocyte and lymphocyte counts were lower at day+50 and +100 after rituximab. Acute GVHD was reduced after preemptive treatment (20 vs 38 %, $p = 0.02$). Finally, in 70 pediatric stem cell transplant recipients, preemptive strategy was prospectively applied in case of EBV high load (>40,000/mL whole blood) (Worth et al. 2011). When CD3 count was <300 × 10^6/L, classical doses rituximab was given (13/70), if not, IS was decreased (6/70). Incidence of PTLD was decreased as compared to historical controls (1.4 vs 21.7 %, $p = 0.003$), even more in patients with high viral load (2.7 vs 62.5 %, $p < 0.0001$). There were few side effects of rituximab, the main being longer time for B-cell recovery (27.6 vs 8.9 months). All PTLD occurred in patients with low CD3 count. The possibility to further improve this prevention in high-risk patients by using sirolimus as

GVHD prophylaxis was proposed but should be demonstrated (Reddy et al. 2011; Cen and Longnecker 2011).

Monitoring of EBV-specific immunity in correlation with EBV load is proposed to better differentiate patients with asymptomatic high EBV load from those at risk of PTLD. This strategy was efficient in children after liver transplantation for which an inadequate ratio between high viral load and low specific EBV T lymphocytes (EBV–CTL) at the time of EBV primary infection was 100 % predictive of following PTLD (Smets et al. 2002). EBV–CTL were detected by the enzyme-linked immunospot assay (Elispot) through their specific IFNγ secretion after stimulation with an autologous EBV + cell line. No PTLD occurred when EBV–CTL were >2/mm^3 of blood, independently of the viral load. The same profile was described in 14 healthy adults followed prospectively (Vogl et al. 2012). EBV reactivation was related to lower Elispot-detected EBV-TL in 8/12 subjects. However, the Elispot monitoring is time-consuming, requires the production of autologous EBV cell line for all the patients, and is therefore difficult to implement for routine follow-up. An alternative would be the tetramer technique that allows easy detection of EBV-peptide-specific HLA-restricted lymphocytes by flow cytometry. No correlation was found with viral reactivation in healthy carriers (Vogl et al. 2012). In 31 prospectively followed liver-transplanted children, level of tetramer EBV-specific CD8 T cells were the same in 11 patients with chronic high viral load and 20 controls (Gotoh et al. 2010). Genetic profile of EBV in these 11 children showed type 0 latency (EBER, BART, and LMP2), suggesting that high viral load can be related to viral immune escaping. No early or late PTLD occurred during the study (follow-up 0.8–7.4 years). In a small series, tetramer EBV-specific CD8 T cells were lower in 10 PTLD patients and 6 healthy controls as compared to the 2 transplanted patients without complication, with only 30 % of them being functional (Sebelin-Wulf et al. 2007). Moreover, the absolute CD4 T-cell count was also lower ($336 \pm 161/\mu L$ vs $1008 \pm 424/\mu L$, $p = 0.00001$). CD4 T cells <230/μL were associated with high viral load >1,000/μg DNA. In 307 patients after allogeneic hematopoietic cell transplant, EBV–CTL at D28 post-transplantation, assessed by flow cytometry after specific peptide stimulation, did not allow us to predict the 25 PTLD (Hoegh-Petersen et al. 2011) although they tend to be higher in patients without complication. In pediatric liver-transplanted children, the PPV of high viral load for PTLD was increased from 29 to 57 % if associated with low IFNγ (A/A) polymorphism (Lee et al. 2006b). No differences were found for polymorphism in TGF-β, TNF-α, and interleukin 2, 6, and 10. Children had 100 % risk of having high viral load in case of low release of ATP by CD3+ cells, stimulated for 24 h with phytohemagglutinin (<125 ng/mL). Following IS tapering, increase in ATP release and decreased viral load were observed (Lee 2006a). Finally, the potential implication of EBV-specific natural killer (NK) cells was recently proposed. Important NK response was described in adults followed prospectively after IM, with particular expansion of a CD56 highly positive sub-population (CD56^{bright}) (Williams et al. 2005). Higher NK cells were correlated with lower viral load. In a retrospective review of adult and pediatric solid organ transplant recipients, polymorphism of Fc-γ receptor 3A (FCGR3A) and killer cell immunoglobulin-like receptor (KIR) genes, involved in NK cell function, was shown to influence significantly survival after PTLD diagnosis but were not risk factors for PTLD (Stem et al. 2010). More recently, both NK cell phenotype and function were different in pediatric patients developing PTLD after thoracic transplantation and those without complication (Wiesmayr et al. 2012). Patients with chronic high viral load but no PTLD displayed intermediate values. The interest to monitor NK cells in transplanted patients should, however, be confirmed.

Antiviral drugs were also proposed as preemptive therapy. Although they are not active on latent EBV cycle involved in PTLD, they can impair or reduce lytic cycle, limiting intercellular EBV transmission. Use of ganciclovir in adults and children after renal transplantation was associated, especially within the first-year post-transplant, with up to 83 % reduction in the risk of PTLD (Funch et al. 2005). One month of treatment in this period gave a 38 % reduced risk. Effect of acyclovir was less marked. Forty-seven children with positive PCR for EBV after liver transplant were treated with valganciclovir but no IS reduction (Hierro et al. 2008). Symptoms were present in 28 % of them. After a short treatment (1 month), negative PCR was observed in 34 % of patients, with 82 % relapse after valganciclovir arrest. After longer therapy (8 months), PCR negativity was found in 47.6 % children and was maintained in 60 % of them off treatment. PTLD was suspected but not confirmed in one child (2.1 %).

2.3   Treatment

Early diagnosis and treatment are mandatory to avoid PTLD-related morbidity and mortality, the latter having been often reported between 30 and 60 % (Bakker et al. 2007). High index of suspicion is necessary due to non-specificity of initial symptoms. Unexplained fever, adenopathies, hepatosplenomegaly, sepsis like syndrome, and extra-nodal involvement (graft, gastrointestinal tract, and sinonasal cavity) are common presentations. Graft localization is more often reported in early PTLD. The interest of fluorodeoxyglucose positron emission tomography for disease staging and follow-up after treatment is described, but its role in precocious diagnosis remains to be demonstrated. Unrecognized PTLD can rapidly lead to graft and multiorgan severe dysfunction and death.

In early PTLD after solid organ transplantation, reducing IS is the first step of treatment, aiming to allow restoration of specific immunity and control of the EBV-infected proliferative cells, with a response rate reported up to 75 % (Heslop 2009). While it could be an option in late PTLD after bone marrow/stem cell transplantation, IS tapering is usually needless in early cases because of the profound immunosuppression of the patient at that time. For those, unmanipulated donor cell injection or in vitro expanded EBV–CTL adoptive transfer would be a better treatment. The sustained response rate to such therapy was demonstrated to be >70 % in 49 PTLD patients after hematopoietic stem cell transplantation

(Doubroniva et al. 2012; Heslop 2012), with same results observed for donor unselected lymphocytes, donor EBV–CTL, or third-party EBV–CTL. Symptoms improved after 5–15 days, with radiological and complete response described after 3 weeks and 3 months, respectively. Cell therapy was also efficient in central nervous system localization of PTLD. Authors showed that treatment failure was associated with mismatched viral antigen or HLA presentation between EBV–CTL and tumor, explaining why the target proliferative cells were not recognized and eliminated. In a PTLD SCID mice model, injection of EBV–CTL resulted in delayed tumor development and tumors were prevented in 40 % of cases ($p = 0.001$). Results were improved by preselection of CD8+ cells or conditioning of CTLS with a combination of IL 2, 7, and 15 (Johannessen et al. 2011). Daily injection of IL2 in addition to EBV–CTL infusion led to PTLD prevention in 78 % of animals. Cell therapy was also shown to be an efficient preemptive strategy in 13 high-risk children receiving partially matched stem cell transplant (Leen et al. 2009). One single dose of EBV–CTL was administered 30 days after transplantation, and neither PTLD nor GVHD was observed. Limitations for cell therapy are GVHD risk (mostly with unmanipulated donor T cells) and need for special facilities to expand EBV–CTL and have them readily available in rapid progressing diseases, as their expansion requires 2–3 months (Heslop 2009; Bollard et al. 2012).

Targeting the proliferative cells with anti-CD20 antibodies (rituximab, 4-weekly injections of 375 mg/m^2) is well recognized as part of initial treatment in case of CD20+ tumor, with response rates around 50 %. In a review of 80 PTLD in adults after solid organ transplantation, 74 % received first-line rituximab treatment in addition to reduced IS, with or without following chemotherapy (Evens et al. 2010). The 3-year progression-free survival and overall survival were 70 and 73 % in patients having received rituximab as compared to 21 and 33 % in others ($p < 0.0001$ and $= 0.0001$). Multivariate analysis showed 3 risk factors associated with disease progression and lower survival: central nervous system or bone marrow involvement and hypoalbuminemia. Overall survival in case of 0, 1, or more than 2 risk factors was, respectively, 93, 68, and 11 %. In case PTLD do not respond to IS reduction and rituximab within 8 weeks, it is now recommended to rapidly start chemotherapy (Montserrat et al. 2012). Seventy adult patients developed B-cell PTLD after solid organ transplant and were prospectively included in a phase 2 international study (Trappe et al. 2012). Subjects that did not respond to IS reduction received 4 courses of rituximab followed after 4 weeks by chemotherapy (cyclophosphamide, doxorubicin, vincristine, and prednisolone = CHOP). PTLD was of late type in 76 %, with monomorphic classification in 96 % and associated with EBV in only 44 % of patients. Fifty-nine patients received the complete treatment and showed partial and complete response in 90 and 68 % of cases. The median survival was 6.6 years. The main adverse events were severe leucopenia (68 %) or infections (41 %), with CHOP-related mortality in 11 % (5/7 being non-rituximab responders). EBV-associated PTLD occurred earlier and in younger patients, involved more often the graft and the kidney, and less frequently lymph nodes, were associated with lower performance status and developed more severe infections. Of the 40 patients with complete response, 8 relapsed and died from PTLD (5 within 6 months). Globally, lower risk of disease progression was observed in EBV-associated PTLD (HR 0.007) and higher risk was seen in patients in partial remission after sequential treatment (HR 20.83). The interest of low-dose chemotherapy was also described in children after liver transplant (Gross et al. 2005). Patients with EBV-related PTLD received six cycles of low-dose cyclophosphamide and prednisone (one cycle every 3 weeks with cyclophosphamide 600 mg/m^2 on day 1, prednisone 1 mg/kg orally twice a day on days 1–5) after a trial of IS reduction. Two-year disease-free survival and overall survival were 67 and 73 %. The same group reported results with rituximab added to the low-dose chemotherapy (375 mg/m^2 on days 1, 8, and 15 of the 2 first cycles) (Gross et al. 2012). Two-year disease-free survival and overall survival were 71 and 83 %. Side effects were similar to the previous study. Of the 55 children, 10 died, 7 of PTLD, and 3 of infections.

Antiviral drugs (ganciclovir, acyclovir) are not described as an efficient treatment for overt PTLD because they are inactive on latent EBV cycle present in these diseases. Surgery or radiotherapy can be useful in localized disease or central nervous system involvement, but often require additional treatment (Heslop 2009).

Outcome of the different treatments for EBV–PTLD in hematopoietic stem cell transplant recipients was extensively reviewed (Styczynski et al. 2009a), showing that preemptive therapy with rituximab of EBV–CTL reduced most significantly the risk of death (survival rates of 89.7 and 94.1 %). The same treatments used for overt PTLD also improved survival rates (63 and 88.2 %, respectively). Reduction in IS approximately allowed a survival rate of 56.6 %. Chemotherapy and antiviral agents did not influence survival of the patients.

According to all the results described, guidelines for PTLD treatment after solid organ or stem cell transplant were defined (Styczynski et al. 2009b; Parker et al. 2010). After solid organ transplantation, recommendations are the following. Reduction in IS should be initiated in all patients (limited disease, 25 % reduction; extensive disease, 50 % reduction in calcineurin inhibitors, stop azathioprine/mycophenolate, keep prednisone at maximum 10 mg/day; extensive disease and critically ill patient, stop all drugs except prednisone). Graft function should be carefully monitored. For patients with low-risk PTLD (age >60 years, normal LDH, performance status ECOG 0–1) and disease progression despite IS reduction, rituximab monotherapy is recommended. Addition of chemotherapy (CHOP-based regimen) should not be delayed in patients with disease progression despite this treatment or immediately in clinically aggressive PTLD. When chemotherapy is given, prophylactic granulocyte colony-stimulating factor and anti-infectious agents should be used. In case of central nervous system involvement, reduced IS should be followed by local radiotherapy and steroids. High-dose methotrexate can be considered in young patients. Surgery or radiotherapy may be adequate for localized diseases. Treatment with EBV–CTL infusions, antiviral drugs, and/or arginine butyrate, interferon or intravenous immunoglobulin should be currently limited to clinical trials. In cases where the graft failed or has been removed, retransplantation can be considered, if possible at least one year after PTLD healing. After bone marrow/stem cell transplantation, recommendations are the

Fig. 1 According to specific guidelines, algorithm for EBV monitoring and prevention, treatment, and follow-up of EBV-related PTLD in patients after solid organ or bone marrow/stem cell transplantation (adapted from Heslop 2009 and Parker et al. 2010)

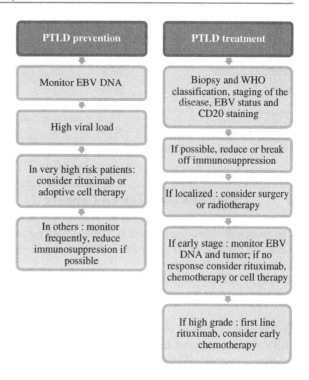

following. Regarding prevention of EBV reactivation, recipients of allogeneic grafts and graft donors should be tested for EBV serology, and antiviral agents and immunoglobulin have no impact and are not recommended. For PTLD preemptive therapy, high-risk patients after allogeneic transplant should be closely monitored for EBV load and PTLD symptoms, and for preemptive treatment with rituximab (1–2 doses), IS reduction or EBV–CTL should be considered in patients with high viral load. Monitoring is not necessary in HLA identical without T-cell depletion or auto transplantation. For PTLD treatment, rituximab (4–8 doses), IS reduction, or EBV–CTL are recommended as first-line therapy. Chemotherapy is a second-line treatment, while antiviral drugs are not recommended.

2.4   Conclusion

EBV-related PTLD are still a common complication after solid organ or stem cell transplantation, with significant morbidity and mortality. Preemptive and curative treatments have been clearly improved in the last decade, leading to specific guidelines that are summarized in Fig. 1. Exact impact of such strategy on PTLD incidence and patient/graft survival remains to be demonstrated in large multi-centre prospective studies.

Table 3 Principal cancers possibly associated with EBV and attributable to new cases per year

Type of cancer	% Attributable to EBV	EBV + new cases/year
Gastric carcinoma	9 %	84.050
Nasopharyngeal carcinoma	98 %	78.100
Hodgkin's lymphoma	46 %	28.600
Burkitt's lymphoma	82 %	6.700

Cohen et al. (2011) and Schiller and Lowy (2010)

3   Other EBV-Associated Cancers

The principal cancers are listed in Table 3 with percentage of those attributable to EBV. At least 200,000 new cases of EBV-related cancers arise every year.

3.1   Gastric Carcinoma

Despite association in less than 10 % of cases, EBV-associated gastric carcinoma (EBVaGC) is the most frequent EBV-related cancer because of high frequency of gastric carcinoma (Fukayama and Ushiku 2011). Two types are histologically described, lymphoepithelioma-like carcinoma and ordinary gastric carcinoma, and are characterized by the presence of EBV latent infection in the stomach neoplastic cells. Clinical features include male predominance, preponderant proximal location in the stomach, lower rate of lymph node involvement in early stages, and relatively good prognosis as compared to EBV-negative gastric carcinoma (Chang and Kim 2005; Chen et al. 2012; Murphy et al. 2009; Truong et al. 2009). Frequency of EBVaGC is described in up to 25 % of remnant cancers after previous gastrectomy (33 % after gastrojejunostomy, 12 % after gastroduodenostomy), and cautious follow-up of these patients would therefore be part of the prevention (Fukayama and Ushiku 2011). The pathogeny being based on methylation of promoter region of many oncogenous genes induced by EBV latent membrane protein 2A, better understanding of epigenetic modifications could lead to more specific prevention and treatment, as recently exemplified with modulation of WNT5A expression with methyltransferase inhibitor (Liu et al. 2012).

Current treatment relies on chemotherapy and surgery. Partial chemoresistance to commonly used drugs (5-fluorouracil and docetaxel) induced by EBV infection was, however, described in vitro (Seo et al. 2011; Shin et al. 2011). As one of the effects of chemotherapy is to induce a switch from latent to lytic EBV cycle, adding antiviral drugs as ganciclovir could improve results (Zhao et al. 2012). Possible interest of endoscopic resection was recently described (Lee et al. 2012).

3.2 Nasopharyngeal Carcinoma

Etiology of NPC is multifactorial and includes genetic predisposition, EBV infection, and environmental carcinogens (salted fish, lack of dietary vegetables, and tobacco). NPC arises in Asia for more than 80 % of cases, with a male predominance and an age peak between 40 and 50 years (Lee et al. 2012; Cao et al. 2011). It is a highly malignant disease with early lymphatic spread and distant metastases. Modernization of lifestyle has allowed a significant decrease in incidence by more than 50 % in the last 3 decades, with a reduction in mortality of about 60 %. Primary prevention remains, however, difficult, and the main objective is to diagnose the disease as early as possible because it could be silent for a long time before overt disease. Indeed, patients with early-stage diseases (I and II) have a 5-year survival rate of up to 94 % compared to 80 % in stages III and IV. EBV serology is the basis for NPC screening, and detection of viral capsid antigen (VCA)-IgA has a sensitivity and specificity of up to 90 % for the diagnosis. Adding another serology, such as EBV nuclear antigen 1 (EBNA-1)-IgA or early antigen (EA)-IgA, can improve prediction. High-risk subjects have VCA-IgA >1/80 or VCA-IgA and EA-IgA >1/5 and should be retested every 6 months to 1 year. Patients with an intermediate risk, meaning VCA-IgA alone >1/5, should be screened every 2–3 years and subjects with negative serology every 5 years (Cao et al. 2011). This strategy allowed diagnosing NPC at stages I or II in more than 80 % of cases in several studies. Adding nasopharyngoscopy can further increase sensitivity. The role of EBV load monitoring remains to be determined.

Treatment relies first on megavoltage radiotherapy that has allowed and increased in 5-year survival from 30 to 70 %. Adding chemotherapy is beneficial, and concurrent chemotherapy is the more potent modality. On this basis, more recent studies showed 5-year survival of 75 % or more, with locoregional control of the disease in at least 90 % of cases. The interest of using induction and concurrent radiotherapy is debated, and 5 randomized trials are ongoing. Finally, the possibility to have personalized treatment based on tumor staging, plasma EBV quantification, and different genes' expression (epidermal growth factor receptor, vascular endothelial growth factor, or excision repair cross-complementation group 1) is under study.

3.3 Hodgkin's and Burkitt's Lymphoma

In both diseases, prevention mostly relies on early diagnosis. Hodgkin's lymphoma should be treated by current chemotherapy/radiotherapy protocols, according to disease staging and risk factors (Ansell 2012; Parker et al. 2010). Localized disease may be successfully treated by radiotherapy alone. High-dose chemotherapy followed by autologous stem cell transplantation is the rescue treatment in case of relapse. Burkitt's lymphoma is often treated with the CODOX–M/IVAC chemotherapy regimen (cyclophosphamide, vincristine, doxorubicin, high-dose methotrexate/Ifosfamide, etoposide, and high-dose cytarabine). Radiotherapy has been eliminated except in case of emergency. New therapeutic approaches include rituximab, targeted cellular immune therapy, and small molecule inhibitors (Miles et al. 2012).

3.4 T/NK Cell EBV-Associated Lymphoproliferative Diseases in Non-immunocompromised Patients

These rare cancers are aggressive diseases, occurring more often in young age and leading to death in about 44 % of cases (Kimura et al. 2012). They are classified into 4 sub-types: chronic active EBV disease, hemophagocytic lymphohistiocytosis, severe mosquito bite allergy, and hydroa vacciniforme. Prevention is until now impossible. Globally, treatment with chemotherapy, immunomodulators, or antiviral agents gives poor results and best outcome is obtained after hematopoietic stem cell transplantation with sustained complete response in 57–63 % of cases. Higher mortality rate is associated with age >8 years and liver dysfunction at diagnosis. Probability of survival after transplantation was higher in patients transplanted at a younger age with inactive disease at the time of transplant. These diseases arise often in young children, and the possibility to treat them with high doses of mother's lymphocyte infusion was studied in 5 patients, with complete response in 3 and partial response in 2 (Wang et al. 2010). The in vitro killing effect of a combination of valproic acid and bortezomib on the T/NK proliferating cells was recently described and could be an option for future treatment (Iwata et al. 2012).

4 EBV Vaccination

As described above, more than 200,000 EBV-related cancers arise every year and EBV vaccination would be a major progress in prevention and public health (Cohen et al. 2011; Schiller and Lowy 2010). The largest study about EBV vaccine was done with a recombinant gp350 vaccine delivered in alum/monophosphoryl A adjuvant (Sokal et al. 2007). Vaccination did not protect against infection in this placebo-controlled study involving 181 EBV-naïve adults, but significantly reduce the incidence of IM by 78 %. A small study of healthy subjects showed immunogenicity of an EBNA-3 peptide with tetanus toxoid and water oil adjuvant with virus specific T-cell response, but was not studied further (Elliott et al. 2008). Therapeutic vaccination is also under investigation, and an EBNA-1/LMP-2 vaccine was able to induce immunity in NPC patients (Long et al. 2011). Until now, no vaccines were able to give sterilizing immunity and to completely prevent EBV infection. However, such a result is perhaps not necessary for prevention of EBV-related malignancies, which are mostly associated with high EBV load. Vaccination reducing IM and/or viral load could *per se* be efficient in prevention of following malignancies. Recommendations are to reinforce research in the field (Cohen et al. 2011; Schiller and Lowy 2010).

References

Ansell SM (2012) Hodgkin lymphoma: 2012 update on diagnosis, risk-stratification, and management. Am J Hematol 87:1096–1103

Bacigalupo A, Socie G, Lanino E et al (2010) Fludarabine, cyclophosphamide antithymocyte globulin, with or without low dose total body irradiation, for alternative donor transplants, in acquired severe aplastic anemia: a retrospective study from the EBMT-SAA working party. Haematol 95:976–982

Bakker NA, van Imhoff GW, Verschuuren EAM et al (2007) Presentation and early detection of post-transplant lymphoproliferative disorder after solid transplantation. Transpl Int 20:207–218

Bingler MA, Feingold B, Miller SA et al (2008) Chronic high Epstein-Barr viral load state and risk for late-onset posttransplant lymphoproliferative disease/lymphoma in children. Am J Transplant 8:442–445

Bollard CM, Rooney CM, Heslop HE (2012) T-cell therapy in the treatment of post-transplant lymphoproliferative disease. Nat Rev Clin Oncol 9:510–519

Cao SM, Simons MJ, Qian CN (2011) The prevalence and prevention of nasopharyngeal carcinoma in China. Chin J Cancer 30:114–119

Cen O, Longnecker R (2011) Rapamycin reverses splenomegaly and inhibits tumor development in a transgenic model of Epstein-Barr virus-related Burkitt's lymphoma. Mol Cancer Ther 10:679–686

Chang MS, Kim WH (2005) Epstein-Barr virus in human malignancy: a special reference to Epstein-Barr virus associated gastric carcinoma. Cancer Res Treat 37:257–267

Chen JN, He D, Tang F et al (2012) Epstein-Barr virus-associated gastric carcinoma: a newly defined entity. J Clin Gastroenterol 46:262–271

Cockfield SM (2001) Identifying the patient at risk for post-transplant lymphoproliferative disorder. Transpl Infect Dis 3:70–78

Cohen JI (1999) The biology of Epstein-Barr virus: lessons learned from the virus and the host. Curr Opin Immunol 11:365–370

Cohen JI, Fauci AS, Varmus H et al (2011) Epstein-Barr virus: an important vaccine target for cancer prevention. Sci Transl Med 3:107fs7

D'Antiga L, Del Rizzo M, Mengoli C et al (2007) Sustained Epstein-Barr virus detection in paediatric liver transplantation. Insights into the occurrence of late PTLD. Liver Transplant 13:343–348

Dominietto A, Tedone E, Soracco M et al (2012) In vivo B-cell depletion with rituximab for alternative donor hemopoietic SCT. Bone Marrow Transplant 47:101–106

Doubroniva E, Oflaz-Sozmen B, Prockop SE et al (2012) Adoptive immunotherapy with unselected or EBV-specific T cells for biopsy-proven EBV + lymphomas after allogeneic hematopoietic cell transplantation. Blood 119:2644–2656

Elliott SL, Suhrbier A, Miles JJ et al (2008) Phase I trial of a CD8+ T-cell peptide epitope-based vaccine for infectious mononucleosis. J Virol 82:1448–1457

Evens AM, David KA, Helenowski I et al (2010) Multicenter analysis of 80 solid organ transplantation recipients with post-transplant lymphoproliferative disease: outcomes and prognostic factors in the modern era. J Clin Oncol 28:1038–1046

Fukayama M, Ushiku T (2011) Epstein-Barr virus-associated gastric carcinoma. Pathol Res Pract 207:529–537

Funch DP, Walker AM, Schneider G et al (2005) Ganciclovir and acyclovir reduce the risk of post-transplant lymphoproliferative disorder in renal transplant recipients. Am J Transplant 5:2894–2900

Gotoh K, Ito Y, Ohta R et al (2010) Immunologic and virologic analyses in pediatric liver transplant recipients with chronic high Epstein-Barr virus loads. J Infect Dis 202:461–469

Gross TG, Bucavals JC, Park JR et al (2005) Low-dose chemotherapy for Epstein-Barr virus-positive post-transplantation lymphoproliferative disease in children after solid organ transplantation. J Clin Oncol 23:6481–6488

Gross TG, Orjuela MA, Perkins SL et al (2012) Low-dose chemotherapy and rituximab for posttransplant lymphoproliferative disease (PTLD): a children's oncology group report. Am J Transplant 12:3069–3075

Gulley ML, Tang W (2010) Using Epstein-Barr load assays to diagnose, monitor, and prevent posttransplant lymphoproliferative disorder. Clin Microbiol Rev 23:350–366

Heslop HE (2009) How I treat EBV lymphoproliferation. Blood 114:4002–4008

Heslop HE (2012) Equal-opportunity treatment of EBV–PTLD. Blood 119:2436–2438

Hierro L, Diez-Dorado R, Diaz C et al (2008) Efficacy and safety of valganciclovir in liver-transplanted children infected with Epstein-Barr virus. Liver Transpl 14:1185–1193

Hoegh-Petersen M, Goodyear D, Geddes MN et al (2011) High incidence of post transplant lymphoproliferative disorder after antithymocyte globulin-based conditioning and ineffective prediction by day 28 EBV-specific T lymphocyte counts. Bone Marrow Transplant 46:1104–1112

Iwata S, Saito T, Ito Y et al (2012) Antitumor activities of valproic acid on Epstein-Barr virus-associated T and natural killer lymphoma cells. Cancer Sci 103:375–381

Johannessen I, Bieleski L, Urquhart G et al (2011) Epstein-Barr virus, B cell lymphoproliferative disease, and SCID mice: modeling T cell immunotherapy in vivo. J Med Virol 83:1585–1596

Kamdar KY, Rooney CM, Heslop HE (2011) Post-transplant lymphoproliferative disease following liver transplantation. Curr Opin Organ Transplant 16:274–280

KDIGO Transplant Work Group (2009) KDIGO clinical practice guideline for the care of kidney transplant recipients. Am J Transplant 9:S1–S155

Kerkar N, Morotti RA, Madan RP et al (2010) The changing face of post-transplant lymphoproliferative disease in the era of molecular EBV monitoring. Pediatr Transplant 14:504–511

Kimura H, Ito Y, Kawabe S et al (2012) EBV-associated T/NK cell lymphoproliferative diseases in nonimmunocompromised hosts: prospective analysis of 108 cases. Blood 119:673–686

Kindwall-Keller TL, Cooper BW, Laughlin MJ et al (2009) Preemptive rituximab treatment may reduce the incidence of post-transplant lymphoproliferative disorders (PTLD) in patients with EBV reactivation after allogeneic stem cell transplantation (Abstract). Biol Blood Marrow Transplant 15:91

Landgren O, Gilbert ES, Rizzo JD et al (2009) Risk factors for lymphoproliferative disorders after allogeneic hematopoietic cell transplantation. Blood 113:4992–5001

Lee AWM, Ng WT, Chan YH et al (2012a) The battle against nasopharyngeal cancer. Radiother Oncol 104:272–278

Lee HL, Kim DC, Lee SP et al (2012b) Treatment of Epstein-Barr virus-associated gastric carcinoma with endoscopic submucosal dissection. Gastrointest Endosc 76:913–915

Lee TC, Savoldo B, Rooney CM et al (2005) Quantitative EBV viral loads and immunosuppression alterations can decrease PTLD incidence in pediatric liver transplant recipients. Am J Transplant 5:2222–2228

Lee TC, Goss JA, Rooney CM et al (2006a) Quantification of a low cellular immune response to aid in identification of pediatric liver transplant recipients at high-risk for EBV infection. Clin Transplant 20:689–694

Lee TC, Savoldo B, Barshes NR et al (2006b) Use of cytokine polymorphisms and Epstein-Barr virus viral load to predict development of post-transplant lymphoproliferative disorder in paediatric liver transplant recipients. Clin Transplant 20:389–393

Leen AM, Christin A, Myers GD et al (2009) Cytotoxic T lymphocyte therapy with donor T cells prevents and treats adenovirus and Epstein-Barr virus infections after haploidentical and matched unrelated stem cell transplantation. Blood 114:4283–4292

Liebowitz D (1998) Epstein-Barr virus and a cellular signaling pathway in lymphomas from immunosuppressed patients. N Engl J Med 338:1413–1421

Liu X, Wang Y, Wang X et al (2012) Epigenetic silencing of WNT5A in Epstein-Barr virus-associated gastric carcinoma. Arch Virol. doi:10.1007/s00705-012-1481-x

Long HM, Taylor GS, Rickinson AB (2011) Immune defence against EBV and EBV-associated disease. Curr Opin Immunol 23:258–264

Lynch BA, Vasef MA, Comito M et al (2003) Post transplant lymphoproliferative disorder. Effect of in vivo lymphocyte-depleting strategies on development of lymphoproliferative disorders in children post allogeneic bone marrow transplantation. Bone Marrow Transplant 32:527–533

Miles RR, Arnold S, Cairo MS (2012) Risk factors and treatment of childhood and adolescent Burkitt lymphoma/leukaemia. Br J Haematol 156:730–743

Montserrat E (2012) PTLD treatment: a step forward, a long way to go. Lancet Oncol 13:120–121

Murphy G, Pfeiffer R, Camargo MC et al (2009) Meta-analysis shows that prevalence of Epstein-Barr virus-positive gastric cancer differs based on sex and anatomic location. Gastroenterology 137:824–833

Okano M, Gross TG (2012) Acute or chronic life treatening diseases associated with Epstein-Barr virus infection. Am J Med Sci 343:483–489

Parker A, Bowles K, Bradley JA et al (2010) Management of post-transplant lymphoproliferative disorder in adult solid organ transplant recipients—BCSH and BTS guidelines. Br J Haematol 149:693–705

Reddy N, Rezvani K, Barrett AJ et al (2011) Strategies to prevent EBV reactivation and posttransplant lymphoproliferative disorders (PTLD) after allogeneic stem cell transplantation in high-risk patients. Biol Blood Marrow Transplant 17:591–597

Rickinson AB (1998) Epstein-Barr virus in action in vivo. N Engl J Med 338:1461–1462

Savani BN, Pohlmann PR, Jagasia M et al (2009) Does peritransplant use of rituximab reduce the risk of EBV reactivation and PTLPD? Blood 113:6263–6264

Schiller JT, Lowy DR (2010) Vaccines to prevent infections by oncoviruses. Annu Rev Microbiol 64:23–41

Sebelin-Wulf K, Nguyen TD, Oertel S et al (2007) Quantitative analysis of EBV-specific CD4/CD8 T cell numbers, absolute CD4/CD8 T cell numbers and EBV load in solid organ transplant recipients with PTLD. Transplant Immunol 17:203–210

Seo JS, Kim TG, Hong YS et al (2011) Contribution of Epstein-Barr virus infection to chemoresistance of gastric carcinoma cells to 5-fluorouracil. Arch Pharm Res 34:635–643

Shin HJ, Kim DN, Lee SK (2011) Association between Epstein-Barr virus infection and chemoresistance to docetaxel in gastric carcinoma. Mol Cells 32:173–179

Smets F, Sokal EM (2002a) Lymphoproliferation in children after liver transplantation. J Pediatr Gastroenterol Nutr 34:499–505

Smets F, Sokal EM (2002b) Epstein-Barr virus-related lymphoproliferation in children after liver transplant: role of immunity, diagnosis and management. Pediatr Transplant 6:280–287

Smets F, Bodeus M, Goubau P et al (2000) Characteristics of Epstein-Barr virus primary infection in pediatric liver transplant recipients. J Hepatol 32:100–104

Smets F, Latinne D, Bazin H et al (2002) Ratio between Epstein-Barr viral load and anti-Epstein-Barr virus specific T-cell response as a predictive marker of posttransplant lymphoproliferative disease. Transplantation 73:1603–1610

Sokal EM, Hoppenbrouwers K, Vandermeulen C et al (2007) Recombinant gp350 vaccine for infectious mononucleosis: a phase 2, randomized, double-blind, placebo-controlled trial to evaluate the safety, immunogenicity and efficacy of an Epstein-Barr virus vaccine in healthy young adults. J Infect Dis 196:1749–1753

Stern M, Opelz G, Döhler B et al (2010) Natural killer-cell receptor polymorphisms and posttransplantation non-Hodgkin lymphoma. Blood 115:3960–3965

Steven NM (1997) Epstein-Barr virus latent infection in vivo. Rev Med Virol 7:97–106

Styczynski J, Einsele H, Gil L et al (2009a) Outcome of treatment of Epstein-Barr virus-related post-transplant lymphoproliferative disorder in hematopoietic stem cell recipients: a comprehensive review of reported cases. Transpl Infect Dis 11:383–392

Styczynski J, Reusser P, Einsele H et al (2009b) Management of HSV, VZV and EBV infections in patients with hematological malignancies and after SCT: guidelines from the second European conference on infections in leukemia. Bone Marrow Transplant 43:757–770

Swerdlow SH, Campo E, Jaffe ES et al (2008) WHO classification of tumours of haematopoietic and lymphoid tissues, 4th edn. International Agency for Research on Cancer, Lyon

Trappe R, Oertel S, Leblond V et al (2012) Sequential treatment with rituximab followed by CHOP chemotherapy in adult B-cell post-transplant lymphoproliferative disorder (PTLD): the prospective international multicentre phase 2 PTLD-1 trial. Lancet Oncol 13:196–206

Truong CD, Feng W, Li W et al (2009) Characteristics of Epstein-Barr virus-associated gastric cancer: a study of 235 cases at a comprehensive cancer center in USA. J Exp Clin Cancer Res 28:14–22

Vianna RM, Mangus RS, Fridell JA et al (2008) Induction immunosuppression with thymoglobulin and rituximab in intestinal and multivisceral transplantation. Transplantation 85:1290–1293

Vogl BA, Fagin U, Nerbas L et al (2012) Longitudinal analysis of frequency and reactivity of Epstein-Barr virus-specific T lymphocytes and their association with intermittent viral reactivation. J Med Virol 84:119–131

Wagner HJ, Wessel M, Jabs W et al (2001) Patients at risk for development of posttransplant lymphoproliferative disorder: plasma versus peripheral blood mononuclear cells as material for quantification of Epstein-Barr viral load by using real-time quantitative polymerase chain reaction. Transplantation 72:1012–1019

Wang Q, Liu H, Zhang X et al (2010) High doses of mother's lymphocyte infusion to treat EBV-positive T-cell lymphoproliferative disorders in childhood. Blood 116:5941–5947

Weiner C, Weintraub L, Wistinghausen B et al (2012) Graft rejection in pediatric liver transplant patients with Epstein-Barr viremia and post-transplant lymphoproliferative disease. Pediatr Transplant 16:458–464

Wiesmayr S, Webber SA, Macedo C (2012) Decreased NKp46 and NKG2D and elevated PD-1 are associated with altered NK-cell function in pediatric transplant patients with PTLD. Eur J Immunol 42:541–550

Williams H, McAuley K, Macsween KF et al (2005) The immune response to primary EBV infection: a role for natural killer cells. Br J Haematol 129:266–274

Worth A, Conyers R, Cohen J et al (2011) Pre-emptive rituximab based on viraemia and T cell reconstitution: a highly effective strategy for the prevention of Epstein-Barr virus-associated lymphoproliferative disease following stem cell transplantation. Br J Haematol 155:377–385

Zhao J, Jin H, Cheung KF et al (2012) Zinc finger E-box binding factor 1 plays a central role in regulating Epstein-Barr virus (EBV) latent-lytic switch and acts as a therapeutic target in EBV-associated gastric cancer. Cancer 118:924–936

HTLV-1 and Leukemogenesis: Virus–Cell Interactions in the Development of Adult T-Cell Leukemia

Linda Zane and Kuan-Teh Jeang

Abstract

Human T-cell lymphotropic virus type 1 (HTLV-1) was originally discovered in the early 1980s. It is the first retrovirus to be unambiguously linked causally to a human cancer. HTLV-1 currently infects approximately 20 million people worldwide. In this chapter, we review progress made over the last 30 years in our understanding of HTLV-1 infection, replication, gene expression, and cellular transformation.

Contents

L. Zane · K.-T. Jeang (✉)
Molecular Virology Section, Laboratory of Molecular Microbiology, The National Institutes of Allergy and Infectious Diseases, The National Institutes of Health, Bethesda, MD 20892-0460, USA
e-mail: KJEANG@nih.gov

M. H. Chang and K.-T. Jeang (eds.), *Viruses and Human Cancer*,
Recent Results in Cancer Research 193, DOI: 10.1007/978-3-642-38965-8_11,
© Springer-Verlag Berlin Heidelberg 2014

1   Introduction

Human T-cell lymphotropic virus type 1 (HTLV-1) is the first identified human retrovirus. This virus belongs to the Deltaretrovirus genera of the Orthoretrovirinae subfamily which includes HTLV-2, HTLV-3, HTLV-4 (Mahieux and Gessain 2005; Mahieux and Gessain 2009), Bovine Leukemia Virus (BLV), and Simian T-cell lymphotropic virus (STLV). The virus was discovered in 1980–1981 by analyzing T cells from a patient suffering T-cell leukemias (ATL) (Poiesz et al. 1980; Hinuma et al. 1981; Miyoshi et al. 1981; Yoshida 1982; Watanabe et al. 1983; Gallo 2005). ATL is a rapidly fatal disease first described in Japan (Takatsuki 2005). Since then, a causal association between HTLV-1 and ATL has become firmly established (Gallo 2005). To date, HTLV-1 is the only known retrovirus that is directly linked to a human cancer. In addition to this cancer link, the virus can also cause inflammatory diseases such as HTLV-1-associated Myelopathy (HAM)/tropical spastic paraparesis (TSP), uveitis, infective dermatitis, and myositis (Gessain 2011; Goncalves et al. 2010).

2   HTLV-1 Infection

2.1   Epidemiology

Approximately 20 million people worldwide are infected with HTLV-1 (Proietti et al. 2005). However, HTLV-1 is not evenly distributed throughout the world. Indeed, the areas of highest prevalence of HTLV-1 are mainly southern Japan, the Caribbean islands, parts of South America and Central Africa, with foci in the Middle East, and Australia (Goncalves et al. 2010). This geographic distribution of HTLV-1 with some clustering of regions with high prevalence is still not understood (Proietti et al. 2005). Among HTLV-1-infected people, 2–5 % will develop ATL after a long latency period of 30–60-year post-infection; by comparison, approximately 0.25–5 % of the infected individuals will develop HAM/TSP. The development of ATL or TSP/HAM is not influenced by the subtype of HTLV-1 infection (Watanabe 2011; Ono et al. 1994). Indeed, while six subtypes of HTLV-1 (subtypes A-F) have been reported, the great majority of infections are caused by the cosmopolitan subtype A.

HTLV-1 has 3 modes of transmission: (1) mother to child, mainly through prolonged breastfeeding (>6 months); sexual, (2) mainly but not exclusively occurring from male to female; and (3) by blood products contaminated with infected lymphocytes (Goncalves et al. 2010; Matsuura et al. 2010). Male individuals and those infected in their early childhood are at the highest risk of developing ATL (Goncalves et al. 2010; Matsuura et al. 2010).

2.2 Tropism and Receptors

In vitro, HTLV-1 can infect many cell types including several non-lymphoid tumor cell lines such as human osteogenic sarcoma cells, lung cells, cervical carcinoma cells (HeLa), human gastric HGC-27 cells, human promyelocytic leukemia HL60 cells, as well as primary endothelial cells, monocytes, microglial cells, and mammary epithelial cells (Clapham et al. 1983; Hayami et al. 1984; Ho et al. 1984; Hiramatsu et al. 1986; Akagi et al. 1988; LeVasseur et al. 1998). However, in vivo, HTLV-1 is found primarily in CD4+ and CD8+ T lymphocytes (Nagai et al. 2001) and less frequently in other cell types such as monocytes, endothelial cells, myeloid, and plasmacytoid dendritic cells (Macatonia et al. 1992; Koyanagi et al. 1993; Jones et al. 2008), and CD34+ hematopoietic progenitor cells (Banerjee et al. 2008, 2010; Feuer et al. 1996; Grant et al. 2002; Tripp et al. 2003, 2005). Until the discovery of the glucose transporter GLUT1 as a receptor for HTLV-1 in 2003, little was known about the entry receptors for HTLV-1 (Manel et al. 2003). Currently, the published data from different laboratories support the idea of a multireceptor model for HTLV-1 entry (Fig. 1). Three cell surface proteins have been found to be involved in HTLV-1 entry: glucose transporter 1 (GLUT1), neuropilin-1 (NRP-1), and heparan sulfate proteoglycans (HSPG) (Jones et al. 2011). The following steps possibly explain HTLV-1 entry into cells. First, the surface subunit (SU) of the virally encoded envelope glycoprotein interacts with the heparan sulfate proteoglycans/neuropilin-1 complexes. Next, these interactions trigger conformational changes of the SU which are followed by the binding of the SU to GLUT1, and finally membrane fusion occur to allow the entry of the virus into the target cell (Jones et al. 2005, 2011; Pinon et al. 2003; Ghez et al. 2006; Lambert et al. 2009).

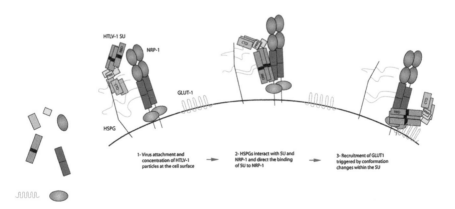

Fig. 1 A multireceptor model for HTLV-1 entry. *HSPG* = heparan sulfate proteoglycans; *SU* = the subunit of HTLV-1 envelope glycoprotein; *NRP-1* = neuropilin-1; *GLUT-1* = glucose transporter 1; *CTD* = C-terminal domain; *RBD* = receptor-binding domain; *PRR* = proline-rich region. This drawing is modified after Jones et al. (2012) (Jones et al. 2011)

2.3   Viral Replication

At the cellular level, HTLV-1 is transmitted via two major routes: through cell-to-cell contact (horizontal transmission) and via clonal expansion of HTLV-1-infected cells (vertical transmission).

2.3.1   Cell-to-Cell Transmission

Naturally infected T lymphocytes produce little to no free viral particles, and the infectivity of these cell-free particles is very low. In vivo, HTLV-1 intercellular transmission, *i.e.,* horizontal, reverse-transcription-based replication, requires close cell-to-cell contact. To date, three mechanisms have been reported in the literature (Fig. 2). First, in 2003, Igakura et al. showed the formation of a "virological synapse," composed of viral and cellular molecules, at the point of contact between the HTLV-1-infected and recipient target cells (Igakura et al. 2003; Nejmeddine et al. 2005). Second, Pais-Correia et al. described that after viral budding, HTLV-1 virions are retained on the cell surface of infected cells in extracellular viral assemblies composed of collagen, agrin, and linker proteins

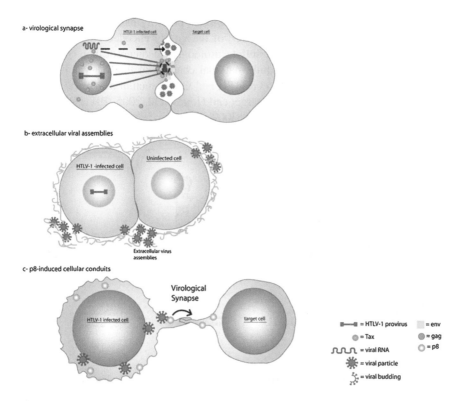

Fig. 2 Mechanisms of cell-to-cell transmission of HTLV-1. This drawing is modified after Yasunaga and Matsuoka (2011)

such as tetherin and galectin-3 (Pais-Correia et al. 2010). When HTLV-1-infected cells attach to uninfected cells, the viral particles contained in these extracellular biofilm-like structures are rapidly transferred to the surface of the target cells, resulting in infection (Pais-Correia et al. 2010). Third, it was recently demonstrated by Franchini and colleagues that HTLV-1 encodes a protein, $p12^I$, in its pX region. The processing of $p12^I$ generates $p8^I$. This protein increases T-cell contact by clustering lymphocyte function-associated antigen-1 (LFA-1); it promotes T-cell conjugation through LFA-1 and intercellular adhesion molecule 1 (ICAM-1) interaction; and it enhances HTLV-1 cell-to-cell transmission by inducing the formation of cellular conduits (Van Prooyen et al. 2010; Fukumoto et al. 2009).

2.3.2 Clonal Expansion

HTLV-1 infection is associated with an elevated proviral load, very low cell-to-cell transmission rate, and high viral genetic stability. This high genetic stability of HTLV-1 (and other deltaretroviruses) is due to its replication in vivo via "the clonal expansion of infected cells" (Wattel et al. 1995; Cavrois et al. 1996; Cavrois et al. 1996; Wattel et al. 1996; Zane et al. 2009). Indeed, instead of using the error-prone viral RT, the HTLV-1 genome is propagated as an integrated provirus that is replicated during cellular DNA synthesis. Since HTLV-1 mostly integrates randomly into the host genome, sequential analyses of integration sites have verified that the proliferation of HTLV-1-infected cells is clonal and persistent (Etoh et al. 1997; Cavrois et al. 1998). In some animal models [e.g., experimentally infected squirrel monkeys (*Saimiri sciureus*) and sheep with HTLV-1 and BLV, respectively], it has been shown that deltaretrovirus infection is a two-step process that includes an early (primo-infection) and transient phase of reverse transcription, before the establishment of an immune response, followed by the persistent multiplication of infected cells by clonal expansion (Mortreux et al. 2001; Pomier et al. 2008). The clonal cells survive over time, and it has been found that ATL originates from one of these clones present during the primo-infection (Moules et al. 2005).

2.4 Viral Expression

As shown in Fig. 3, the HTLV-1 proviral genome contains retroviral structural and non-structural genes. The viral *gag*, *pro*, *pol*, and *env* genes are flanked by the long terminal repeats (LTR) at both ends, and a region named pX is located between *env* and the 3' LTR. The 5' LTR serves as the viral promoter for transcription. The Pol open reading frame encodes reverse transcriptase, protease, and integrase. Gag provides the virion core proteins, and Env is used for viral infectivity. The pX region contains four partially overlapping open reading frames (ORFs); and through the use of alternative splicing and internal initiation codons, it encodes several regulatory proteins. Orf-I produces the $p12^I$ protein which can be proteolytically cleaved at the amino terminus to generate the $p8^I$ protein, while differential splicing of mRNAs from orf-II results in the production of the $p13^{II}$ and

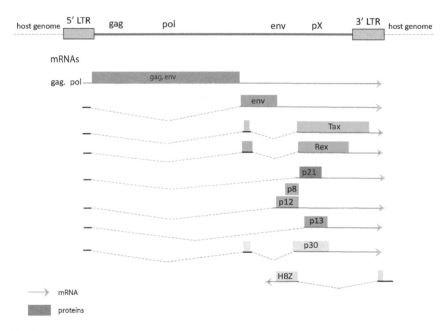

Fig. 3 The HTLV-1 proviral genome showing the expression of various spliced transcripts and the open reading frames (ORFs) that they encode. This drawing is modified from Matsuoka and Jeang (2007)

p30[II] proteins. Orf-III and orf-IV encode the Rex and Tax proteins, respectively; and an antisense mRNA transcribed from the 3′ LTR generates the HTLV-1 basic leucine zipper (HBZ) protein. Below, we will discuss in brief the roles of Tax and HBZ on the induction and the maintenance of leukemogenesis, respectively (Matsuoka and Jeang 2007).

3   Tax Expression Dictates the Fate of HTLV-1-infected Cells

Expression of the viral Tax oncoprotein is sufficient to immortalize T cells (Grassmann et al. 1992), transform rodent cells (Tanaka et al. 1990), and induce tumorigenesis in mouse models (Hinrichs et al. 1987; Nerenberg et al. 1987; Green et al. 1989; Iwakura et al. 1991; Kwon et al. 2005; Hasegawa et al. 2006; Fu et al. 2011). Recently, Banerjee et al. have reported on the transformation of human cells into leukemic cells. Using immune-deficient NOD/SCID mice, they showed that CD4+ lymphomas can arise from mice that are injected with CD34+hematopoietic progenitor stem cells transduced to express Tax (Banerjee et al. 2010). These data raise the notion that a target of Tax transformation may be the CD34+ hematopoietic progenitor stem cells, instead of and perhaps in addition to the currently considered mature CD4+ or CD8+ T lymphocytes.

To become tumorigenic, cells have to grow more rapidly than non-transformed cells. The tumorigenic cells accumulate genetic changes (clastogenic damage or aneuploidy) and enforce the propagation of these aberrant changes by neutralizing the cell cycle checkpoints. To be effective, tumorigenic cells must also evade the host's immune responses (Hanahan and Weinberg 2000, 2011).

Data from multiple laboratories over the past 25 years have begun to shed light on how Tax confers growth advantage to HTLV-1-infected cells, and how this viral oncoprotein triggers DNA damage accumulation and inhibits the cell cycle checkpoints during its transformation of a normal cell into a leukemic cell (Fig. 4).

3.1 Tax Promotes the Survival and the Proliferation of HTLV-1-infected Cells

3.1.1 Tax and Apoptosis and Senescence

Like other oncogenes, Tax confers pro-proliferative and pro-survival properties to cells (Schmitt et al. 1998; Xiao et al. 2001; Iwanaga et al. 2008). Curiously, its expression also has been reported to trigger apoptosis (Yamada et al. 1994; Chlichlia et al. 1995; Fujita and Shiku 1995; Chen et al. 1997; Hall et al. 1998; Kao et al. 2000; Nicot and Harrod 2000) and senescence (Kinjo et al. 2010; Yang et al. 2011; Zhi et al. 2011). These apparently contradictory findings are reconciled if one realizes that Tax performs a single signaling event that differentially elicits either a growth or death/senescence response depending on the context of the cell. Thus, the Tax signal for cells to grow capably stimulates cellular to proliferation under physiologically conditions favorable for growth conditions. On the other hand, under austere conditions that are non-permissive for cellular growth, the same Tax proliferative signal presumably attempts to initiate an increased metabolic program that cannot be consummated and instead the cells react by committing apoptosis or entering senescence (Kasai and Jeang 2004). Stated another way, Tax signaling is always intended to promote cell division. Cells, depending on context, can respond to that dictate to proliferate by growing or by executing apoptosis/senescence. Thus, Tax does not have two countervailing and contradictory functions; rather, it is the same function that elicits two different cellular outcomes (proliferation *versus* apoptosis/senescence) depending on the status of the infected cell (Jeang 2010; Boxus and Willems 2012). In vivo, because HTLV-1 infection ultimately leads to leukemogenesis and T-cell proliferation in some individuals, in these persons, it is clear that the prevailing effect of Tax is pro-proliferative and anti-apoptotic (Copeland et al. 1994; Kishi et al. 1997; Arai et al. 1998; Mulloy et al. 1998; Kawakami et al. 1999); in others who do not develop ATL, it is possible that apoptotic/senescent cellular responses predominate. For understanding the process of leukemogenesis, Tax's activity on factors such as p53 (Mulloy et al. 1998; Haoudi and Semmes 2003; Jung et al. 2008) and NF-kB is

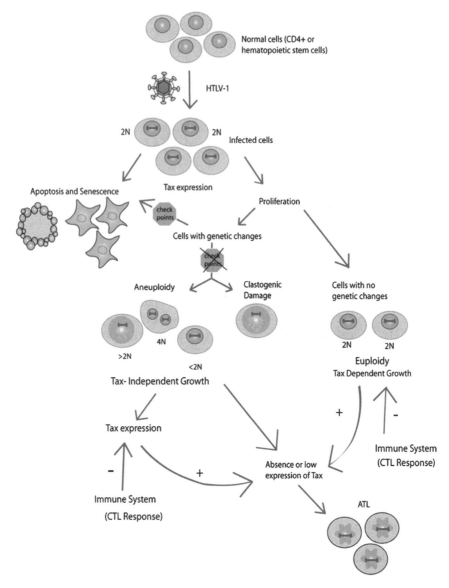

Fig. 4 Multistep processes that lead to the transformation of normal hematopoietic cells into ATL cells. The scheme incorporates the concept that ATL leukemogenesis is induced by Tax. This drawing is modified from Matsuoka and Jeang (2011)

consistent with the requirements in transformed cells of activating anti-apoptotic genes and suppressing pro-apoptotic genes (Kawakami et al. 1999; Tsukahara et al. 1999; Nicot et al. 2000; Mori et al. 2001; Pise-Masison et al. 2002; Krueger et al. 2006; Okamoto et al. 2006; Waldele et al. 2006).

3.1.2   Tax and NF-κB

NF-κB is a major survival factor engaged by HTLV-1. NF-κB is constitutively active in most tumor cells, and its suppression inhibits the growth of tumor (Chaturvedi et al. 2011; Perkins 2012). Although tightly controlled in normal cells, including T cells, NF-κB is constitutively activated in both transformed and untransformed HTLV-1-infected cells (Watanabe et al. 2005; Qu and Xiao 2011).

The NF-κB family of transcription factors has five closely related DNA-binding proteins (RelA (p65), RelB, c-Rel, NF-κB1/p50, and NF-κB2/p52) that can form various homodimers and heterodimers to regulate the transcription of genes containing κB motifs in their promoters. Latent or unstimulated cells sequester NF-κB dimers in the cytoplasm using inhibitors of kappa B (IκBs) proteins such as IκBα and p100. Upon activation, IκBs are degraded (canonical pathway) or p100 is processed to generate p52 (non-canonical pathway) leading to the translocation of active NF-κB proteins into the nucleus to activate transcription (Qu and Xiao 2011; Rauch and Ratner 2011). In the canonical pathway, IκBα degradation requires its phosphorylation by a specific IκB kinase (IKK) complex composed of two catalytic subunits IKKα (or IKK1) and IKKβ (or IKK2), and a regulatory subunit IKKγ (or NEMO). This phosphorylation results in rapid ubiquitination and proteasomal degradation of IκBα, allowing RelA (or p65), and other NF-κB members to localize to the nucleus in order to induce gene expression. In the non-canonical NF-κB pathway, IKKγ is specifically recruited into the p100 complex to phosphorylate p100, leading to p100 ubiquitination and processing to p52 which then associates with NF-κB-binding partners and translocates into the nucleus to induce or repress gene expression (Qu and Xiao 2011).

Work from many investigators has shown that Tax activates both canonical and non-canonical NF-κB signaling pathways in HTLV-1-infected cells (Xiao et al. 2001; Iha et al. 2003; Qu and Xiao 2011). Tax persistently activates IKK through binding to IKKγ, leading to the degradation of IκBα (canonical pathway) (Chu et al. 1999; Harhaj and Sun 1999; Jin et al. 1999; Xiao et al. 2000); and Tax promotes the formation of an IKKα-IKKγ-p100 complex followed by the processing of the NF-κB p100 precursor protein to its active p52 form (non-canonical pathway) (Xiao et al. 2001). Tax also binds to and increases the stability and activity of NF-κB (Hirai et al. 1992; Suzuki et al. 1993; Suzuki et al. 1994) and inactivates NF-κB inhibitors (Maggirwar et al. 1995; Suzuki et al. 1995; Good and Sun 1996; McKinsey et al. 1996; Petropoulos et al. 1996).

Recently, two independent studies using two different Tax transgenic mouse models have revealed that Tax-induced tumorigenesis is dependent on the NF-κB pathway and that both canonical and non-canonical NF-κB pathways are involved in this process (Kwon et al. 2005; Fu et al. 2011). The first study used mice expressing a wild-type Tax or a mutant form of Tax that is unable to activate the NF-κB pathway. A lethal cutaneous disease that shares several features in common with the skin disease that occurs during the preleukemic stage in HTLV-1-infected patients developed in the wild-type Tax–expressing mice (Kwon et al. 2005). In the second study, the investigators found that the genetic knockout of the NF-κB2 gene alone dramatically delayed tumor onset in Tax-expressing transgenic mice (Fu et al. 2011).

3.1.3   Tax and the Cell Cycle

Progression through the cell cycle is a tightly controlled process regulated by interactions between cyclins and cyclin-dependent kinases (CDKs). Tax deregulates the progression of infected cells through different phases of the cell cycle, especially the progression through G1.

Tax propels the cell through G1 by increasing the formation of complexes of cyclin D/CDK4, cyclin D/CDK6, and cyclin E/CDK6 via several mechanisms (Marriott and Semmes 2005). First, Tax can transcriptionally activate the expression of cyclins D2 (Akagi et al. 1996; Santiago et al. 1999; Iwanaga et al. 2001) and E (Iwanaga et al. 2001), CDK2 and 4 (Iwanaga et al. 2001), and transcriptionally repress CKIs such as $p18^{INK4c}$, $p19^{INK4D}$, and $p27^{KIP1}$ (Suzuki et al. 1999; Iwanaga et al. 2001). Additionally, Tax can directly bind CDK4 (Haller et al. 2002; Fraedrich et al. 2005) and $p16^{INK4a}$, thereby preventing the inhibitory $p16^{INK4a}$ molecule from binding to CDK4 and CDK6 (Low et al. 1997). Finally, Tax directly binds the retinoblastoma (RB) protein, which is a target substrate of cyclin D/CDK4/CDK6 and cyclin E/CDK2 complexes, and triggers proteosomal degradation of the RB protein; this then leads to the release of the E2F1 transcription factor from RB and the transcription of E2F1-responsive genes whose products are necessary for passage of the cells through G1 into S phase (Kehn et al. 2005). Moreover, it has also been reported that Tax expression activates the transcription of the *E2F1* gene (Mori 1997; Lemasson et al. 1998; Ohtani et al. 2000).

Another fundamental property of Tax is that it can inhibit the G1/S checkpoint to allow cell cycle progression to happen even with the presence of DNA damage (Marriott and Semmes 2005). Accordingly, Tax can inhibit p53 activity which functions to monitor DNA structure integrity at the G1/S transition (Tabakin-Fix et al. 2006).

3.2   Tax-Expressing Cells Accumulate DNA Damage

Genetic instability of HTLV-1-infected cells generates the acquisition of eight biological changes predicted to be needed for the multistep development of ATL (Hanahan and Weinberg 2000, 2011). Two major types of genetic instability include the loss of DNA repair capabilities and the loss of euploidy. Indeed, Tax is able to disrupt normal cellular processes of DNA repair and chromosomal segregation (Majone et al. 1993; Saggioro et al. 1994, 1996; Lemoine and Marriott 2002).

3.2.1   Tax and Clastogenic Damage

The chromosomes in ATL cells show clastogenic damage (Marriott et al. 2002). Tax engenders direct DNA damage by increasing reactive oxygen species (Kinjo et al. 2010) and/or by inhibiting p53 checkpoint function (Tabakin-Fix et al. 2006).

Two major mechanisms have been hypothesized to explain Tax-abrogation of p53 function. One model suggests that there is a competition between p53 and Tax for binding to the transcription coactivator CREB–binding protein (CBP)/p300 (Ariumi et al. 2000); a second model suggests that Tax activation of NF-κB is required for its inactivation of p53 (Miyazato et al. 2005). More recent data suggest that neither model satisfactorily explain Tax-p53 functional interaction, leaving incompletely answered the question of how Tax disables p53 function.

3.2.2 Tax and Aneuploidy

The majority of cancer cells including ATL cells are aneuploid. Aneuploidy has been proposed to be a cause of transformation. It has been shown that Tax can induce aneuploidy via several mechanisms. Tax can directly trigger chromosomal separation errors in two ways. First, Tax has been shown to cause multipolar mitosis (Peloponese et al. 2005; Ching et al. 2006; Nitta et al. 2006). Tax can also induce aberrant centrosomal multiplication by targeting the cellular TAX1BP2 protein, which normally blocks centriole over-duplication (Ching 2006). Second, during mitosis, Tax engages RANBP1 and fragments spindle poles, provoking multipolar segregation (Peloponese et al. 2005). Moreover, Tax has also been shown to promote premature mitotic exit by binding and activating the anaphase-promoting complex/cyclosome (APC/C). Finally, Tax-expressing cells are lost for the "aneuploidy" checkpoint in mitosis because of Tax-mediated inactivation of the critical spindle assembly checkpoint protein, Mad1 (Liu et al. 2005; Jin et al. 1998).

---

4 ATL

4.1 Absence of Tax Expression and Evasion of the Host's Immune Surveillance

HTLV-1 chronic infection arises when an equilibrium is established between viral virulence and the host immunity. HTLV-1 requires Tax expression to transform cells, but Tax is also the main target of the host's cytotoxic T Lymphocytes (CTLs) (Jacobson et al. 1990; Kannagi et al. 1991; Elovaara et al. 1993; Yamano et al. 2002). Thus, biologically, the virus has to evolve a process to control Tax expression to evade the host's immune surveillance. Early after infection, the current view is that Tax is needed to initiate the cascade of events leading to transformation. On the other hand, Tax-expressing cells immediately become recognized as foreign entities and are targeted by the host's immune system (cytotoxic T cells, CTL) for elimination. Accordingly, a balance has to be reached between growth advantage conferred by Tax to the cell and the susceptibility of the same cell to CTL killing. Early in virus infection when growth advantages conferred by Tax outweigh CTL killing, Tax expression is maintained in virus-infected cells; later in infection, the opposite may be the case which then explains why most HTLV-1 transformed cells become silenced for Tax expression. Thus, it is currently considered that although Tax is needed early to initiate transformation, when cells become transformed, Tax is no longer needed for maintenance of transformation. Given that situation and the need to evade CTL killing, it is not surprising that in ATLs cells late in the course of virus infection, more than 60 % of such cells show no detectable Tax transcripts (Takeda et al. 2004; Taniguchi et al. 2005; Miyazaki et al. 2007). While it is still not fully understood how Tax expression is silence, some of this likely occurs from genetic changes in the Tax gene (Furukawa et al. 2001; Takeda et al. 2004), epigenetic changes in the viral promoter in the 5′LTR (DNA hypermethylation and histone modifications) (Koiwa et al. 2002; Takeda et al. 2004; Taniguchi et al. 2005), and/or deletion of 5′LTR sequences (Tamiya et al. 1996).

4.2   HBZ Expression

The mechanism of how cells acquire Tax-independent proliferation is not completely understood. One explanation is that the genetic host chromosomal changes accumulated over time in HTLV-1-infected cells may have conferred sufficient virus-independent transformation/growth properties to those cells. An additional explanation may be the expression of the viral HBZ transcript/protein. Indeed, HBZ mRNA is highly expressed in ATL cells (Murata et al. 2006; Satou et al. 2006; Miyazaki et al. 2007). Using in vivo models, it has been shown that HBZ is expressed later than Tax in the infected cell, and its expression increases over time (Li et al. 2009). In contrast to Tax, HBZ sequence is not mutated in ATL cells (Fan et al. 2010), and the 3′LTR containing its promoter remains intact (Taniguchi et al. 2005; Fan et al. 2010). Moreover, although HBZ is an immunogenic protein, HBZ-specific CTLs seem unable to efficiently eliminate HTLV-1-infected cells (Suemori et al. 2009). HBZ further promotes virus-infected cell to proliferate late in infection (Satou et al. 2006), and its silencing of viral expression appears to enhance virus-infected cells to escape the host's immune response (Gaudray et al. 2002). The complementary expression patterns of Tax and HBZ suggest that Tax and HBZ may act early and late, respectively, in virus infection with the former used to initiate transformation and the latter utilized to maintain the transformed phenotype of ATL cells.

5   Concluding Remarks

Despite robust progress, several questions regarding ATL leukemogenesis remain unresolved. First, what is the true cellular target of virus/Tax transformation? To date, only human CD34+ hematopoietic progenitor stem cells have been successfully transformed by Tax while other differentiated human primary cells have been refractory to Tax-mediated transformation. Thus, it is unclear what cellular factor differences between progenitor versus differentiated human cells govern Tax-induced transformation? Second, how does Tax fully inactivate p53 function? As mentioned above, current hypotheses on how Tax inactivates p53 appears to be unsatisfactory. Third, what factors are needed for the initiation of ATL versus those needed for maintenance of ATL? One anticipates that progress will be made on these and other questions in the coming years.

Acknowledgments Work in our laboratory is supported in part by intramural funding from the NIAID and by the IATAP program from the Office of the Director, NIH. We thank Lauren Lee for her assistance in the preparation of the figures.

References

Akagi T, Ono H et al (1996) Expression of cell-cycle regulatory genes in HTLV-I infected T-cell lines: possible involvement of Tax1 in the altered expression of cyclin D2, p18Ink4 and p21Waf1/Cip1/Sdi1. Oncogene 12(8):1645–1652

Akagi T, Yoshino T et al (1988) Isolation of virus-producing transformants from human gastric cancer cell line, HGC-27, infected with human T-cell leukemia virus type I. Jpn J Cancer Res 79(7):836–842

Arai M, Kannagi M et al (1998) Expression of FAP-1 (Fas-associated phosphatase) and resistance to Fas-mediated apoptosis in T cell lines derived from human T cell leukemia virus type 1-associated myelopathy/tropical spastic paraparesis patients. AIDS Res Hum Retroviruses 14(3):261–267

Ariumi Y, Kaida A et al (2000) HTLV-1 tax oncoprotein represses the p53-mediated trans-activation function through coactivator CBP sequestration. Oncogene 19(12):1491–1499

Banerjee P, Sieburg M et al (2008) Human T-cell lymphotropic virus type 1 infection of CD34+ hematopoietic progenitor cells induces cell cycle arrest by modulation of p21(cip1/waf1) and survivin. Stem Cells 26(12):3047–3058

Banerjee P, Tripp A et al (2010) Adult T-cell leukemia/lymphoma development in HTLV-1-infected humanized SCID mice. Blood 115(13):2640–2648

Boxus M, Willems L (2012) How the DNA damage response determines the fate of HTLV-1 Tax-expressing cells. Retrovirology 9:2

Cavrois M, Gessain A et al (1996a) Proliferation of HTLV-1 infected circulating cells in vivo in all asymptomatic carriers and patients with TSP/HAM. Oncogene 12(11):2419–2423

Cavrois M, Leclercq I et al (1998) Persistent oligoclonal expansion of human T-cell leukemia virus type 1-infected circulating cells in patients with Tropical spastic paraparesis/HTLV-1 associated myelopathy. Oncogene 17(1):77–82

Cavrois M, Wain-Hobson S et al (1996b) Adult T-cell leukemia/lymphoma on a background of clonally expanding human T-cell leukemia virus type-1-positive cells. Blood 88(12):4646–4650

Chaturvedi MM, Sung B et al (2011) NF-kappaB addiction and its role in cancer: 'one size does not fit all'. Oncogene 30(14):1615–1630

Chen X, Zachar V et al (1997) Role of the Fas/Fas ligand pathway in apoptotic cell death induced by the human T cell lymphotropic virus type I Tax transactivator. J Gen Virol 78(Pt 12):3277–3285

Ching YP, Chan SF et al (2006) The retroviral oncoprotein Tax targets the coiled-coil centrosomal protein TAX1BP2 to induce centrosome overduplication. Nat Cell Biol 8(7):717–724

Chlichlia K, Moldenhauer G et al (1995) Immediate effects of reversible HTLV-1 tax function: T-cell activation and apoptosis. Oncogene 10(2):269–277

Chu ZL, Shin YA et al (1999) IKKgamma mediates the interaction of cellular IkappaB kinases with the tax transforming protein of human T cell leukemia virus type 1. J Biol Chem 274(22):15297–15300

Clapham P, Nagy K et al (1983) Productive infection and cell-free transmission of human T-cell leukemia virus in a nonlymphoid cell line. Science 222(4628):1125–1127

Copeland KF, Haaksma AG et al (1994) Inhibition of apoptosis in T cells expressing human T cell leukemia virus type I Tax. AIDS Res Hum Retroviruses 10(10):1259–1268

Elovaara I, Koenig S et al (1993) High human T cell lymphotropic virus type 1 (HTLV-1)-specific precursor cytotoxic T lymphocyte frequencies in patients with HTLV-1-associated neurological disease. J Exp Med 177(6):1567–1573

Etoh K, Tamiya S et al (1997) Persistent clonal proliferation of human T-lymphotropic virus type I-infected cells in vivo. Cancer Res 57(21):4862–4867

Fan J, Ma G et al (2010) APOBEC3G generates nonsense mutations in human T-cell leukemia virus type 1 proviral genomes in vivo. J Virol 84(14):7278–7287

Feuer G, Fraser JK et al (1996) Human T-cell leukemia virus infection of human hematopoietic progenitor cells: maintenance of virus infection during differentiation in vitro and in vivo. J Virol 70(6):4038–4044

Fraedrich K, Muller B et al (2005) The HTLV-1 Tax protein binding domain of cyclin-dependent kinase 4 (CDK4) includes the regulatory PSTAIRE helix. Retrovirology 2:54

Fu J, Qu Z et al (2011) The tumor suppressor gene WWOX links the canonical and noncanonical NF-kappaB pathways in HTLV-I Tax-mediated tumorigenesis. Blood 117(5):1652–1661

Fujita M, Shiku H (1995) Differences in sensitivity to induction of apoptosis among rat fibroblast cells transformed by HTLV-I tax gene or cellular nuclear oncogenes. Oncogene 11(1):15–20

Fukumoto R, Andresen V et al (2009) In vivo genetic mutations define predominant functions of the human T-cell leukemia/lymphoma virus p12I protein. Blood 113(16):3726–3734

Furukawa Y, Kubota R et al (2001) Existence of escape mutant in HTLV-I tax during the development of adult T-cell leukemia. Blood 97(4):987–993

Gallo RC (2005) History of the discoveries of the first human retroviruses: HTLV-1 and HTLV-2. Oncogene 24(39):5926–5930

Gaudray G, Gachon F et al (2002) The complementary strand of the human T-cell leukemia virus type 1 RNA genome encodes a bZIP transcription factor that down-regulates viral transcription. J Virol 76(24):12813–12822

Gessain A (2011) Human retrovirus HTLV-1: descriptive and molecular epidemiology, origin, evolution, diagnosis and associated diseases. Bull Soc Pathol Exot 104(3):167–180

Ghez D, Lepelletier Y et al (2006) Neuropilin-1 is involved in human T-cell lymphotropic virus type 1 entry. J Virol 80(14):6844–6854

Goncalves DU, Proietti FA et al (2010) Epidemiology, treatment, and prevention of human T-cell leukemia virus type 1-associated diseases. Clin Microbiol Rev 23(3):577–589

Good L, Sun SC (1996) Persistent activation of NF-kappa B/Rel by human T-cell leukemia virus type 1 tax involves degradation of I kappa B beta. J Virol 70(5):2730–2735

Grant C, Barmak K et al (2002) Human T cell leukemia virus type I and neurologic disease: events in bone marrow, peripheral blood, and central nervous system during normal immune surveillance and neuroinflammation. J Cell Physiol 190(2):133–159

Grassmann R, Berchtold S et al (1992) Role of human T-cell leukemia virus type 1 X region proteins in immortalization of primary human lymphocytes in culture. J Virol 66(7):4570–4575

Green JE, Hinrichs SH et al (1989) Exocrinopathy resembling Sjogren's syndrome in HTLV-1 tax transgenic mice. Nature 341(6237):72–74

Hall AP, Irvine J et al (1998) Tumours derived from HTLV-I tax transgenic mice are characterized by enhanced levels of apoptosis and oncogene expression. J Pathol 186(2):209–214

Haller K, Wu Y et al (2002) Physical interaction of human T-cell leukemia virus type 1 Tax with cyclin-dependent kinase 4 stimulates the phosphorylation of retinoblastoma protein. Mol Cell Biol 22(10):3327–3338

Hanahan D, Weinberg RA (2000) The hallmarks of cancer. Cell 100(1):57–70

Hanahan D, Weinberg RA (2011) Hallmarks of cancer: the next generation. Cell 144(5):646–674

Haoudi A, Semmes OJ (2003) The HTLV-1 tax oncoprotein attenuates DNA damage induced G1 arrest and enhances apoptosis in p53 null cells. Virology 305(2):229–239

Harhaj EW, Sun SC (1999) IKKgamma serves as a docking subunit of the IkappaB kinase (IKK) and mediates interaction of IKK with the human T-cell leukemia virus Tax protein. J Biol Chem 274(33):22911–22914

Hasegawa H, Sawa H et al (2006) Thymus-derived leukemia-lymphoma in mice transgenic for the Tax gene of human T-lymphotropic virus type I. Nat Med 12(4):466–472

Hayami M, Tsujimoto H et al (1984) Transmission of adult T-cell leukemia virus from lymphoid cells to non-lymphoid cells associated with cell membrane fusion. Gann 75(2):99–102

Hinrichs SH, Nerenberg M et al (1987) A transgenic mouse model for human neurofibromatosis. Science 237(4820):1340–1343

Hinuma Y, Nagata K et al (1981) Adult T-cell leukemia: antigen in an ATL cell line and detection of antibodies to the antigen in human sera. Proc Natl Acad Sci USA 78(10):6476–6480

Hirai H, Fujisawa J et al (1992) Transcriptional activator Tax of HTLV-1 binds to the NF-kappa B precursor p105. Oncogene 7(9):1737–1742

Hiramatsu K, Masuda M et al (1986) Mode of transmission of human T-cell leukemia virus type I (HTLV I) in a human promyelocytic leukemia HL60 cell. Int J Cancer 37(4):601–606

Ho DD, Rota TR et al (1984) Infection of human endothelial cells by human T-lymphotropic virus type I. Proc Natl Acad Sci USA 81(23):7588–7590

Igakura T, Stinchcombe JC et al (2003) Spread of HTLV-I between lymphocytes by virus-induced polarization of the cytoskeleton. Science 299(5613):1713–1716

Iha H, Kibler KV et al (2003) Segregation of NF-kappaB activation through NEMO/IKKgamma by Tax and TNFalpha: implications for stimulus-specific interruption of oncogenic signaling. Oncogene 22(55):8912–8923

Iwakura Y, Tosu M et al (1991) Induction of inflammatory arthropathy resembling rheumatoid arthritis in mice transgenic for HTLV-I. Science 253(5023):1026–1028

Iwanaga R, Ohtani K et al (2001) Molecular mechanism of cell cycle progression induced by the oncogene product Tax of human T-cell leukemia virus type I. Oncogene 20(17):2055–2067

Iwanaga R, Ozono E et al (2008) Activation of the cyclin D2 and cdk6 genes through NF-kappaB is critical for cell-cycle progression induced by HTLV-1 Tax. Oncogene 27(42):5635–5642

Jacobson S, Shida H et al (1990) Circulating CD8+ cytotoxic T lymphocytes specific for HTLV-I pX in patients with HTLV-I associated neurological disease. Nature 348(6298):245–248

Jeang KT (2010) HTLV-1 and adult T-cell leukemia: insights into viral transformation of cells 30 years after virus discovery. J Formos Med Assoc 109(10):688–693

Jin DY, Spencer F et al (1998) Human T cell Leukemia virus type 1 oncoprotein Tax targets the human mitotic checkpoint protein MAD1. Cell 93(1):81–91

Jin DY, Giordano V et al (1999) Role of adapter function in oncoprotein-mediated activation of NF-kappaB. Human T-cell leukemia virus type I Tax interacts directly with IkappaB kinase gamma. J Biol Chem 274(25):17402–17405

Jones KS, Akel S et al (2005) Induction of human T cell leukemia virus type I receptors on quiescent naive T lymphocytes by TGF-beta. J Immunol 174(7):4262–4270

Jones KS, Lambert S et al (2011) Molecular aspects of HTLV-1 entry: functional domains of the HTLV-1 surface subunit (SU) and their relationships to the entry receptors. Viruses 3(6):794–810

Jones KS, Petrow-Sadowski C et al (2008) Cell-free HTLV-1 infects dendritic cells leading to transmission and transformation of CD4(+) T cells. Nat Med 14(4):429–436

Jung KJ, Dasgupta A et al (2008) Small-molecule inhibitor which reactivates p53 in human T-cell leukemia virus type 1-transformed cells. J Virol 82(17):8537–8547

Kannagi M, Harada S et al (1991) Predominant recognition of human T cell leukemia virus type I (HTLV-I) pX gene products by human CD8+ cytotoxic T cells directed against HTLV-I-infected cells. Int Immunol 3(8):761–767

Kao SY, Lemoine FJ et al (2000) HTLV-1 Tax protein sensitizes cells to apoptotic cell death induced by DNA damaging agents. Oncogene 19(18):2240–2248

Kasai T, Jeang KT (2004) Two discrete events, human T-cell leukemia virus type I Tax oncoprotein expression and a separate stress stimulus, are required for induction of apoptosis in T-cells. Retrovirology 1:7

Kawakami A, Nakashima T et al (1999) Inhibition of caspase cascade by HTLV-I tax through induction of NF-kappaB nuclear translocation. Blood 94(11):3847–3854

Kehn K, Fuente Cde L et al (2005) The HTLV-I Tax oncoprotein targets the retinoblastoma protein for proteasomal degradation. Oncogene 24(4):525–540

Kinjo T, Ham-Terhune J et al (2010) Induction of reactive oxygen species by human T-cell leukemia virus type 1 tax correlates with DNA damage and expression of cellular senescence marker. J Virol 84(10):5431–5437

Kishi S, Saijyo S et al (1997) Resistance to fas-mediated apoptosis of peripheral T cells in human T lymphocyte virus type I (HTLV-I) transgenic mice with autoimmune arthropathy. J Exp Med 186(1):57–64

Koiwa T, Hamano-Usami A et al (2002) 5′-long terminal repeat-selective CpG methylation of latent human T-cell leukemia virus type 1 provirus in vitro and in vivo. J Virol 76(18):9389–9397

Koyanagi Y, Itoyama Y et al (1993) In vivo infection of human T-cell leukemia virus type I in non-T cells. Virology 196(1):25–33

Krueger A, Fas SC et al (2006) HTLV-1 Tax protects against CD95-mediated apoptosis by induction of the cellular FLICE-inhibitory protein (c-FLIP). Blood 107(10):3933–3939

Kwon H, Ogle L et al (2005) Lethal cutaneous disease in transgenic mice conditionally expressing type I human T cell leukemia virus Tax. J Biol Chem 280(42):35713–35722

Lambert S, Bouttier M et al (2009) HTLV-1 uses HSPG and neuropilin-1 for entry by molecular mimicry of VEGF165. Blood 113(21):5176–5185

Lemasson I, Thebault S et al (1998) Activation of E2F-mediated transcription by human T-cell leukemia virus type I Tax protein in a p16(INK4A)-negative T-cell line. J Biol Chem 273(36):23598–23604

Lemoine FJ, Marriott SJ (2002) Genomic instability driven by the human T-cell leukemia virus type I (HTLV-I) oncoprotein, Tax. Oncogene 21(47):7230–7234

LeVasseur RJ, Southern SO et al (1998) Mammary epithelial cells support and transfer productive human T-cell lymphotropic virus infections. J Hum Virol 1(3):214–223

Li M, Kesic M et al (2009) Kinetic analysis of human T-cell leukemia virus type 1 gene expression in cell culture and infected animals. J Virol 83(8):3788–3797

Liu B, Hong S et al (2005) HTLV-I Tax directly binds the Cdc20-associated anaphase-promoting complex and activates it ahead of schedule. Proc Natl Acad Sci USA 102(1):63–68

Low KG, Dorner LF et al (1997) Human T-cell leukemia virus type 1 Tax releases cell cycle arrest induced by p16INK4a. J Virol 71(3):1956–1962

Macatonia SE, Cruickshank JK et al (1992) Dendritic cells from patients with tropical spastic paraparesis are infected with HTLV-1 and stimulate autologous lymphocyte proliferation. AIDS Res Hum Retroviruses 8(9):1699–1706

Maggirwar SB, Harhaj E et al (1995) Activation of NF-kappa B/Rel by Tax involves degradation of I kappa B alpha and is blocked by a proteasome inhibitor. Oncogene 11(5):993–998

Mahieux R, Gessain A (2005) New human retroviruses: HTLV-3 and HTLV-4. Med Trop (Mars) 65(6):525–528

Mahieux R, Gessain A (2009) The human HTLV-3 and HTLV-4 retroviruses: new members of the HTLV family. Pathol Biol (Paris) 57(2):161–166

Majone F, Semmes OJ et al (1993) Induction of micronuclei by HTLV-I Tax: a cellular assay for function. Virology 193(1):456–459

Manel N, Kim FJ et al (2003) The ubiquitous glucose transporter GLUT-1 is a receptor for HTLV. Cell 115(4):449–459

Marriott SJ, Lemoine FJ et al (2002) Damaged DNA and miscounted chromosomes: human T cell leukemia virus type I tax oncoprotein and genetic lesions in transformed cells. J Biomed Sci 9(4):292–298

Marriott SJ, Semmes OJ (2005) Impact of HTLV-I Tax on cell cycle progression and the cellular DNA damage repair response. Oncogene 24(39):5986–5995

Matsuoka M, Jeang KT (2007) Human T-cell leukaemia virus type 1 (HTLV-1) infectivity and cellular transformation. Nat Rev Cancer 7(4):270–280

Matsuoka M, Jeang KT (2011) Human T-cell leukaemia virus type 1 (HTLV-1) and leukemic transformation: viral infectivity, Tax, HBZ and therapy. Oncogene 30(12):1379–1389

Matsuura E, Yamano Y et al (2010) Neuroimmunity of HTLV-I Infection. J Neuroimmune Pharmacol 5(3):310–325

McKinsey TA, Brockman JA et al (1996) Inactivation of IkappaBbeta by the tax protein of human T-cell leukemia virus type 1: a potential mechanism for constitutive induction of NF-kappaB. Mol Cell Biol 16(5):2083–2090

Miyazaki M, Yasunaga J et al (2007) Preferential selection of human T-cell leukemia virus type 1 provirus lacking the 5' long terminal repeat during oncogenesis. J Virol 81(11):5714–5723

Miyazato A, Sheleg S et al (2005) Evidence for NF-kappaB- and CBP-independent repression of p53's transcriptional activity by human T-cell leukemia virus type 1 Tax in mouse embryo and primary human fibroblasts. J Virol 79(14):9346–9350

Miyoshi I, Kubonishi I et al (1981) Type C virus particles in a cord T-cell line derived by co-cultivating normal human cord leukocytes and human leukaemic T cells. Nature 294(5843):770–771

Mori N (1997) High levels of the DNA-binding activity of E2F in adult T-cell leukemia and human T-cell leukemia virus type I-infected cells: possible enhancement of DNA-binding of E2F by the human T-cell leukemia virus I transactivating protein, Tax. Eur J Haematol 58(2):114–120

Mori N, Fujii M et al (2001) Human T-cell leukemia virus type I tax protein induces the expression of anti-apoptotic gene Bcl-xL in human T-cells through nuclear factor-kappaB and c-AMP responsive element binding protein pathways. Virus Genes 22(3):279–287

Mortreux F, Kazanji M et al (2001) Two-step nature of human T-cell leukemia virus type 1 replication in experimentally infected squirrel monkeys (Saimiri sciureus). J Virol 75(2):1083–1089

Moules V, Pomier C et al (2005) Fate of premalignant clones during the asymptomatic phase preceding lymphoid malignancy. Cancer Res 65(4):1234–1243

Mulloy JC, Kislyakova T et al (1998) Human T-cell lymphotropic/leukemia virus type 1 Tax abrogates p53-induced cell cycle arrest and apoptosis through its CREB/ATF functional domain. J Virol 72(11):8852–8860

Murata K, Hayashibara T et al (2006) A novel alternative splicing isoform of human T-cell leukemia virus type 1 bZIP factor (HBZ-SI) targets distinct subnuclear localization. J Virol 80(5):2495–2505

Nagai M, Brennan MB et al (2001) CD8(+) T cells are an in vivo reservoir for human T-cell lymphotropic virus type I. Blood 98(6):1858–1861

Nejmeddine M, Barnard AL et al (2005) Human T-lymphotropic virus, type 1, tax protein triggers microtubule reorientation in the virological synapse. J Biol Chem 280(33):29653–29660

Nerenberg M, Hinrichs SH et al (1987) The tat gene of human T-lymphotropic virus type 1 induces mesenchymal tumors in transgenic mice. Science 237(4820):1324–1329

Nicot C, Harrod R (2000) Distinct p300-responsive mechanisms promote caspase-dependent apoptosis by human T-cell lymphotropic virus type 1 Tax protein. Mol Cell Biol 20(22):8580–8589

Nicot C, Mahieux R et al (2000) Bcl-X(L) is up-regulated by HTLV-I and HTLV-II in vitro and in ex vivo ATLL samples. Blood 96(1):275–281

Nitta T, Kanai M et al (2006) Centrosome amplification in adult T-cell leukemia and human T-cell leukemia virus type 1 Tax-induced human T cells. Cancer Sci 97(9):836–841

Ohtani K, Iwanaga R et al (2000) Cell type-specific E2F activation and cell cycle progression induced by the oncogene product Tax of human T-cell leukemia virus type I. J Biol Chem 275(15):11154–11163

Okamoto K, Fujisawa J et al (2006) Human T-cell leukemia virus type-I oncoprotein Tax inhibits Fas-mediated apoptosis by inducing cellular FLIP through activation of NF-kappaB. Genes Cells 11(2):177–191

Ono A, Miura T et al (1994) Subtype analysis of HTLV-1 in patients with HTLV-1 uveitis. Jpn J Cancer Res 85(8):767–770

Pais-Correia AM, Sachse M et al (2010) Biofilm-like extracellular viral assemblies mediate HTLV-1 cell-to-cell transmission at virological synapses. Nat Med 16(1):83–89

Peloponese JM Jr, Haller K et al (2005) Abnormal centrosome amplification in cells through the targeting of Ran-binding protein-1 by the human T cell leukemia virus type-1 Tax oncoprotein. Proc Natl Acad Sci USA 102(52):18974–18979

Perkins ND (2012) The diverse and complex roles of NF-kappaB subunits in cancer. Nat Rev Cancer 12(2):121–132

Petropoulos L, Lin R et al (1996) Human T cell leukemia virus type 1 tax protein increases NF-kappa B dimer formation and antagonizes the inhibitory activity of the I kappa B alpha regulatory protein. Virology 225(1):52–64

Pinon JD, Klasse PJ et al (2003) Human T-cell leukemia virus type 1 envelope glycoprotein gp46 interacts with cell surface heparan sulfate proteoglycans. J Virol 77(18):9922–9930

Pise-Masison CA, Radonovich M et al (2002) Transcription profile of cells infected with human T-cell leukemia virus type I compared with activated lymphocytes. Cancer Res 62(12):3562–3571

Poiesz BJ, Ruscetti FW et al (1980) Detection and isolation of type C retrovirus particles from fresh and cultured lymphocytes of a patient with cutaneous T-cell lymphoma. Proc Natl Acad Sci USA 77(12):7415–7419

Pomier C, Alcaraz MT et al (2008) Early and transient reverse transcription during primary deltaretroviral infection of sheep. Retrovirology 5:16

Proietti FA, Carneiro-Proietti AB et al (2005) Global epidemiology of HTLV-I infection and associated diseases. Oncogene 24(39):6058–6068

Qu Z, Xiao G (2011) Human T-cell lymphotropic virus: a model of NF-kappaB-associated tumorigenesis. Viruses 3(6):714–749

Rauch DA, Ratner L (2011) Targeting HTLV-1 activation of NFkappaB in mouse models and ATLL patients. Viruses 3(6):886–900

Saggioro D, Majone F et al (1994) Tax protein of human T-lymphotropic virus type I triggers DNA damage. Leuk Lymphoma 12(3–4):281–286

Santiago F, Clark E et al (1999) Transcriptional up-regulation of the cyclin D2 gene and acquisition of new cyclin-dependent kinase partners in human T-cell leukemia virus type 1-infected cells. J Virol 73(12):9917–9927

Satou Y, Yasunaga J et al (2006) HTLV-I basic leucine zipper factor gene mRNA supports proliferation of adult T cell leukemia cells. Proc Natl Acad Sci USA 103(3):720–725

Schmitt I, Rosin O et al (1998) Stimulation of cyclin-dependent kinase activity and G1- to S-phase transition in human lymphocytes by the human T-cell leukemia/lymphotropic virus type 1 Tax protein. J Virol 72(1):633–640

Semmes OJ, Majone F et al (1996) HTLV-I and HTLV-II Tax: differences in induction of micronuclei in cells and transcriptional activation of viral LTRs. Virology 217(1):373–379

Suemori K, Fujiwara H et al (2009) HBZ is an immunogenic protein, but not a target antigen for human T-cell leukemia virus type 1-specific cytotoxic T lymphocytes. J Gen Virol 90(Pt 8):1806–1811

Suzuki T, Hirai H et al (1993) A trans-activator Tax of human T-cell leukemia virus type 1 binds to NF-kappa B p50 and serum response factor (SRF) and associates with enhancer DNAs of the NF-kappa B site and CArG box. Oncogene 8(9):2391–2397

Suzuki T, Hirai H et al (1995) Tax protein of HTLV-1 destabilizes the complexes of NF-kappa B and I kappa B-alpha and induces nuclear translocation of NF-kappa B for transcriptional activation. Oncogene 10(6):1199–1207

Suzuki T, Hirai H et al (1994) Tax protein of HTLV-1 interacts with the Rel homology domain of NF-kappa B p65 and c-Rel proteins bound to the NF-kappa B binding site and activates transcription. Oncogene 9(11):3099–3105

Suzuki T, Narita T et al (1999) Down-regulation of the INK4 family of cyclin-dependent kinase inhibitors by tax protein of HTLV-1 through two distinct mechanisms. Virology 259(2):384–391

Tabakin-Fix Y, Azran I et al (2006) Functional inactivation of p53 by human T-cell leukemia virus type 1 Tax protein: mechanisms and clinical implications. Carcinogenesis 27(4):673–681

Takatsuki K (2005) Discovery of adult T-cell leukemia. Retrovirology 2:16

Takeda S, Maeda M et al (2004) Genetic and epigenetic inactivation of tax gene in adult T-cell leukemia cells. Int J Cancer 109(4):559–567

Tamiya S, Matsuoka M et al (1996) Two types of defective human T-lymphotropic virus type I provirus in adult T-cell leukemia. Blood 88(8):3065–3073

Tanaka A, Takahashi C et al (1990) Oncogenic transformation by the tax gene of human T-cell leukemia virus type I in vitro. Proc Natl Acad Sci USA 87(3):1071–1075

Taniguchi Y, Nosaka K et al (2005) Silencing of human T-cell leukemia virus type I gene transcription by epigenetic mechanisms. Retrovirology 2:64

Tripp A, Banerjee P et al (2005) Induction of cell cycle arrest by human T-cell lymphotropic virus type 1 Tax in hematopoietic progenitor (CD34+) cells: modulation of p21cip1/waf1 and p27kip1 expression. J Virol 79(22):14069–14078

Tripp A, Liu Y et al (2003) Human T-cell leukemia virus type 1 tax oncoprotein suppression of multilineage hematopoiesis of CD34+ cells in vitro. J Virol 77(22):12152–12164

Tsukahara T, Kannagi M et al (1999) Induction of Bcl-x(L) expression by human T-cell leukemia virus type 1 Tax through NF-kappaB in apoptosis-resistant T-cell transfectants with Tax. J Virol 73(10):7981–7987

Van Prooyen N, Gold H et al (2010) Human T-cell leukemia virus type 1 p8 protein increases cellular conduits and virus transmission. Proc Natl Acad Sci USA 107(48):20738–20743

Waldele K, Silbermann K et al (2006) Requirement of the human T-cell leukemia virus (HTLV-1) tax-stimulated HIAP-1 gene for the survival of transformed lymphocytes. Blood 107(11):4491–4499

Watanabe M, Ohsugi T et al (2005) Dual targeting of transformed and untransformed HTLV-1-infected T cells by DHMEQ, a potent and selective inhibitor of NF-kappaB, as a strategy for chemoprevention and therapy of adult T-cell leukemia. Blood 106(7):2462–2471

Watanabe T (2011) Current status of HTLV-1 infection. Int J Hematol 94(5):430–434

Watanabe T, Seiki M et al (1983) Retrovirus terminology. Science 222(4629):1178

Wattel E, Cavrois M et al (1996) Clonal expansion of infected cells: a way of life for HTLV-I. J Acquir Immune Defic Syndr Hum Retrovirol 13(Suppl 1):S92–S99

Wattel E, Vartanian JP et al (1995) Clonal expansion of human T-cell leukemia virus type I-infected cells in asymptomatic and symptomatic carriers without malignancy. J Virol 69(5):2863–2868

Xiao G, Cvijic ME et al (2001) Retroviral oncoprotein Tax induces processing of NF-kappaB2/p100 in T cells: evidence for the involvement of IKKalpha. EMBO J 20(23):6805–6815

Xiao G, Harhaj EW et al (2000) Domain-specific interaction with the I kappa B kinase (IKK)regulatory subunit IKK gamma is an essential step in tax-mediated activation of IKK. J Biol Chem 275(44):34060–34067

Yamada T, Yamaoka S et al (1994) The human T-cell leukemia virus type I Tax protein induces apoptosis which is blocked by the Bcl-2 protein. J Virol 68(5):3374–3379

Yamano Y, Nagai M et al (2002) Correlation of human T-cell lymphotropic virus type 1 (HTLV-1) mRNA with proviral DNA load, virus-specific CD8(+) T cells, and disease severity in HTLV-1-associated myelopathy (HAM/TSP). Blood 99(1):88–94

Yang L, Kotomura N et al (2011) Complex cell cycle abnormalities caused by human T-lymphotropic virus type 1 Tax. J Virol 85(6):3001–3009

Yasunaga J, Matsuoka M (2011) Molecular mechanisms of HTLV-1 infection and pathogenesis. Int J Hematol 94(5):435–442

Yoshida M, Miyoshi I et al (1982) Isolation and characterization of retrovirus from cell lines of human adult T-cell leukemia and its implication in the disease. Proc Natl Acad Sci USA 79(6):2031–2035

Zane L, Sibon D et al (2009) Clonal expansion of HTLV-1 infected cells depends on the CD4 versus CD8 phenotype. Front Biosci 14:3935–3941

Zhi H, Yang L et al (2011) NF-kappaB hyper-activation by HTLV-1 tax induces cellular senescence, but can be alleviated by the viral anti-sense protein HBZ. PLoS Pathog 7(4):e1002025

Prevention of Human T-Cell Lymphotropic Virus Type 1 Infection and Adult T-Cell Leukemia/Lymphoma

Makoto Yoshimitsu, Yohann White and Naomichi Arima

Abstract

Adult T-cell leukemia/lymphoma (ATLL) is a highly aggressive peripheral T-cell malignancy that develops after long-term chronic infection with human T-cell lymphotropic virus type-1 (HTLV-1). Despite the recent advances in chemotherapy, allogeneic hematopoietic stem cell transplantation (alloHSCT), and supportive care, the prognosis for patients with ATL is one of the poorest among hematological malignancies; overall survival (OS) at 3 years is only 24 % in the more aggressive subtypes of ATLL. HTLV-1 is a human retrovirus infecting approximately 10–20 million people worldwide, particularly in southern and southeastern Japan, the Caribbean, highlands of South America, Melanesia, and Equatorial Africa. Despite this high frequency of human infection, only 2–5 % of HTLV-1-infected individuals develop ATLL. Three major routes of viral transmission have been established: (1) mother-to-child transmission through breast-feeding; (2) sexual transmission, predominantly from men to women; and (3) cellular blood components. Multiple factors (e.g., virus, host cell, and immune factors) have been implicated in the development of ATLL, although the underlying mechanisms of leukemogenesis have not been fully elucidated. No preventive vaccine against HTLV-1 is currently available, and interrupting the well-recognized primary modes of HTLV-1 transmission is the mainstay of ATLL prevention. Prevention of

M. Yoshimitsu (✉) · N. Arima
Department of Hematology and Immunology, Kagoshima University Hospital,
8-35-1 Sakuragaoka, Kagoshima, 890-8520, Japan
e-mail: myoshimi@m.kufm.kagoshima-u.ac.jp

Y. White · N. Arima
Division of Hematology and Immunology, Center for Chronic Viral Diseases, Graduate
School of Medical and Dental Sciences, 8-35-1 Sakuragaoka, Kagoshima, 890-8544, Japan

M. H. Chang and K.-T. Jeang (eds.), *Viruses and Human Cancer*,
Recent Results in Cancer Research 193, DOI: 10.1007/978-3-642-38965-8_12,
© Springer-Verlag Berlin Heidelberg 2014

mother-to-child transmission through the replacement of breast-feeding has been shown to have the most significant impact on the incidence of HTLV-1 infection, and public health policies should consider the risk of malnutrition, especially in developing countries where malnutrition is the significant cause of infant mortality.

Keywords
HTLV-1 · ATLL · ATL · Breast-feeding · Transfusion

Contents

1   Introduction of Prevention of HTLV-1 and Adult T-Cell Leukemia/Lymphoma

Human T-cell lymphotropic virus type 1 (HTLV-1) was first discovered as the human retrovirus causally linked to the T-cell hematological malignancy and adult T-cell leukemia/lymphoma (Poiesz et al. 1980; Yoshida et al. 1982). The virus is transmitted through contact with body fluids containing HTLV-1-infected cells, mostly from mother-to-child transmission through breast-feeding or through blood transfusions. Adult T-cell leukemia/lymphoma (ATLL) develops after a prolonged incubation period in a minority of individuals infected with HTLV-1, and strategies aimed at preventing ATLL are based on interrupting HTLV-1 transmission. Firstly, interrupting HTLV-1 transmission by screening for HTLV-1 among blood donors and restricting breast-feeding by mothers who are HTLV-1 carriers have been the primary public health approaches in HTLV-1 endemic areas. The second strategy, although intuitive but that has yet to be realized due to a lack of effective modalities, is the prevention of progression to ATLL among HTLV-1 carriers. Approximately 90 % of HTLV-1 carriers remain as healthy uninfected individuals throughout their lifetime, and there are no clear predictors of progression to ATLL. In addition, preventive strategies such as vaccination could theoretically lead to preventive strategies based on our current understanding of immune-mediated disorders also known to be linked to HTLV-1, including HTLV-1-associated myelopathy/tropical spastic paraparesis (HAM/TSP).

2 Epidemiology of HTLV-1 Infection

2.1 Worldwide

Nearly 20 million people worldwide are estimated to be infected with HTLV-1 (de The and Kazanji 1996). Among them, only less than 10 % develops HTLV-1-related disorders, including adult T-cell leukemia/lymphoma throughout their lifetime. A number of studies investigating the geographical and ethno-epidemiological distribution of the virus have been conducted over the last 3 decades (Goncalves et al. 2010; Sonoda et al. 2011) and have revealed that southwestern Japan, parts of Africa, the Caribbean Islands, and Central and South America are the endemic areas. In Europe and North America, infection with HTLV-1 is predominantly found in immigrant populations originating from endemic countries.

2.2 Japan

A recent seroprevalence study has revealed that the estimated number of HTLV-1 carriers in Japan is at least 1.08 million (Satake et al. 2012). This is 10 % lower than that reported in 1988. The estimated prevalence rates were 0.66 % in men and 1.02 % in women.

3 Mode of Transmission and Clinical Outcome

The primary modes of infection with HTLV-1 have been well described; namely (1) mother-to-child transmission, mainly via breast-feeding (Kinoshita et al. 1987; Yamanouchi et al. 1985); (2) sexual transmission, predominantly from men to women (Murphy et al. 1989; Tajima et al. 1982); and (3) through cellular blood components (Okochi and Sato 1984). Studies suggest an association between the mode of infection and the type of HTLV-1-associated disease seen. ATLL has been mainly associated with acquiring infection through breast-feeding, and HAM/TSP with acquiring infection through blood transfusions (Osame et al. 1990). Reports of ATLL cases occurring in patients infected through blood transfusion are few (Chen et al. 1989). The risk of transmission from mother to child during breast-feeding has been estimated to be 20 % (Hino et al. 1985), while the risk of transmission during pregnancy or the peripartum period was estimated to be less than 5 % (Fujino and Nagata 2000). ATLL development may be linked to a prolonged period of infection with HTLV-1 acquired through vertical transmission.

4 Epidemiology of Adult T-Cell Leukemia/Lymphoma

Only a small proportion of HTLV-1 carriers develop ATLL after a long latency period. Despite the wide geographical distribution, data on the incidence and prevalence of ATLL, except for Japan, are scarce. Furthermore, it is likely that existing reports may be underestimating the prevalence of lymphoma subtype especially due to similarities in clinical presentation compared with other T-cell lymphomas and limited diagnostic capabilities in resource-poor settings. Among the Japanese population, the incidence of ATLL among carriers is estimated to be between 4.5 and 7.3 % in men and 2.6 and 3.5 % in women (Koga et al. 2010; Kondo et al. 1989; Tokudome et al. 1989). ATLL is reported to develop among individuals predominantly in their fifth decade of life in Japan (Takatsuki et al. 1996), whereas in Jamaica and Brazilian series, patients tend to present with the disease in the fourth decade of life, suggesting that other immunological or host genetic factors may play a role in the pathogenesis of ATLL (Gibbs et al. 1987; Pombo de Oliveira et al. 1995).

5 Mechanisms of HTLV-1 Transmission

HTLV-1 can infect a wide variety of human cell types in vitro (Koyanagi et al. 1993; Sommerfelt et al. 1988), and its presumed receptor is therefore thought to be a widely expressed molecule. Glucose transporter-1 (GLUT1), heparin sulfate proteoglycan (HSPG), and neuropilin-1 have been reported to be involved in the interaction between the viral envelop and the host cell membrane, and for viral entry into the target cells (Jones et al. 2005; Lambert et al. 2009; Manel et al. 2003). The current model postulates that HTLV-1 particles first come into contact with HSPG, followed by recruitment of the HTLV-1/HSPG complex by neuropilin-1, finally interacting with GLUT1. Formation of the HSPG/neuropilin-1/GLUT1 complex appears to be essential for the fusion of the viral envelope and host cell membrane and viral entry.

Cell-free HTLV-1 virions are poorly infectious in vitro for most of cell types, including their primary target cells, CD4 T-cells. Direct cell-to-cell contact appears to be essential for HTLV-1 infection, except for myeloid and plasmacytoid dendritic cells (DCs), which appear to be susceptible to infection by cell-free HTLV-1 virions (Jones et al. 2008). DCs may therefore play an important role in transmission, possibly facilitating spread during contact between breast milk and the infant's gastrointestinal mucosa.

Three major mechanisms of cell-to-cell transmission of HTLV-1 have been proposed: (1) HTLV-1-infection of lymphocytes results in polarization of their microtubules and viral components upon contact with other T-cells, forming a so-called virological synapse (Igakura et al. 2003); (2) HTLV-1-infected cells produce and transiently store viral particles in extracellular adhesive structures rich in extracellular matrix components, including collagen and agrin, and cellular linker proteins, such as tetherin and galectin-3, similar to bacterial biofilms. Extracellular viral assemblies then rapidly adhere to other cells upon contact, allowing viral spread and infection of target cells (Pais-Correia et al. 2010); and (3) the HTLV-1–pX region-encoded p8 protein increases T-cell conjugation through lymphocyte function-associated antigen-1 clustering. In addition, p8 induces cellular conduits among T-cells and increases viral transmission (Van Prooyen et al. 2010).

6 Prevention of Transmission of HTLV-1

The prognosis for ATLL remains one of the worst among hematological malignancies, even with the best available therapies, and no preventive vaccine against HTLV-1 is currently available. Prevention of transmission of HTLV-1 is therefore an important strategy in preventing ATLL.

6.1 Prevention of Vertical Transmission

Based on retrospective and prospective epidemiological studies, the mother-to-child transmission rate is estimated to be 20 % (Hino et al. 1985). Prevention of mother-to-child transmission by restricting breast-feeding has the most significant impact on the incidence of HTLV-1 infection and associated diseases. In a prefecture-wide intervention study in Nagasaki, southern Japan, in which mothers with HTLV-1 infection were counseled to avoid breast-feeding, there was a marked reduction of mother-to-child transmission from 20.3 to 2.5 %. Thus, prenatal screening for HTLV-1 may be an important public health strategy in endemic areas, in conjunction with counseling of mothers with HTLV-1 infection to avoid breast-feeding. Although children breast-fed for less than 6 months have significantly lower incidence of HTLV-1 infection than those breast-fed for more than 6 months, the risk of transmission is significantly higher for the former group compared with formula-fed infants (Hino 2011).

Even with exclusive bottle-feeding, 2.5 % of infants born to carrier mothers become infected with HTLV-1. As intrauterine transmission of HTLV-1 is rare, transplacental transmission during delivery seems to be the probable mode of transmission, as has been reported for hepatitis B and hepatitis C viruses.

Although exclusive formula-feeding reduces the risk of mother-to-child transmission of HTLV-1, risk of malnutrition is a significant concern in developing countries, where malnutrition remains a significant contributor to infant mortality.

6.2 Prevention of Horizontal Transmission

HTLV-1 can also spread through contact with body fluids, whole blood or blood components. As ATLL is associated with prolonged infection acquired during vertical transmission, and with infection through blood transfusions, the purpose of the prevention of horizontal transmission is mainly to reduce the general pool of HTLV-1 carriers.

6.3 Transfusion and Sexual Transmission

HTLV-1 screening programs aimed at preventing transfusion-related transmission of HTLV-1 through systematic screening of all blood donors as a public health control measure have been implemented in many endemic areas, since 1986 (Inaba et al. 1989; Osame et al. 1990). Restricting breast-feeding and blood donor screening resulted in a decrease in HTLV-1 carriers from 2.79 to 0.44 % in Kagoshima Prefecture, southern Japan (Table 1). In HTLV-1 non-endemic areas, reports indicate that HTLV-1 infection may be concentrated in select donor populations, especially among immigrants from endemic areas. For developing countries, the cost of imported screening test kits may be prohibitive, necessitating the development of more cost-effective tools and programs for blood donor screening. In most African countries, transfusion remains a significant contributor to HTLV-1 transmission.

Table 1 Changes in the number of HTLV-1 carrier among blood transfusion donor at Kagoshima prefecture, southern part of Japan (1999–2008)

Year	The number of blood transfusion donor	The number of HTLV-1 carrier	%
1999	98,644	2,751	2.79
2000	91,456	1,368	1.50
2001	92,281	1,048	1.14
2002	89,458	827	0.92
2003	86,000	686	0.80
2004	82,310	565	0.69
2005	73,792	435	0.59
2006	69,133	388	0.56
2007	69,741	360	0.52
2008	71,226	313	0.44

Sexual transmission of HTLV-1 is primarily due to the transmission from men to women. Recommendations to prevent sexually transmitted infections should be emphasized, including condom use and avoiding multiple and anonymous sexual partners. Access to accurate information about HTLV-1 infection and appropriate counseling are important preventive strategies, as blood donor candidates and sexually active persons are usually asymptomatic and are primarily of reproductive age.

7   Development of Adult T-cell leukemia lymphoma

7.1   Pathogenesis of Adult T-Cell Leukemia/Lymphoma

The pathogenesis of ATLL is not completely understood. Extensive studies have revealed that HTLV-1 transactivator/transcriptional activator (Tax) plays a critical role in the transformation of virus-infected cells. Tax is thought to be a potent oncoprotein, as it results in immortalization of human primary T-cells and Tax transgenic mice malignancy. Tax enhances viral replication through transactivation of the viral promoter, the $5'$ long tandem repeat (LTR), results in activation of the nuclear factor kappa-B (NF-kB) pathway, interferes with cell cycle regulators, induces aneuploidy and DNA damage, and impairs DNA repair. Thus, Tax is thought to play a key role in the pathogenesis of ATLL (Matsuoka and Jeang 2011).

HTLV-1 bZIP factor (HBZ) is coded for by the minus strand of the HTLV-1 provirus and can be found in all ATLL cells (Satou et al. 2006). HBZ protein was originally reported to suppress Tax-mediated viral transcription; however, HBZ RNA has also been shown to promote cell proliferation. Importantly, HBZ transgenic mice developed CD4/forkhead box protein-3 (Foxp3)-positive T-cell lymphoma, resembling the immunophenotype and clinical features of human ATLL. These findings suggest that HBZ is a critical factor in leukemogenesis. The proposed model for the interplay between Tax and HBZ is that Tax is needed to initiate the transformation of HTLV-1-infected cells, while HBZ is required to maintain the transformed phenotype in ATLL (Matsuoka and Jeang 2011).

7.2   Determinants of Progression from Asymptomatic Carrier Status to ATLL

The determinants of ATLL progression in HTLV-1 carriers have been investigated in many epidemiological and clinical studies. In Japanese cohorts, the average age at diagnosis is about 65 years (Yamada et al. 2011), significantly greater than in the Jamaican cohort, who present in their mid-forties, suggesting that other host and environmental factors may also be involved in ATLL pathogenesis (Hanchard 1996). The age at the time of HTLV-1 infection is also a critical factor in ATLL development, as ATLL rarely develops in HTLV-1 carriers who acquired infection through horizontal transmission. Several studies have examined host genetic factors, including HLA haplotypes, due to the observation that patients with ATLL were more likely to have a family history of ATLL when compared with the general population. The frequency of HLA-A*26, HLA-B*4002, HLA-B*4006, and HLA-B*4801 alleles was significantly higher in ATLL patients than in HTLV-1 asymptomatic carriers in Japan (Yashiki et al. 2001). In the Miyazaki cohort, HTLV-1 carriers with a higher anti-HTLV-1 titer and lower anti-Tax reactivity were at greatest risk of developing ATLL (Hisada et al. 1998), and higher HTLV-1 proviral load was a significant risk factor for progression from asymptomatic HTLV-1 carrier status to ATLL. A nationwide prospective study of HTLV-1 carriers in Japan was initiated to identify the determinants of ATLL development. Fourteen subjects out of 1,218 asymptomatic carriers developed ATLL, and all of the 14 subjects had higher baseline proviral loads, whereas there were no cases of ATLL among those with a baseline proviral load of less than 4 copies/100 peripheral blood mononuclear cells (Iwanaga et al. 2010).

8 Prognosis for Patients with Adult T-Cell Leukemia/ Lymphoma

8.1 Acute and Lymphoma Sub-Types

The prognosis for patients with acute and lymphoma subtypes of ATLL remains poor, even with chemotherapy or allogeneic hematopoietic stem cell transplantation (alloHSCT). With currently best available chemotherapy in one series (Tsukasaki et al. 2007), the rate of complete response (CR) was 40 % and overall survival (OS) at 3 years was 24 %. The median survival time (MST) is 13 months.

8.2 Chronic and Smoldering Sub-Types

In a previous study, in which Japanese patients with ATLL were followed for a total duration of 7 years, the 4-year survival rates for chronic and smoldering sub-types were 26.9 and 62.8 %, respectively, with a MST of 24.3 months for the chronic sub-type (Shimoyama 1991). Therefore, the chronic and smoldering subtypes of ATLL are characterized by an indolent clinical course and are usually managed by observation or "watchful waiting" until disease progression to acute crisis, which is similar to the approach to the management of chronic lymphoid leukemia or smoldering myeloma. However, a recent report with long-term follow-up of these indolent sub-types of ATLL (chronic and smoldering) revealed that the MST was 4.1 years and the estimated 5-, 10-, and 15-year survival rates were 47.2, 25.4, and 14.1 %, respectively (Takasaki et al. 2010), which were poorer than expected. These findings suggest that even patients with indolent forms of ATLL should be carefully observed in clinical practice, and further research is needed to improve the management of these patients.

9 Current Treatment Options

9.1 Conventional Chemotherapy

The results of a phase III randomized control trial suggest that the vincristine, cyclophosphamide, doxorubicin, and prednisone (VCAP); doxorubicin, ranimustine, and prednisone (AMP); and vindesine, etoposide, carboplatin, and prednisone (VECP) regimens show no benefit over biweekly cyclophosphamide, doxorubicin, vincristine, and prednisone (CHOP) in newly diagnosed acute, lymphoma, or unfavorable chronic subtypes of ATLL in terms of OS, primary study endpoint, or progression-free survival (Tsukasaki et al. 2007). However, the rate of CR was higher in the VCAP-AMP-VECP arm than the biweekly CHOP arm (40 vs 25 %, respectively; $P = 0.020$). OS at 3 years was 24 % in the VCAP-AMP-VECP arm and 13 % in the CHOP arm ($P = 0.085$). Nonetheless, the MST of 13 months still compares unfavorably to other hematological malignancies.

9.2 Allogeneic Hematopoietic Stem Cell Transplantation

Allogeneic HSCT (alloHSCT) has been explored as a promising alternative therapeutic modality that can provide long-term remission in a proportion of patients with ATLL (Choi et al. 2011; Hishizawa et al. 2010; Utsunomiya et al. 2001). In a recent large nationwide retrospective analysis, investigators compared outcomes of 386 patients with ATL who underwent alloHSCT. After a median follow-up of 41 months, 3-year OS for the entire cohort was 33 % (Hishizawa et al. 2010). Another retrospective study based on 294 ATLL patients who received alloHSCT revealed that the development of mild-to-moderate acute GVHD confers a lower risk of disease progression and a beneficial influence on survival (Kanda et al. 2012), which is indicative of a graft-versus-ATLL effect. Another large retrospective analysis of alloHSCT for ATLL ($n = 586$) in Japan revealed no significant difference in OS between myeloablative conditioning (MAC) and reduced intensity conditioning (RIC). There was a tendency toward better OS in older patients receiving RIC (Ishida et al. 2012). The number of ATLL patients eligible for allogeneic transplantation is few because of older age at presentation and the low rate of CR. Selection criteria for alloHSCT for patients with ATLL remain to be determined.

9.3 Interferon-α (IFN-α) and Zidovudine (AZT)

Results of a recent meta-analysis on the use of AZT/IFN for 254 ATLL patients worldwide showed that the treatment of ATLL patients with AZT and IFN resulted in better response and prolonged OS (Bazarbachi et al. 2010). Two hundred and seven patients received AZT/IFN therapy. In these patients, 5-year OS rates were

46 % for 75 patients who received antiviral therapy ($P = 0.004$). In acute ATLL, achievement of complete remission with antiviral therapy resulted in 82 % 5-year survival. These results suggest that the treatment of ATLL using AZT/IFN results in high response and CR rates except for lymphoma type of ATLL, resulting in prolonged survival in a significant proportion of patients. Although this is a retrospective analysis, the results seem to be promising, and further studies comparing AZT/IFN-α and conventional chemotherapy or alloHSCT are warranted.

9.4 Prevention of ATLL

The prevention of ATLL mostly relies on the prevention of HTLV-1 transmission as previously described. Another strategy could be the prevention of ATLL development among HTLV-1 carriers. Despite the prolonged carrier status before ATLL development, there are no interventions exploiting this window of opportunity to treat ATLL. This is partly because only approximately 10 % of HTLV-1 carriers develop HTLV-1-related disease in their lifetime. Careful risk–benefit analysis including the acceptability of side effects during interventions is needed.

9.5 Future Directions for the Prevention of ATLL

9.5.1 Immunological Impairment of HTLV-1-Specific T-Cells

Vertical transmission, high proviral loads, and suppression of HTLV-1-specific T-cell immune responses are important risk factors for ATLL development. It has been reported that Tax-specific cytotoxic T lymphocytes (CTLs) detected in chronic and smoldering ATLL and a subset of asymptomatic carriers are anergic to antigen stimulation (Takamori et al. 2011). Such functional impairment of CTLs seems specific to HTLV-1, as cytomegalovirus-specific CTLs, for example, remain intact.

In animal models, oral inoculation of HTLV-1 virions induces T-cell tolerance against HTLV-1 (Hasegawa et al. 2003). As breast-feeding is the main route of vertical transmission in HTLV-1 infection, this may induce neonatal T-cell tolerance against HTLV-1.

In addition to immunological tolerance, T-cell exhaustion may be another mechanism of antigen-specific T-cell suppression. We have reported on the upregulation of programmed death-1 (PD-1) expression on Tax-specific CTLs, suggesting Tax-specific T-cell exhaustion (Kozako et al. 2009).

9.5.2 Vaccine

Vaccination of uninfected individuals against HTLV-1 is not a sophisticated feasible strategy for the prevention of ATLL, as ATLL develops after a long latency period in individuals vertically transmitted HTLV-1 carriers within the first 6 months of life, and vertical transmission is almost completely prevented by avoiding breast-feeding. Thus, the purpose of vaccination should be to augment HTLV-1-specific T-cell responses in asymptomatic carriers, enhancing clearance of infected and transformed cells, thereby protecting against ATLL.

HTLV-1 Tax-targeted vaccines in a rat model of HTLV-1-induced lymphomas showed promising antitumor effects (Ohashi et al. 2000). In addition, HTLV-1-immunized monkeys developed a strong cellular immune response with HTLV-1-derived peptide vaccines, and a significant reduction in HTLV-1 proviral load was observed in these immunized monkeys after challenge (Kazanji et al. 2006). Therefore, these results provide the scientific rationale for clinical use of such a vaccine for preventing ATLL. There remain, however, several obstacles to be overcome before clinical application can be realized. HTLV-1 synthetic peptides are poorly immunogenic, with inefficient induction of antigen-specific CTLs. We have shown in previous reports that oligomannose-coated liposomes (OMLs) encapsulating the HLA-A*0201-restricted HTLV-1 Tax-epitope (OML/Tax) resulted in the efficient induction of HTLV-1-specific T-cell responses (Kozako et al. 2011). Further, immunization of HLA-A*0201 transgenic mice with OML/Tax resulted in the efficient induction of HTLV-1-specific IFN-γ producing T-cells, and DCs exposed to OML/Tax showed increased expression of DC maturation markers. In addition, HTLV-1-Tax-specific CD8+ T-cells were efficiently induced by OML/Tax derived from HTLV-1 carriers ex vivo. OML/Tax increased the number of HTLV-1-specific CD8+ T-cells by an average 170-fold. Furthermore, these HTLV-1-specific CD8+ cells efficiently lysed HTLV-1 epitope peptide-pulsed T2-A2 cells. These results suggest that OML/Tax induces antigen-specific cellular immune responses without the need for adjuvants and may be an effective vaccine candidate to reduce progression to ATLL.

Better prognosticators would help identify individuals most at risk for progression to ATLL, allowing us to limit the exposure of lower-risk individuals to unwanted immunological responses to vaccination, including autoimmune-like conditions such as HAM/TSP.

10 Conclusion

To date, restricting breast-feeding by mothers with HTLV-1 infection has been the mainstay of HTLV-1 prevention thereby extending ATLL. Antenatal screening for HTLV-1 should be implemented in the endemic areas, with provision of accurate information and counseling. In addition, screening of blood donor candidates has been shown to be effective in preventing HTLV-1 transmission. Recommendations to prevent sexual transmission should be emphasized, including condom use and adopting safe sexual behavior. The development of an effective and safe vaccine could be an important tool in protecting HTLV-1-infected carriers against ATLL.

References

Bazarbachi A, Plumelle Y, Carlos Ramos J, Tortevoye P, Otrock Z, Taylor G, Gessain A, Harrington W, Panelatti G, Hermine O (2010) Meta-analysis on the use of zidovudine and interferon-alfa in adult T-cell leukemia/lymphoma showing improved survival in the leukemic subtypes. J Clin Oncol 28:4177–4183

Chen YC, Wang CH, Su IJ, Hu CY, Chou MJ, Lee TH, Lin DT, Chung TY, Liu CH, Yang CS (1989) Infection of human T-cell leukemia virus type I and development of human T-cell leukemia lymphoma in patients with hematologic neoplasms: a possible linkage to blood transfusion. Blood 74:388–394

Choi I, Tanosaki R, Uike N, Utsunomiya A, Tomonaga M, Harada M, Yamanaka T, Kannagi M, Okamura J (2011) Long-term outcomes after hematopoietic SCT for adult T-cell leukemia/ lymphoma: results of prospective trials. Bone Marrow Transplant 46:116–118

de The G, Kazanji M (1996) An HTLV-I/II vaccine: from animal models to clinical trials? J Acquir Immune Defic Syndr Hum Retrovirol 13(Suppl 1):S191–S198

Fujino T, Nagata Y (2000) HTLV-I transmission from mother to child. J Reprod Immunol 47:197–206

Gibbs WN, Lofters WS, Campbell M, Hanchard B, LaGrenade L, Cranston B, Hendriks J, Jaffe ES, Saxinger C, Robert-Guroff M et al (1987) Non-hodgkin lymphoma in Jamaica and its relation to adult T-cell leukemia-lymphoma. Ann Intern Med 106:361–368

Goncalves DU, Proietti FA, Ribas JG, Araujo MG, Pinheiro SR, Guedes AC, Carneiro-Proietti AB (2010) Epidemiology, treatment, and prevention of human T-cell leukemia virus type 1- associated diseases. Clin Microbiol Rev 23:577–589

Hanchard B (1996) Adult T-cell leukemia/lymphoma in Jamaica: 1986–1995. J Acquir Immune Defic Syndr Hum Retrovirol 13(Suppl 1):S20–S25

Hasegawa A, Ohashi T, Hanabuchi S, Kato H, Takemura F, Masuda T, Kannagi M (2003) Expansion of human T-cell leukemia virus type 1 (HTLV-1) reservoir in orally infected rats: inverse correlation with HTLV-1-specific cellular immune response. J Virol 77:2956–2963

Hino S (2011) Establishment of the milk-borne transmission as a key factor for the peculiar endemicity of human T-lymphotropic virus type 1 (HTLV-1): the ATL prevention program Nagasaki. Proc Jpn Acad Ser B Phys Biol Sci 87:152–166

Hino S, Yamaguchi K, Katamine S, Sugiyama H, Amagasaki T, Kinoshita K, Yoshida Y, Doi H, Tsuji Y, Miyamoto T (1985) Mother-to-child transmission of human T-cell leukemia virus type-I. Jpn J Cancer Res 76:474–480

Hisada M, Okayama A, Shioiri S, Spiegelman DL, Stuver SO, Mueller NE (1998) Risk factors for adult T-cell leukemia among carriers of human T-lymphotropic virus type I. Blood 92:3557–3561

Hishizawa M, Kanda J, Utsunomiya A, Taniguchi S, Eto T, Moriuchi Y, Tanosaki R, Kawano F, Miyazaki Y, Masuda M et al (2010) Transplantation of allogeneic hematopoietic stem cells for adult T-cell leukemia: a nationwide retrospective study. Blood 116:1369–1376

Igakura T, Stinchcombe JC, Goon PK, Taylor GP, Weber JN, Griffiths GM, Tanaka Y, Osame M, Bangham CR (2003) Spread of HTLV-I between lymphocytes by virus-induced polarization of the cytoskeleton. Science 299:1713–1716

Inaba S, Sato H, Okochi K, Fukada K, Takakura F, Tokunaga K, Kiyokawa H, Maeda Y (1989) Prevention of transmission of human T-lymphotropic virus type 1 (HTLV-1) through transfusion, by donor screening with antibody to the virus. One-year experience. Transfusion 29:7–11

Ishida T, Hishizawa M, Kato K, Tanosaki R, Fukuda T, Taniguchi S, Eto T, Takatsuka Y, Miyazaki Y, Moriuchi Y et al (2012) Allogeneic hematopoietic stem cell transplantation for adult T-cell leukemia-lymphoma with special emphasis on preconditioning regimen: a nationwide retrospective study. Blood 120:1734–1741

Iwanaga M, Watanabe T, Utsunomiya A, Okayama A, Uchimaru K, Koh KR, Ogata M, Kikuchi H, Sagara Y, Uozumi K et al (2010) Human T-cell leukemia virus type I (HTLV-1) proviral load and disease progression in asymptomatic HTLV-1 carriers: a nationwide prospective study in Japan. Blood 116:1211–1219

Jones KS, Petrow-Sadowski C, Bertolette DC, Huang Y, Ruscetti FW (2005) Heparan sulfate proteoglycans mediate attachment and entry of human T-cell leukemia virus type 1 virions into CD4+ T cells. J Virol 79:12692–12702

Jones KS, Petrow-Sadowski C, Huang YK, Bertolette DC, Ruscetti FW (2008) Cell-free HTLV-1 infects dendritic cells leading to transmission and transformation of CD4(+) T cells. Nat Med 14:429–436

Kanda J, Hishizawa M, Utsunomiya A, Taniguchi S, Eto T, Moriuchi Y, Tanosaki R, Kawano F, Miyazaki Y, Masuda M et al (2012) Impact of graft-versus-host disease on outcomes after allogeneic hematopoietic cell transplantation for adult T-cell leukemia: a retrospective cohort study. Blood 119:2141–2148

Kazanji M, Heraud JM, Merien F, Pique C, de The G, Gessain A, Jacobson S (2006) Chimeric peptide vaccine composed of B- and T-cell epitopes of human T-cell leukemia virus type 1 induces humoral and cellular immune responses and reduces the proviral load in immunized squirrel monkeys (*Saimiri sciureus*). J Gen Virol 87:1331–1337

Kinoshita K, Amagasaki T, Hino S, Doi H, Yamanouchi K, Ban N, Momita S, Ikeda S, Kamihira S, Ichimaru M et al (1987) Milk-borne transmission of HTLV-I from carrier mothers to their children. Jpn J Cancer Res 78:674–680

Koga Y, Iwanaga M, Soda M, Inokuchi N, Sasaki D, Hasegawa H, Yanagihara K, Yamaguchi K, Kamihira S, Yamada Y (2010) Trends in HTLV-1 prevalence and incidence of adult T-cell leukemia/lymphoma in Nagasaki, Japan. J Med Virol 82:668–674

Kondo T, Kono H, Miyamoto N, Yoshida R, Toki H, Matsumoto I, Hara M, Inoue H, Inatsuki A, Funatsu T et al (1989) Age- and sex-specific cumulative rate and risk of ATLL for HTLV-I carriers. Int J Cancer 43:1061–1064

Koyanagi Y, Itoyama Y, Nakamura N, Takamatsu K, Kira J, Iwamasa T, Goto I, Yamamoto N (1993) In vivo infection of human T-cell leukemia virus type I in non-T cells. Virology 196:25–33

Kozako T, Yoshimitsu M, Fujiwara H, Masamoto I, Horai S, White Y, Akimoto M, Suzuki S, Matsushita K, Uozumi K et al (2009) PD-1/PD-L1 expression in human T-cell leukemia virus type 1 carriers and adult T-cell leukemia/lymphoma patients. Leukemia 23:375–382

Kozako T, Hirata S, Shimizu Y, Satoh Y, Yoshimitsu M, White Y, Lemonnier F, Shimeno H, Soeda S, Arima N (2011) Oligomannose-coated liposomes efficiently induce human T-cell leukemia virus-1-specific cytotoxic T lymphocytes without adjuvant. FEBS J 278:1358–1366

Lambert S, Bouttier M, Vassy R, Seigneuret M, Petrow-Sadowski C, Janvier S, Heveker N, Ruscetti FW, Perret G, Jones KS et al (2009) HTLV-1 uses HSPG and neuropilin-1 for entry by molecular mimicry of VEGF165. Blood 113:5176–5185

Manel N, Kim FJ, Kinet S, Taylor N, Sitbon M, Battini JL (2003) The ubiquitous glucose transporter GLUT-1 is a receptor for HTLV. Cell 115:449–459

Matsuoka M, Jeang KT (2011) Human T-cell leukemia virus type 1 (HTLV-1) and leukemic transformation: viral infectivity, Tax, HBZ and therapy. Oncogene 30:1379–1389

Murphy EL, Figueroa JP, Gibbs WN, Brathwaite A, Holding-Cobham M, Waters D, Cranston B, Hanchard B, Blattner WA (1989) Sexual transmission of human T-lymphotropic virus type I (HTLV-I). Ann Intern Med 111:555–560

Ohashi T, Hanabuchi S, Kato H, Tateno H, Takemura F, Tsukahara T, Koya Y, Hasegawa A, Masuda T, Kannagi M (2000) Prevention of adult T-cell leukemia-like lymphoproliferative disease in rats by adoptively transferred T cells from a donor immunized with human T-cell leukemia virus type 1 Tax-coding DNA vaccine. J Virol 74:9610–9616

Okochi K, Sato H (1984) Transmission of ATLV (HTLV-I) through blood transfusion. Princess Takamatsu Symp 15:129–135

Osame M, Janssen R, Kubota H, Nishitani H, Igata A, Nagataki S, Mori M, Goto I, Shimabukuro H, Khabbaz R et al (1990) Nationwide survey of HTLV-I-associated myelopathy in Japan: association with blood transfusion. Ann Neurol 28:50–56

Pais-Correia AM, Sachse M, Guadagnini S, Robbiati V, Lasserre R, Gessain A, Gout O, Alcover A, Thoulouze MI (2010) Biofilm-like extracellular viral assemblies mediate HTLV-1 cell-to-cell transmission at virological synapses. Nat Med 16:83–89

Poiesz BJ, Ruscetti FW, Gazdar AF, Bunn PA, Minna JD, Gallo RC (1980) Detection and isolation of type C retrovirus particles from fresh and cultured lymphocytes of a patient with cutaneous T-cell lymphoma. Proc Natl Acad Sci USA 77:7415–7419

Pombo de Oliveira MS, Matutes E, Schulz T, Carvalho SM, Noronha H, Reaves JD, Loureiro P, Machado C, Catovsky D (1995) T-cell malignancies in Brazil. Clinico-pathological and molecular studies of HTLV-I-positive and -negative cases. Int J Cancer 60:823–827

Satake M, Yamaguchi K, Tadokoro K (2012) Current prevalence of HTLV-1 in Japan as determined by screening of blood donors. J Med Virol 84:327–335

Satou Y, Yasunaga J, Yoshida M, Matsuoka M (2006) HTLV-I basic leucine zipper factor gene mRNA supports proliferation of adult T cell leukemia cells. Proc Natl Acad Sci USA 103:720–725

Shimoyama M (1991) Diagnostic criteria and classification of clinical subtypes of adult T-cell leukaemia-lymphoma. A report from the Lymphoma Study Group (1984–87). Br J Haematol 79:428–437

Sommerfelt MA, Williams BP, Clapham PR, Solomon E, Goodfellow PN, Weiss RA (1988) Human T cell leukemia viruses use a receptor determined by human chromosome 17. Science 242:1557–1559

Sonoda S, Li HC, Tajima K (2011) Ethnoepidemiology of HTLV-1 related diseases: ethnic determinants of HTLV-1 susceptibility and its worldwide dispersal. Cancer Sci 102:295–301

Tajima K, Tominaga S, Suchi T, Kawagoe T, Komoda H, Hinuma Y, Oda T, Fujita K (1982) Epidemiological analysis of the distribution of antibody to adult T-cell leukemia-virus-associated antigen: possible horizontal transmission of adult T-cell leukemia virus. Gann 73:893–901

Takamori A, Hasegawa A, Utsunomiya A, Maeda Y, Yamano Y, Masuda M, Shimizu Y, Tamai Y, Sasada A, Zeng N et al (2011) Functional impairment of Tax-specific but not cytomegalovirus-specific CD8+ T lymphocytes in a minor population of asymptomatic human T-cell leukemia virus type 1-carriers. Retrovirology 8:100

Takasaki Y, Iwanaga M, Imaizumi Y, Tawara M, Joh T, Kohno T, Yamada Y, Kamihira S, Ikeda S, Miyazaki Y et al (2010) Long-term study of indolent adult T-cell leukemia-lymphoma. Blood 115:4337–4343

Takatsuki K, Matsuoka M, Yamaguchi K (1996) Adult T-cell leukemia in Japan. J Acquir Immune Defic Syndr Hum Retrovirol 13(Suppl 1):S15–S19

Tokudome S, Tokunaga O, Shimamoto Y, Miyamoto Y, Sumida I, Kikuchi M, Takeshita M, Ikeda T, Fujiwara K, Yoshihara M et al (1989) Incidence of adult T-cell leukemia/lymphoma among human T-lymphotropic virus type I carriers in Saga, Japan. Cancer Res 49:226–228

Tsukasaki K, Utsunomiya A, Fukuda H, Shibata T, Fukushima T, Takatsuka Y, Ikeda S, Masuda M, Nagoshi H, Ueda R et al (2007) VCAP-AMP-VECP compared with biweekly CHOP for adult T-cell leukemia-lymphoma: Japan Clinical Oncology Group Study JCOG9801. J Clin Oncol 25:5458–5464

Utsunomiya A, Miyazaki Y, Takatsuka Y, Hanada S, Uozumi K, Yashiki S, Tara M, Kawano F, Saburi Y, Kikuchi H et al (2001) Improved outcome of adult T cell leukemia/lymphoma with allogeneic hematopoietic stem cell transplantation. Bone Marrow Transplant 27:15–20

Van Prooyen N, Gold H, Andresen V, Schwartz O, Jones K, Ruscetti F, Lockett S, Gudla P, Venzon D, Franchini G (2010) Human T-cell leukemia virus type 1 p8 protein increases cellular conduits and virus transmission. Proc Natl Acad Sci USA 107:20738–20743

Yamada Y, Atogami S, Hasegawa H, Kamihira S, Soda M, Satake M, Yamaguchi K (2011) Nationwide survey of adult T-cell leukemia/lymphoma (ATL) in Japan. Rinsho Ketsueki 52:1765–1771

Yamanouchi K, Kinoshita K, Moriuchi R, Katamine S, Amagasaki T, Ikeda S, Ichimaru M, Miyamoto T, Hino S (1985) Oral transmission of human T-cell leukemia virus type-I into a common marmoset (*Callithrix jacchus*) as an experimental model for milk-borne transmission. Jpn J Cancer Res 76:481–487

Yashiki S, Fujiyoshi T, Arima N, Osame M, Yoshinaga M, Nagata Y, Tara M, Nomura K, Utsunomiya A, Hanada S et al (2001) HLA-A*26, HLA-B*4002, HLA-B*4006, and HLA-B*4801 alleles predispose to adult T cell leukemia: the limited recognition of HTLV type 1 tax peptide anchor motifs and epitopes to generate anti-HTLV type 1 tax CD8(+) cytotoxic T lymphocytes. AIDS Res Hum Retroviruses 17:1047–1061

Yoshida M, Miyoshi I, Hinuma Y (1982) Isolation and characterization of retrovirus from cell lines of human adult T-cell leukemia and its implication in the disease. Proc Natl Acad Sci USA 79:2031–2035

Molecular Biology of Human Herpesvirus 8: Novel Functions and Virus–Host Interactions Implicated in Viral Pathogenesis and Replication

Emily Cousins and John Nicholas

Abstract

Human herpesvirus 8 (HHV-8), also known as Kaposi's sarcoma-associated herpesvirus (KSHV), is the second identified human gammaherpesvirus. Like its relative Epstein-Barr virus, HHV-8 is linked to B-cell tumors, specifically primary effusion lymphoma and multicentric Castleman's disease, in addition to endothelial-derived KS. HHV-8 is unusual in its possession of a plethora of "accessory" genes and encoded proteins in addition to the core, conserved herpesvirus and gammaherpesvirus genes that are necessary for basic biological functions of these viruses. The HHV-8 accessory proteins specify not only activities deducible from their cellular protein homologies but also novel, unsuspected activities that have revealed new mechanisms of virus–host interaction that serve virus replication or latency and may contribute to the development and progression of virus-associated neoplasia. These proteins include viral interleukin-6 (vIL-6), viral chemokines (vCCLs), viral G protein–coupled receptor (vGPCR), viral interferon regulatory factors (vIRFs), and viral antiapoptotic proteins homologous to FLICE (FADD-like IL-1β converting enzyme)-inhibitory protein (FLIP) and survivin. Other HHV-8 proteins, such as signaling membrane receptors encoded by open reading frames K1 and K15, also interact with host mechanisms in unique ways and have been implicated in viral pathogenesis. Additionally, a set of micro-RNAs encoded by HHV-8 appear to modulate expression of multiple host proteins to provide conditions

E. Cousins · J. Nicholas (✉)
Department of Oncology, Sidney Kimmel Comprehensive Cancer Center at Johns Hopkins,
1650 Orleans Street, Baltimore, MD 21287, USA
e-mail: nichojo@jhmi.edu

E. Cousins
e-mail: ecousins@jhmi.edu

M. H. Chang and K.-T. Jeang (eds.), *Viruses and Human Cancer*,   227
Recent Results in Cancer Research 193, DOI: 10.1007/978-3-642-38965-8_13,
© Springer-Verlag Berlin Heidelberg 2014

conducive to virus persistence within the host and could also contribute to HHV-8-induced neoplasia. Here, we review the molecular biology underlying these novel virus–host interactions and their potential roles in both virus biology and virus-associated disease.

Contents

1   Introduction

Human herpesvirus 8 (HHV-8) is classified as a gamma-2 herpesvirus and is related to Epstein-Barr virus (EBV), a member of the gamma-1 subfamily. One important aspect of the gammaherpesviruses is their association with neoplasia, either naturally or in animal model systems. HHV-8 is associated with B-cell–derived primary effusion lymphoma (PEL) and multicentric Castleman's disease (MCD) as well as endothelial-derived Kaposi's sarcoma (KS) (Arvanitakis et al. 1996; Carbone et al. 2000; Chang and Moore 1996; Gaidano et al. 1997). EBV is associated with a number of B-cell malignancies, such as Burkitt's lymphoma, Hodgkin's lymphoma, and posttransplant lymphoproliferative disease, in addition to epithelial nasopharyngeal and gastric carcinomas, T-cell lymphoma, and muscle tumors (Kawa 2000; Okano 2000; Young and Murray 2003). Despite the similarities between the viruses and their associated malignancies, the particular protein functions and activities involved in the relevant aspects of virus biology and neoplasia appear to be quite distinct. Indeed, HHV-8 specifies a number of proteins that had not previously been identified in gammaherpesviruses, herpesviruses, or even viruses in general, and these proteins are believed to play vital functions in virus biology and to be centrally involved in viral pathogenesis.

One such gene is viral interleukin-6 (vIL-6), which was immediately upon its discovery implicated as a candidate contributor to HHV-8 pathogenesis (Moore et al. 1996; Neipel et al. 1997a; Nicholas et al. 1997). Previous reports had indicated that IL-6 was produced by and supported the growth of KS cells, promoted inflammation and angiogenesis typical of KS, served as an important B-cell growth factor, and was found at elevated levels in MCD patient sera (Burger et al. 1994; Ishiyama et al. 1994; Miles et al. 1990; Roth 1991; Yoshizaki et al. 1989). Similarly, the discovery of viral chemokines, vCCLs 1-3, and demonstration of their pro-angiogenic activities in experimental systems suggested that these proteins also could contribute to disease, in addition to their suspected roles in immune evasion during HHV-8 productive replication (Boshoff et al. 1997; Stine et al. 2000). The chemokine receptor homologue, vGPCR, was found to induce angiogenic cellular cytokines of the type produced in and suspected to promote the growth of KS lesions (Cannon et al. 2003; Pati et al. 2001; Schwarz and Murphy 2001). The constitutively active membrane receptors encoded by HHV-8 open reading frames (ORFs) K1 and K15 could function similarly (Brinkmann et al. 2007; Caselli et al. 2007; Samaniego et al. 2001; Wang et al. 2006). vGPCR and K1 also acted as oncogenes, promoting cell transformation and inducing tumorigenesis in animal models (Bais et al. 1998; Lee et al. 1998b; Yang et al. 2000). However, like the v-cytokines, vGPCR and K1 are expressed predominantly or exclusively during productive, lytic replication; therefore, any contributions to malignant pathogenesis are likely to be mediated through paracrine signaling. There is ample evidence that cytokine-mediated paracrine signal transduction plays a role in KS, and B-cell growth can also be influenced by this route, as discussed below. Apart from the likely involvement of these viral proteins in HHV-8-associated pathogenesis, the roles of some of these "unique" viral products in virus biology are only beginning to be appreciated. For example, the pro-survival signaling induced by vCCLs and vGPCR and the antiapoptotic activities of vIRF-1 have been demonstrated to enhance productive replication. Therefore, functions that serve normal virus biology, such as inhibiting infection-induced apoptosis, may have the "side effect" of promoting virus-associated neoplasia. This concept is familiar to viral oncologists, but the precise mechanisms deployed by HHV-8 are novel.

Classical oncogene and tumor suppressor activities are mediated in an autocrine fashion, and viral genes expressed during latency are potential contributors to malignant disease. Chief among these for HHV-8 is the latency-associated nuclear antigen, LANA, which specifies essential replication and genome segregation activities in dividing cells and impacts several host pathways to promote cell survival and proliferation (Verma et al. 2007). These activities have obvious connections to processes involved in malignant transformation. Likewise, the viral homologue of cellular FLICE-inhibitory protein, vFLIP, is both latently expressed and crucially important for maintaining cell viability. vFLIP acts via induction of NF-κB activity and associated antiapoptotic mechanisms, rather than via inhibition of receptor-mediated caspase activation (Chugh et al. 2005; Guasparri et al. 2004). Latency genes v-cyclin (ORF72) and micro-RNAs (miRNAs) have also been implicated in viral pathogenesis (Gottwein et al. 2011; Liang et al. 2012; Suffert et al. 2011; Verschuren et al. 2004). In addition to these latency products, vIL-6, vIRF-1, vIRF-3, K1, and K15, while expressed maximally during productive replication, have also been detected in latently infected cells (of some types) and may contribute in a direct, autocrine fashion to viral neoplasia. The interplay between lytic and latent activities is likely to be important for KS, in which cytokine dysregulation is believed to drive the disease, and this interplay may also be significant in PEL and MCD. These issues and details of the molecular biology of virus–host interactions involving these various HHV-8-encoded factors are the topic of this review.

2   HHV-8 Latency Products and Autocrine Dysregulation

2.1   Latency-Associated Nuclear Antigen (LANA)

LANA is specified by ORF73 of HHV-8, and homologues are encoded by every other sequenced member of the gamma-2 herpesvirus subfamily. The basic functions of each of these proteins are to serve as a latency origin-binding protein and to tether viral genomes to host chromosomes for appropriate segregation to daughter cell nuclei during cell division (Barbera et al. 2006; Verma et al. 2007). These activities are equivalent to those of EBNA1 of gamma-1 subfamily Epstein-Barr virus (Lindner and Sugden 2007). However, LANA has further activities that are likely to play roles in viral pathogenesis in addition to contributing to the maintenance of HHV-8 latency.

One such property reported for LANA is its association with and inhibition of the cell cycle checkpoint protein and tumor suppressor p53 (Friborg et al. 1999). However, while the presence of wild-type p53 in most PEL cell lines suggests that inactivation of p53 could be biologically relevant, the susceptibility of PEL cells to p53 activation indicates that LANA is not fully able to inhibit the tumor suppressor (Chen et al. 2010; Petre et al. 2007). LANA also interacts with retinoblastoma protein (Rb) to mediate activation of E2F-responsive targets and can transform rat embryo fibroblasts in combination with transduced H-Ras (Radkov et al. 2000). Additionally, LANA was found to suppress cyclin-dependent kinase inhibitor p16INK4a-mediated cell cycle arrest and to induce E2F-mediated S-phase entry in lymphoid cells (An et al. 2005). However, as in the case of p53, the actual relevance of this experimental finding has been questioned because Rb function appears to be fully intact in PEL cells (Platt et al. 2002).

LANA also interacts with GSK3β, a kinase that targets various proteins involved in cell cycle regulation. GSK3β targets include the transcriptional regulator β-catenin and proto-oncoprotein c-Myc; phosphorylation of these proteins by GSK3β promotes their proteolytic degradation (Karim et al. 2004; Sears et al. 2000). β-catenin, in combination with the transcription factor TCF, induces expression of various genes, including c-myc, c-jun, and cyclin D1; these genes are involved in cell cycle promotion and are dysregulated in oncogenesis. LANA binding of GSK3β leads to nuclear sequestration and inactivation of the kinase, removing its negative regulation of β-catenin and promoting cell proliferation (Fujimuro and Hayward 2003; Fujimuro et al. 2003; Liu et al. 2007a, b). c-Myc residue T-58, the target of GSK3β phosphorylation, was found to be hypophosphorylated in PEL cells, and this underphosphorylation and consequent stabilization of c-Myc were dependent on LANA expression (Bubman et al. 2007). In addition, LANA interacts directly with c-Myc and induces ERK-mediated activation of c-Myc via phosphorylation of residue S-62 (Liu et al. 2007b). LANA binding of c-Myc and activation of ERK activity occur independently of LANA interaction with GSK3β. Thus, through these various interactions, LANA can activate proliferative pathways of likely significance to both HHV-8 latency and pathogenesis.

Recently, LANA has been reported to induce and interact with angiogenin (ANG), a mediator of angiogenesis, which itself upregulates LANA expression and appears to play a role in the establishment of latency and promotion of cell viability (Paudel et al. 2012; Sadagopan et al. 2011). Furthermore, interaction of LANA and ANG with annexin A2 has been identified in both HHV-8 latently infected-telomerase immortalized endothelial (TIME) cells and BCBL-1 PEL cells. Based on results from confocal analyses, it appears that these proteins colocalize and can form complexes together in addition to establishing separate ANG-LANA and ANG-annexin A2 interactions (Paudel et al. 2012). Annexin A2 is involved in regulation of cell proliferation, apoptosis, and cytoskeletal reorganization, among other activities (Shim et al. 2007; Thomas and Augustin 2009). Suppression of annexin A2 or ANG expression in PEL cells was found to increase cell death, and depletion of annexin A2 led to decreased expression of ANG and LANA (Paudel et al. 2012; Sadagopan et al. 2011). Thus, there appears to be an integrated and functionally important relationship between LANA, ANG, and annexin A2 that promotes viability of latently infected cells. Furthermore, the increased level of ANG in HHV-8-infected cells may contribute to KS pathogenesis via induction of endothelial cell activation, migration, and angiogenesis (Sadagopan et al. 2009). While the mechanisms involved in LANA, ANG, and annexin A2 mutual regulation and functional interactions remain to be elucidated, the recently reported ANG interaction with and destabilization of p53 may be significant with respect to pro-survival effects mediated via ANG (Sadagopan et al. 2012).

In addition to the individual LANA-cellular protein interactions outlined above, LANA can mediate transcriptional regulation of cellular genes via more general mechanisms. One such mechanism involves regulation of transcriptionally suppressive DNA methylation. LANA interacts with DNA methyltransferase DNMT3a, leading to its recruitment to and methylation of LANA-targeted promoters (Shamay et al. 2006). LANA also associates with histone methyltransferase SUV39H1 and transcriptional histone deacetylase-associated corepressors mSin3, SAP30, and CIR; these interactions implicate additional mechanisms of direct, promoter-specific repression by LANA (Krithivas et al. 2000; Sakakibara et al.

2004). Such mechanisms are believed to be important for the suppression of both viral lytic and cellular gene expressions to promote latency and long-term cell viability (Verma et al. 2007). LANA-targeted epigenetic repression of specific cellular genes that are silenced in various cancers could contribute to HHV-8-associated malignancies in addition to viral latency (Shamay et al. 2006; Ziech et al. 2010).

2.2 Viral FLICE-Inhibitory Protein (vFLIP)

HHV-8-encoded vFLIP is specified by ORF K13, and the protein is often referred to simply as K13. vFLIP/K13 is related structurally to death effector domain (DED)-containing and death receptor-interacting vFLIPs of other viruses, such as MC159L from mulloscum contagiosum virus and equine herpesvirus-2 E8; these vFLIPs are protective against Fas/CD95- and TNF receptor-induced apoptosis (Bertin et al. 1997; Hu et al. 1997; Thome et al. 1997). It was reported that HHV-8 vFLIP/K13 mediated protection of mouse lymphoma and rat pheochromocytoma cell lines from Fas- and TNFα-induced apoptosis (Belanger et al. 2001; Djerbi et al. 1999). However, the unique ability of HHV-8 vFLIP/K13 to induce NF-κB signaling and its inability to effectively suppress Fas-induced apoptosis suggest that HHV-8 vFLIP/K13 functions primarily through activation of NF-κB rather than via death receptor/caspase inhibition (Chaudhary et al. 1999; Chugh et al. 2005; Matta and Chaudhary 2004). HHV-8 vFLIP/K13 activates NF-κB classical and alternative pathways by interacting directly with the inhibitory κ-kinase (IKK) complexes (IKKα:IKKβ $\pm$ IKKγ/Nemo) to stimulate kinase activity, leading to disruption of IκB interaction with p50/p65-subunit NF-κB and to protease-mediated release of RelB/p52 (active form) from RelB/p100 (Field et al. 2003; Liu et al. 2002; Matta et al. 2007). Thus, vFLIP/K13 is able to activate NF-κB independently of upstream receptor-associated mechanisms involving signaling adaptors and kinases such as TRAFs and RIP; in doing so, vFLIP/K13 can avoid activation of JNK stress signaling (Matta et al. 2007). While it was reported that vFLIP/K13 interaction with TRAF2 is required for vFLIP/K13 binding to IKKγ in BC-3 PEL cells (Guasparri et al. 2006), TRAF2-dependent interaction between vFLIP/K13 and IKKγ and activation of JNK/AP1 signaling by vFLIP/K13 was not evident in subsequent studies (Matta et al. 2007).

NF-κB activation by vFLIP is significant because NF-κB is a suppressor of lytic reactivation in latently infected PEL cells, and vFLIP and NF-κB promote survival of these cells (Brown et al. 2003; Godfrey et al. 2005; Grossmann and Ganem 2008; Guasparri et al. 2004; Keller et al. 2000; Zhao et al. 2007). These effects have clear implications for the maintenance of long-term latency and for the potential contribution of vFLIP/K13 to HHV-8 malignancies. In addition to these biological effects, NF-κB signaling also induces pro-inflammatory and angiogenic cytokines, such as IL-6 and IL-8. These cytokines are produced by KS lesions and are predicted to promote KS pathogenesis and mediate vFLIP/K13-induced cellular proliferation and transformation in experimental systems (An et al. 2003; Grossmann et al. 2006; Sun et al. 2006). Together, these findings suggest that vFLIP/K13 contributes to both viral latency maintenance and to PEL and KS pathogenesis through the constitutive activation of pro-survival and generally anti-lytic NF-κB signaling.

2.3   Kaposins

The K12 transcription unit comprises the K12 ORF and two sets of GC-rich repeat units (DR1, DR2) positioned upstream of K12 (Sadler et al. 1999). Three proteins, kaposins A, B, and C, are produced from this locus by virtue of alternative transcriptional and translational initiation. Kaposin A, corresponding to the K12 translation product, is initiated from a conventional AUG codon in a transcript originating proximal to K12. Kaposins B and C initiate from CUG codons in different reading frames in transcripts containing the upstream repeat elements. Kaposin B is translated from DR1 and DR2 in "frame 1," while frame 2-translated kaposin C contains DR1/2-translated sequences fused to K12. A larger, spliced transcript initiating 5-kb upstream of K12 has also been identified, and this sequence has the potential to encode non-AUG-initiated protein(s) with novel N-terminal sequences derived from codons upstream of DR2 (Li et al. 2002; Pearce et al. 2005). K12-locus transcripts are found in high abundance in latently infected cells but are induced during lytic replication (Li et al. 2002; Sadler et al. 1999; Staskus et al. 1997; Sturzl et al. 1997; Zhong et al. 1996). The relative expression of kaposins A, B, and C in different cell types and tissues varies. For example, kaposins A and C are predominant in primary PEL cells, whereas kaposin B is most abundant in the BCBL-1 cell line (Li et al. 2002). While there have been no functional studies of kaposin C, activities of kaposins A and B have been reported.

As the direct product of ORF K12, kaposin A was the first identified and studied protein of this locus. In transfection experiments, the 6-kDa protein was found to transform immortalized rat-3 and NIH3T3 cells, forming cell colony foci in culture and tumors in athymic mice (Muralidhar et al. 1998). Transformation was dependent on cytohesin-1, a guanine nucleotide exchange factor, which binds to kaposin A. This interaction promotes membrane recruitment and activity of cytohesin-1, which acts on membrane-associated target GTPases such as ARF1 (Kliche et al. 2001). Increased activities of kinases, such as cdc2, PKC, ERK, and CAM kinase II, have been detected in kaposin A-transduced cells, but the underlying mechanisms have not been established (Muralidhar et al. 2000). Studies using gene arrays and signaling assays have implicated activation of MEK/ERK, PI3K/AKT, and STAT3 pathways by kaposin A (Chen et al. 2009b). Initial immunofluorescence studies indicated possible Golgi localization of kaposin A (Muralidhar et al. 2000). However, subsequent confocal fluorescence microscopy, cell fractionation, and flow cytometry analyses detected mostly perinuclear kaposin A with some plasma membrane localization also detectable (Tomkowicz et al.

2002, 2005). A LXXLL motif resembling ligand-interacting regions of nuclear receptors was required for immortalized cell transformation, and mutation of this motif led to greatly diminished nuclear association and newly acquired cytoplasmic localization of kaposin A (Tomkowicz et al. 2005). However, both wild-type and motif-mutated kaposin A were equally capable of activating an AP1 reporter, indicating that transformation occurs via a mechanism distinct from AP1 activation. More recently, kaposin A was identified as interacting with a variant of GTP-binding protein septin-4, a protein which localizes to mitochondria and promotes apoptosis (Lin et al. 2007; Mandel-Gutfreund et al. 2011). Co-expression of kaposin A with the septin-4 variant led to suppression of septin-4 variant-induced apoptosis in transfected cells (Lin et al. 2007). Therefore, inhibition of septin-4 function may be one mechanism by which kaposin A can influence malignant pathogenesis by promoting cell survival. Normally, this activity would be expected to serve virus latency. This mechanism would be biologically significant if septin-4 was expressed and functional in HHV-8 latently infected cell types.

Kaposin B, translated from the DR repeats and K12, interacts via DR2-encoded sequences with MK2 kinase and enhances its activity (McCormick and Ganem 2005). Kaposin B binds the "C-lobe" region of MK2, a region also targeted by p38 kinase. Kaposin B binding of the C-lobe, like its phosphorylation by p38, prevents inhibitory intramolecular association of C-lobe and C-tail sequences of MK2, which results in activation of the kinase. A single DR1 together with a single DR2 repeat, but neither element alone, is sufficient to mediate MK2 activation (McCormick and Ganem 2006). MK2 activity leads to stabilization of high-turn-over mRNAs containing AU-rich elements (AREs), and many of these mRNAs specify cytokines, such as pro-inflammatory and angiogenic IL-6. Thus, kaposin B has the potential to influence viral pathogenesis. Kaposin B is predicted to function in the maintenance of latently infected cell populations and/or expansion of latent cell pools through pro-survival and mitogenic activities of induced cellular proteins. Importantly, kaposin B stabilizes the mRNA encoding PROX1, the "master regulator" of lymphatic endothelial cell differentiation. PROX1 is targeted by ARE-binding protein HuR, and kaposin B-activated p38 kinase promotes nucleo-cytoplasmic export of HuR (Yoo et al. 2010). Reprogramming of blood endothelial cells to cells expressing lymphatic markers is induced by HHV-8 infection and is believed to be a key process in KS development (Pyakurel et al. 2006; Wang et al. 2004). Stabilization of PROX1 mRNA by kaposin B represents a mechanism by which blood-to-lymphatic endothelial reprogramming is induced by HHV-8. Thus, in addition to kaposin A, kaposin B is likely to contribute significantly to HHV-8-associated disease.

2.4 Viral Interleukin-6 (vIL-6) in PEL

Viral IL-6 (vIL-6) shares approximately 25% sequence identity with its cellular counterpart, human IL-6 (hIL-6); this viral cytokine was independently discovered by multiple groups (Moore et al. 1996; Neipel et al. 1997a; Nicholas et al. 1997).

Though sequence identity between human and viral homologues is low, the cytokines adopt equivalent 4-alpha helical bundle structures and have similar receptor interactions and signaling activities (Boulanger et al. 2003; Chow et al. 2001; Heinrich et al. 2003; Kishimoto et al. 1995). Signaling by hIL-6 requires interaction with gp130 signal transducer and gp80 receptor subunits, which leads to Janus kinase (JAK) activation and phosphorylation, dimerization, and nuclear translocation of STATs 1 and 3 (Heinrich et al. 1998). Several groups have shown that vIL-6 utilizes the same signaling components employed by human IL-6 but that it does not require gp80 for active complex formation and can signal through tetrameric ($gp130_2$:$vIL-6_2$) or hexameric ($gp130_2$:$gp80_2$:$vIL-6_2$) complexes (Aoki et al. 2001; Chen and Nicholas 2006; Chow et al. 2001). Ultimately, both cytokines share functional characteristics, such as the ability to sustain the growth of IL-6-dependent cell lines (Burger et al. 1998; Nicholas et al. 1997). vIL-6, however, is distinct in its ability to signal not only through gp130 complexes located on the plasma membrane but also intracellularly within the endoplasmic reticulum (ER); vIL-6, unlike hIL-6, is secreted inefficiently and localizes in large part to the ER. These unique properties of the viral homologue are likely involved in maintenance of viral latency and important for HHV-8 pathogenicity.

Several studies have shown that vIL-6 is critical for the growth of PEL cells. While vIL-6, IL-10, and vascular endothelial growth factor (VEGF) were detected in PEL cell supernatants, only vIL-6 and IL-10 induced PEL cell proliferation (Aoki and Tosato 1999; Jones et al. 1999). The detection of vIL-6 in these cultures was initially assumed to be the result of spontaneous lytic reactivation, either full or abortive, in a small proportion of cells because vIL-6 expression is induced during productive replication. However, vIL-6 is now known to be expressed at low levels in latent PEL cells (Chandriani and Ganem 2010; Chen et al. 2009a). Depletion of vIL-6 in these cells induces apoptosis and slows the rate of cell growth (Chen et al. 2009a). Similar growth effects were observed with intracellularly delivered single-chain antibody and peptide-conjugated phosphorodiamidate morpholino oligomers (PPMO) directed to vIL-6 and its transcript (Kovaleva et al. 2006; Zhang et al. 2008). Fully ER-retained vIL-6 (cloned to include an ER-targeting KDEL motif) is capable of rescuing the growth effects mediated by vIL-6 depletion (Chen et al. 2009a). These data indicate a prominent role of vIL-6 intracellular, autocrine signaling in support of growth and survival of latently infected PEL cells.

The mechanism through which vIL-6 acts in the ER to promote growth and viability of PEL cells is not entirely clear, but early evidence suggests that a novel interaction with VKORC1v2 (vitamin K epoxide reductase complex subunit 1 variant 2) is critical. VKORC1v2, a protein present in the ER, was identified as a novel binding partner of vIL-6 and was found to be required for PEL cell survival (Chen et al. 2012). Depletion of VKORC1v2 yielded similar growth effects to those observed in vIL-6-depleted cultures. Furthermore, a small peptide inhibitor capable of disrupting the VKORC1v2:vIL-6 interaction recapitulated growth and apoptosis effects observed upon vIL-6 or VKORC1v2 depletion, confirming the biological relevance of the vIL-6:VKORC1v2 interaction (Chen et al. 2012).

These studies suggest that the activities of this viral cytokine within the ER, acting at least in part through VKORC1v2, are important for the maintenance of the virus within latently infected B cells and that these activities contribute to viral pathogenicity.

Increased levels of phosphorylated (active) STAT3 have been detected in several PEL cell lines (Aoki et al. 2003). STAT3 is activated upon vIL-6 signaling through gp130 complexes. Depletion of STAT3 in PEL cells leads to an increase in apoptosis and a decrease in the levels of survivin (Aoki et al. 2003). Survivin is a member of the IAP (inhibitors of apoptosis) family that has been shown to inhibit apoptosis in several cancer cell lines (Ambrosini et al. 1997). These results are significant because they link antiapoptotic activities of survivin to STAT3 signaling and potentially to vIL-6. It is noteworthy, in this regard, that gp130 depletion in PEL cells leads to diminished growth and increased apoptosis in culture (E. Cousins and J. Nicholas, unpublished). In addition to vIL-6:gp130 signaling, STAT3 can also be activated by VEGF (Bartoli et al. 2000). Importantly, vIL-6 has been found to induce VEGF in experimentally transduced cell lines and to play a significant role in PEL growth and dissemination in a xenograft model (Aoki et al. 1999; Aoki and Tosato 1999). Therefore, vIL-6 is involved in a complex set of activities in PEL cells and is not only capable of initiating pro-growth and survival signaling through the ER compartment (Fig. 1) but may contribute to PEL pathogenesis through activation of STAT3/survivin and VEGF signaling.

2.5   Viral Interferon Regulatory Factor-3 (vIRF-3) in PEL

HHV-8 specifies four viral interferon regulatory factor homologues, vIRFs 1-4 (Cunningham et al. 2003; Lee et al. 2009a), which serve to counter the effects of cellular IRFs and to inhibit innate responses of the cell to virus infection and productive replication (see below). While all of the vIRFs are expressed during lytic replication, consistent with their presumed primary functions in evasion of antiviral host cell defenses, vIRF-3 is expressed as a *bona fide* latent product in PEL cells (Jenner et al. 2001; Paulose-Murphy et al. 2001; Rivas et al. 2001). As such, vIRF-3 has been referred to by some investigators as latency-associated nuclear antigen-2 (LANA2), despite its partial localization in the cytoplasm (Munoz-Fontela et al. 2005) and the absence of demonstrable latent expression in any other cell type examined. Nonetheless, in the context of PEL/B cells, vIRF-3 has the potential to impact cellular pathways that may be of biological relevance to viral latency and virus-associated pathogenesis. In common with other vIRFs, vIRF-3 can inhibit cellular IRF function. vIRF-3 does so by interfering with the transcriptional activities of IRFs 3, 5, and 7 in addition to inhibiting PKR, a pro-apoptotic kinase that is activated by interferon and dsRNA (Esteban et al. 2003; Joo et al. 2007; Lubyova and Pitha 2000; Wies et al. 2009). Importantly, vIRF-3 has also been found to interact directly with p53 to inhibit the tumor suppressor

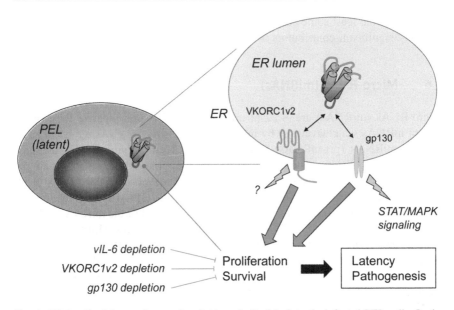

Fig. 1 ER-localized interactions and activities of vIL-6 in latently infected PEL cells. In the context of PEL latency, vIL-6 is expressed at low but functional levels and is largely sequestered in the ER compartment. Latently expressed vIL-6 supports PEL cell growth and viability. In vIL-6-depleted PEL cells, these activities can be complemented by ER-restricted (KDEL motif-tagged) transduced vIL-6, demonstrating sufficiency of ER-localized vIL-6 activity. The vIL-6 signal transducer, gp130, and a novel splice-variant protein, VKORC1v2, each bind vIL-6 within the ER, and depletion of each inhibits PEL cell growth and viability. Available evidence indicates that vIL-6 activity via each of these ER receptors occurs independently. While gp130–mediated activation of STAT and MAPK signaling has been detected in PEL cells, gp130 signaling (in contrast to effects on growth and survival) is not affected by VKORC1v2 depletion or by peptide-mediated disruption of vIL-6:VKORC1v2 interaction. Expression and pro-growth and pro-survival activities of vIL-6 in latently infected PEL cells suggest that the viral cytokine may contribute to latency maintenance in B cells and to PEL disease

and to induce c-Myc-directed transcription (Lubyova et al. 2007; Rivas et al. 2001). vIRF-3 can also interact with and inhibit FOXO3a, a transcription factor targeting pro-apoptotic genes (Munoz-Fontela et al. 2007). These activities indicate that vIRF-3 plays a significant role in promoting cell survival and proliferation in the context of PEL and potentially in general B-cell latency. Indeed, vIRF-3 has been demonstrated to be critically important for PEL cell viability in culture, as its depletion triggers apoptosis (Wies et al. 2008). Pro-survival effects of vIRF-3 may in part be the result of its inhibition of PML-mediated repression of survivin (Marcos-Villar et al. 2009). Furthermore, vIRF-3 has recently been reported to repress CIITA transcription factor-directed expression of interferon-γ and class-II major histocompatibility complex in PEL cells (Schmidt et al. 2011). This immune evasion activity of vIRF-3 is probably vital for the long-term survival of these cells in vivo. Therefore, vIRF-3 activities are likely to be important for latency persistence in at least some cell types (where vIRF-3 is latently expressed) and are probably significant contributors to PEL malignancy.

2.6   Micro-RNAs (miRNAs)

Micro-RNAs (miRNAs) are ~ 22 nt non-coding RNAs that regulate the expression of mRNAs via cleavage or by inhibiting translation. miRNAs are encoded as primary miRNAs (pri-mRNAs), synthesized by RNA Polymerase II, and processed to pre-miRNAs (stem-loop structures) by the RNAse III domain of DROSHA prior to nuclear export by exportin 5 (Lee et al. 2003b; Lund et al. 2004). In the cytoplasm, pre-miRNAs are cleaved into 21–24 nt double-stranded RNAs by an RNAse III domain of DICER, and one strand of the miRNA duplex is then incorporated into RISC (RNA-induced silencing complex) (Bartel 2004; Lee et al. 2003b). The incorporated miRNA guides the loaded RISC to the mRNA target (Schwarz et al. 2003). Generally, the mRNA target is degraded if the miRNA is perfectly complementary to the targeted sequence. Alternatively, binding of a miRNA lacking perfect complementarity inhibits translation of the mRNA (Zeng et al. 2003).

While miRNAs have been detected in all metazoans, virus-encoded miRNAs were discovered only recently (Pfeffer et al. 2004). To date, 12 HHV-8-encoded miRNAs have been identified, termed miR-K12-1 to miR-K12-12. A total of ten of the 12 miRNAs are located between latently expressed ORFs 71 and K12 in the HHV-8 genome; miR-K12-10 is located within the ORF K12, and miR-K12-12 is within the 3′ UTR of K12 (Pfeffer et al. 2005; Samols et al. 2005). All 12 of the HHV-8 miRNAs are oriented "in sense" with ORFs 71 to K12 and are expressed during latency (Cai et al. 2005). Many of the miRNAs can be detected (and some are even induced) during lytic infection (Cai et al. 2005; Umbach and Cullen 2010). Bioinformatic approaches have utilized miRNA seed sequences (nt 2-7/8 of the miRNA) to search for miRNA gene targets (Gottwein and Cullen 2010; Lu et al. 2010b; Nachmani et al. 2009; Qin et al. 2010), but these approaches yield large numbers of potential candidates. Additionally, targets with less than perfect seed sequence complementarity may be overlooked, and experimental validation of identified candidates is necessary to assess authenticity. Functional approaches involving transduction of recombinant viruses containing single or a combination of HHV-8 miRNAs and monitoring potential changes in mRNA levels via microarray have also been used (Ziegelbauer et al. 2009). More recently, immunoprecipitation of RISC and/or associated Argonaute proteins has been conducted, and microarray analysis of co-precipitated mRNAs (RIP-ChIP) has been employed (Dolken et al. 2010). These methods have identified multiple mRNA targets within the host cell. From these data, the virally-encoded miRNAs have been deduced to: (1) limit NK cell recognition of virally infected cells (Nachmani et al. 2009), (2) repress RTA expression to inhibit latent-lytic switching (Bellare and Ganem 2009; Lin et al. 2011; Lu et al. 2010a), (3) interfere with TGFβ signaling (Liu et al.

2012), (4) alter NF-κB signaling pathways (Lei et al. 2010), (5) modulate cytokine production (Abend et al. 2010), and (6) modify cell cycle progression (Gottwein and Cullen 2010). These targets and pathways are summarized in Table 1 and are discussed further below.

HHV-8 employs many strategies to remain undetected by the host immune response, and viral miRNAs are believed to play a vital role. HHV-8-encoded miR-K12-7 targets MICB, a viral infection-induced cell-surface marker that functions to induce natural killer (NK) cell recognition and killing via engagement with the NK-expressed NKG2D receptor (Glas et al. 2000; Nachmani et al. 2009). MICB targeting by virus-encoded microRNAs is conserved in human cytomegalovirus (Stern-Ginossar et al. 2007; Stern-Ginossar et al. 2008) and EBV infection (Nachmani et al. 2009). In addition to immune evasion, several HHV-8 miRNAs target RTA, thereby functioning to maintain viral latency and inhibit viral replication. Several research groups have identified miR-K12-3, miR-K12-5, miR-K12-9, and miR-K12-11 as directly or indirectly targeting RTA expression (Bellare and Ganem 2009; Lin et al. 2011; Lu et al. 2010b). Viral replication is also limited through inhibition of the NF-κB inhibitor IκBα by miR-K12-1; upregulation of NF-κB signaling abrogates lytic replication of HHV-8 (Lei et al. 2010). By reducing virion production, the virus is able to further circumvent the host immune response.

In addition to immune evasion, the virus counteracts pro-apoptotic pathways induced by the host cell upon viral infection. Caspase 3 is a target of multiple viral miRNAs (miR-K12-1, miR-K12-3, and miR-K12-4-3p), and its inhibition desensitizes HHV-8-infected cells to caspase-induced apoptosis (Suffert et al. 2011). Similarly, miR-K12-10a suppresses TWEAK receptor (TWEAKR) expression, which limits TWEAK-induced caspase activation and apoptosis (Abend et al. 2010). TWEAKR inactivation also reduces production of pro-inflammatory cytokines IL-8/CXCL8 and MCP-1/CCL2 (Abend et al. 2010). BCLAF1, a transcriptional repressor, can be targeted by several of the HHV-8 microRNAs (miR-K12-5, miR-K12-9, and miR-K12-10a/b), leading to decreased etoposide-induced apoptosis in PEL cells (Ziegelbauer et al. 2009). Finally, miR-K12-11, a homologue to cellular miR-155, alters expression of BACH1, which leads to a variety of phenotypic changes including upregulation of HMOX1 (increased cell survival), upregulation of xCT (increased infection permissivity in macrophages and endothelial cells), and protection against apoptosis mediated by reactive nitrogen species (Gottwein et al. 2007; Qin et al. 2010; Skalsky et al. 2007).

MicroRNAs encoded by HHV-8 also play significant roles in growth signaling and angiogenesis. Transforming growth factor beta (TGFβ) signaling can be downregulated via the direct targeting of SMAD5 by miR-K12-11 (Liu et al. 2012); downregulation of TGFβ signaling induces cell proliferation. It has been noted that HHV-8⁺ B-cell lines have decreased levels of miR-155, the cellular homologue of miR-K12-11; miR-K12-11 may compensate for the limited levels of miR-155 in these cells (Skalsky et al. 2007). Inhibition of miR-K12-11 was found to derepress TGFβ signaling in HHV-8⁺ B cells (Liu et al. 2012). TGFβ signaling can be modulated by thrombospondin 1 (THBS1), a target of HHV-8 miRNAs

Table 1 HHV-8-encoded microRNAs and their functions

Functional Class	Viral microRNA	Target	Activity/ Function	References
Cell Cycle Regulation	miR-K12-1	p21	Inhibits cell cycle arrest	Gottwein and Cullen (2010)
Apoptosis	miR-K12-10a	TWEAKR	Reduced caspase activation	Abend et al. (2010)
	miR-K12-5, 9, 10	BCLAF1	Inhibits apoptosis	Ziegelbauer et al. (2009)
	miR-K12-1, 3, 4-3p	Caspase 3	Inhibits apoptosis	Suffert et al. (2011)
	miR-K12-11	BACH1	Pro-survival	Gottwein et al. (2007), Skalsky et al. (2007), Qin et al. (2010)
Latency Maintenance	miR-K12-3	NFIB	Inhibits RTA, Stabilizes latency	Lu et al. (2010a, b)
	miR-K12-5, 7-5p, 9	RTA	Stabilizes latency	Bellare and Ganem (2009), Lin et al. (2011), Lu et al. (2010a, b)
Growth Signaling	miR-K12-11	SMAD5	Limits TGFß signaling	Liu et al. (2012)
	miR-K12-6, 11	MAF	Cell fate reprogramming	Hansen et al. (2010)
Immune Evasion	miR-K12-7	MICB	Evasion from NK cells	Nachmani et al. (2009)
Angiogenesis	miR-K12-1, 3, 6, 11	THBS1	Increases angiogenesis	Samols et al. (2007)
Viral Replication	miR-K12-1	IkBalpha	Limits viral replication	Lei et al. (2010)
Chromatin Modification	miR-K12-4-5p	Rbl2	Global epigenetic modification	Lu et al. (2010a, b)
Unvalidated Targets	miR-K12-3	NHP2L1	Splicesome assembly?	Dolken et al. (2010)
	mIR-K12 cluster	LRRC8D	Unknown	Dolken et al. (2010)
	miR-K12 cluster	SEC15L	Vesicle trafficking?	Dolken et al. (2010)
	miR-K12-4-3	GEMIN8	mRNA splicing?	Dolken et al. (2010)
	miR-K12 cluster	ZNF684	Transcriptional regulation?	Dolken et al. (2010)
	miR-K12 cluster	CDK5RAP1	CDK5 inhibitor?	Dolken et al. (2010)

miR-K12-1, miR-K12-3, miR-K12-6, and miR-K12-11 (Samols et al. 2007). Thrombospondin 1 is an antiangiogenic factor, and its downregulation leads to repressed TGFβ signaling (Samols et al. 2005). Thus, multiple viral microRNAs are capable of altering growth signaling and increasing angiogenesis to support the establishment of HHV-8-associated neoplasias. Additionally, viral miRNAs can alter the transcriptomes of endothelial cells and assist in cellular reprogramming. Both blood vessel endothelial cell (BEC) and lymphatic endothelial cell (LEC) expression markers are found in KS tissue, and the tissue can be reprogrammed toward the LEC or BEC fate under the appropriate stimuli. Specifically, miR-K12-6 and miR-K12-11 target the cellular transcription factor MAF (musculoaponeurotic fibrosarcoma oncogene homologue) (Hansen et al. 2010), which is found in LECs but not in BECs. Silencing of MAF by the microRNAs increases expression of BEC marker genes within the KS tissue. Targeting of MAF by miR-155 in CD4$^+$ T cells has also been described (Rodriguez et al. 2007).

Several other studies have identified potential targets of viral microRNAs, though the validation of these targets is still ongoing. For example, Rbl2 (retinoblastoma-like protein 2) has been identified as a target of miR-K12-4-5p (Lu et al. 2010b). Rbl2 is an inhibitor of specific DNA methyltransferases (DNMT3a and 3b). Epigenetic changes have been observed following inhibition of Rbl2, but the consequences of these alterations are unclear (Lu et al. 2010b). Lastly, a large-scale study utilizing a luciferase assay approach has identified several potential targets of various HHV-8 miRNAs. These targets include proteins with roles in vesicle trafficking, spliceosome assembly, transcription regulation, and cell cycle progression (Dolken et al. 2010). Validation of these protein targets is needed and will further enhance the understanding of the roles of miRNAs in HHV-8 biology and pathogenesis.

3   Novel Virus–Host Interactions via Lytic Activities

3.1   Viral Interleukin-6 (vIL-6)

In contrast to its direct autocrine role in PEL pathogenesis, vIL-6 is believed to contribute to KS and MCD predominantly via paracrine signaling. Newly infected cells and those undergoing lytic replication express vIL-6 as an early gene product, and vIL-6 is rapidly induced following RTA expression in these cells (Sun et al. 1999). The majority of HHV-8-infected cells in the KS lesion remain latently infected, but small subsets of cells are lytically active. This minority of cells expresses lytic proteins, including vIL-6, vGPCR, and K1; these proteins ultimately enhance the expression of cellular inflammatory and angiogenic cytokines (Mesri et al. 2010). For example, vIL-6 can induce expression of VEGF (Aoki and Tosato 1999), considered to be a key contributor to KS development. VEGF and cytokines such as IL-6, CXCL8, and bFGF are secreted from lytically active cells and can further activate nearby latently infected and uninfected cells in a paracrine fashion. The secreted proteins modulate survival of HHV-8-infected cells, angiogenesis (predominantly through VEGF), and recruitment of uninfected cells to the lesion site. Inflammatory cytokines (including IL-6) and angiogenic factors have been proposed to play a role in the initial development of KS (Ensoli and Sturzl 1998).

In MCD patients, levels of IL-6 are substantially higher in affected lymph nodes when compared to unaffected nodes of the same patient (Yoshizaki et al. 1989). Additionally, disease severity correlates with IL-6 levels, and when affected lymph nodes are resected, IL-6 levels decrease (Yoshizaki et al. 1989). Similarly, cell lines derived from HIV positive KS patients express IL-6, and KS tissue produces increased amounts of IL-6 compared to normal tissue (Miles et al. 1990). IL-6 antisense oligonucleotides were found to suppress the growth of KS cells from HIV positive patients as well as the production of IL-6, and addition of exogenous, recombinant IL-6 was able to restore growth and proliferation (Miles et al. 1990). An additional study observed substantially elevated levels of IL-6 produced by malignant plasma cells from MCD patients though not from other B-cell tumors (Burger et al. 1994). At the time of these discoveries, the mechanism of disease-associated IL-6 dysregulation was not understood. More recent reports have demonstrated that vIL-6 can induce the expression of IL-6 and VEGF in some cell types (Aoki et al. 1999; Mori et al. 2000). It is likely that vIL-6 also contributes directly to MCD. Additionally, vIL-6 downregulates CCL2 and inhibits the infiltration of neutrophils during acute infection of B cells (Fielding et al. 2005).

3.2 Viral CC-Chemokine Ligands (vCCLs)

The three HHV-8 chemokines, vCCL-1, vCCL-2, and vCCL-3, are encoded by ORFs K6, K4, and K4.1, respectively (Moore et al. 1996; Neipel et al. 1997b; Nicholas et al. 1997; Russo et al. 1996). All are expressed early during the lytic cycle (Jenner et al. 2001; Paulose-Murphy et al. 2001). vCCL-1 and vCCL-2 are most closely related structurally to cellular chemokines CCL3 and CCL4, while vCCL-3 shares significant primary sequence similarity with a number of CC-chemokines. However, the properties of the v-chemokines are distinct from those of their cellular counterparts. With respect to receptor usage, vCCL-1 is an agonist for CCR8; vCCL-2 signals through CCR3, CCR8, and potentially also CCR5; vCCL-3 functionally targets CCR4 and XCR1 (Boshoff et al. 1997; Dairaghi et al. 1999; Luttichau 2008; Luttichau et al. 2007; Nakano et al. 2003; Stine et al. 2000). In addition, vCCL-2 binds several CCR- and CXCR-type chemokine receptors and CX_3CR1 as a neutral (non-signaling) ligand and effectively inhibits cellular chemokine activity through these receptors (Chen et al. 1998; Crump et al. 2001; Dairaghi et al. 1999; Kledal et al. 1997; Luttichau et al. 2001). The nature of the v-chemokine-targeted receptors suggests that they may mediate immune evasion via Th2 polarization and blocking of leukocyte trafficking, as has been demonstrated for vCCL-2 in in vivo experiments (Chen et al. 1998; Weber et al. 2001).

Apart from these immune evasion properties, each of the v-chemokines has been demonstrated to promote angiogenesis, in part via induction of VEGF (Boshoff et al. 1997; Liu et al. 2001; Stine et al. 2000). Like vIL-6, HHV-8 vCCLs have the potential to promote KS pathogenesis via paracrine signaling and may also play roles in PEL, where VEGF has been implicated as an important factor based on data from murine studies (Aoki and Tosato 1999; Ensoli and Sturzl 1998; Haddad et al. 2008). Additional contributions of vCCL-1 and vCCL-2 to pathogenesis may include pro-survival signaling via CCR8, demonstrated in uninfected and HHV-8-infected endothelial cells (Choi and Nicholas 2008). vCCL-1 and vCCL-2 were also found to promote survival of PEL and murine cell lines (Liu et al. 2001; Louahed et al. 2003). Unlike most (non-secreted) viral proteins, the v-chemokines have the potential to function in a paracrine manner. Therefore, these chemokines may promote cell survival of latently infected and uninfected cells surrounding those supporting lytic replication, thus contributing to viral pathogenesis. Nonetheless, an important aspect of the pro-survival activities of vCCL-1 and vCCL-2 is the positive contribution to productive replication via autocrine signaling. The endogenously produced v-chemokines inhibit lytic cycle-induced apoptosis and increase virus yields in HHV-8-infected endothelial cultures (Choi and Nicholas 2008). This activity involves CCR8 signaling-dependent suppression of lytic cycle stress-induced pro-apoptotic protein Bim, a powerful inhibitor of productive replication.

3.3 Viral G Protein-Coupled Receptor (vGPCR)

While not unique to HHV-8, the vGPCR encoded by this virus is structurally and functionally diverged from other gamma-2 herpesvirus vGPCRs and is strongly implicated as a paracrine contributor to KS development (Cannon 2007; Nicholas 2005; Rosenkilde et al. 2001; Verzijl et al. 2004). HHV-8 vGPCR is unusual in its promiscuous functional association with three classes of Gα proteins (i, q, and 13) in addition to its direct association with and activation of the signaling protein SHP2 (Couty et al. 2001; Liu et al. 2004; Philpott et al. 2011; Shepard et al. 2001). Initial reports that vGPCR can function as a classical "autocrine" oncogene in in vitro and in vivo experimental systems employing vGPCR-transduced cell lines (Bais et al. 1998) implied that vGPCR may be expressed as a latent protein (enabling it to contribute directly to HHV-8 oncogenesis). However, no evidence of vGPCR latent expression has been forthcoming. Nonetheless, vGPCR can induce KS-like tumors in receptor-transduced mice, and this phenotype is supported despite only a minority of cells expressing vGPCR (Montaner et al. 2003; Yang et al. 2000). This can be explained by angiogenic, mitogenic, and inflammatory cellular cytokine induction via vGPCR signaling, leading to local endothelial activation, proliferation, and tumorigenesis (Montaner et al. 2004). As stated above, cytokine dysregulation is considered to be the principle driver of KS disease. vGPCR is known to activate key factors, such as VEGF, bFGF, CXCL8, and IL-6, that are found in KS lesions and believed to promote and be required for KS development and progression (Bais et al. 1998; Cannon et al. 2003; Pati et al. 2001; Schwarz and Murphy 2001). Thus, vGPCR produced in a small minority of spontaneously reactivating endothelial cells may induce levels of cellular cyto-kines sufficient to promote KS. It should be noted that in animal models, vGPCR-expressing cells are able to cooperate with cells expressing latency genes v-cyclin and/or vFLIP to increase the frequency of KS achieved with inoculated vGPCR$^+$ cells alone (Montaner et al. 2003). This supports the notion that both autocrine latent and paracrine lytic activities can function together in HHV-8-associated neoplasia.

3.4   Viral Interferon Regulatory Factors (vIRFs)

HHV-8 vIRFs 1-4 are specified by the genomic region encompassing ORFs K9 to K11; an equivalent locus (encoding 8 vIRFs) has been identified only in the closely related rhesus rhadinovirus (Alexander et al. 2000; Cunningham et al. 2003; Moore et al. 1996; Searles et al. 1999). All four of the vIRFs are expressed during lytic replication, but vIRF-1 and vIRF-3 are also expressed in PEL latency (Po-zharskaya et al. 2004; Rivas et al. 2001). Additionally, vIRF-1 transcripts have been detected in KS cells using reverse transcription-polymerase chain reaction techniques (Dittmer 2003). As outlined above (Sect. 2.5), vIRF-3 is required for PEL cell viability (Wies et al. 2008); however, vIRF-1 depletion has no detectable influence on PEL cell growth or survival in culture (Y. Choi and J. Nicholas, unpublished data). The vIRFs appear to function primarily to evade host cell defenses against *de novo* infection and virus productive replication, which trigger cellular IRF and interferon signaling cascades leading to cell cycle arrest and pro-apoptotic signaling (Lee et al. 2009a; Offermann 2007). The vIRFs counter these cellular signals in several ways. IRF5 and IRF7 functional dimerization and pro-moter association are antagonized by direct binding of vIRF-3 to these cellular factors, and vIRF-3 also inhibits IRF3 activity (Joo et al. 2007; Wies et al. 2009). vIRF-1 mediates transcriptional repression of IRF-targeted genes by blocking IRF-directed promoter recruitment of p300/CBP via competitive binding to the tran-scriptional coactivators (Burysek et al. 1999a; Li et al. 2000; Lin et al. 2001; Seo et al. 2000; Zimring et al. 1998). Transcriptional repression of IRF1-, IRF3-, and ISGF3-targeted genes is mediated by vIRF-2, in part by activation of caspase 3-mediated destabilization of IRF3 (Areste et al. 2009; Burysek et al. 1999b; Fuld et al. 2006). In addition, vIRF-2 and vIRF-3 directly target and/or inhibit dsRNA-activated PKR kinase activity; this suppresses PKR promotion of protein trans-lation and inhibits interferon signaling (Burysek and Pitha 2001; Esteban et al. 2003). Interestingly, vIRF-3 associates with and stabilizes the pro-angiogenic transcription factor HIF-1α (Shin et al. 2008), and this could serve not only to promote endothelial cell survival but may also contribute to KS and PEL patho-genesis via induction of cytokines such as VEGF. vIRFs 1, 3, and 4 have been shown to inhibit p53 activity via (1) direct binding to the tumor suppressor (vIRF-1 and vIRF-3), (2) through interaction with p53-phosphorylating and p53-activating ATM kinase (vIRF-1), or (3) via stabilization of MDM2 (vIRF-4), which promotes ubiquitination and proteasomal degradation of p53 (Lee et al. 2009b; Nakamura et al. 2001; Rivas et al. 2001; Seo et al. 2001; Shin et al. 2006). Inhibitory interactions of vIRF-4 with deubiquitinase HAUSP have been reported to be involved in p53 destabilization (Lee et al. 2011). Therefore, p53 represents a major vIRF target that is important to control for efficient virus productive replication to occur. vIRF-4 also binds to CSL, the target of Notch, but the significance of this interaction is unclear (Heinzelmann et al. 2010).

vIRF-1 associates with and inhibits the activities of other cellular proteins involved in innate cellular responses to infection and the promotion of apoptosis. These targets include retinoic acid/interferon induced protein GRIM19 and TGFβ receptor-activated transcription factors Smad3 (tumor suppressor) and Smad4 (co-Smad) (Angell et al. 2000; Ma et al. 2007; Seo et al. 2002, 2005). More recently, vIRF-1 has been found to bind directly to members of the so-called BH3-only protein (BOP) family (Choi and Nicholas 2010; Choi et al. 2012). BOPs are Bcl-2-related proteins that function to promote apoptosis either via inhibitory interactions with pro-survival members of the Bcl-2 family or by direct activation of apoptotic executioners Bax and Bak (Kuwana et al. 2005; Willis and Adams 2005). A region of vIRF-1 comprising residues 170-184 (BOP-binding domain, BBD) interacts with BOP BH3 domains, required for pro-apoptotic activities of these proteins, thereby mediating BOP inhibition (Choi and Nicholas 2010; Choi et al. 2012). vIRF-1 BBD, a predicted amphipathic α-helix, resembles the Bid BH3-inhibitory BH3-B domain of Bid and represents only the second example of a BH3-B-type BH3-inhibitory domain and therefore a novel viral mechanism of apoptotic inhibition. BBD-mediated interactions with BOPs are functionally important, as indicated by the following findings: (1) BBD-mutated vIRF-1 is less active than wild-type vIRF-1 in promoting productive replication and inhibiting apoptosis in lytically infected endothelial cells; (2) vIRF-1:BOP-disrupting BBD peptide causes significant inhibition of virus production in these cells; and (3) depletion of vIRF-1-targeted BOPs Bim or Bid leads to substantial increases in replicative titers (Choi and Nicholas 2010; Choi et al. 2012).

In summary, a wealth of published data indicates that the vIRFs of HHV-8 represent an effective panel of antiapoptotic proteins that promote productive replication via interactions with an important group of cellular proteins involved in host cell defenses against infection (summarized in Fig. 2).

3.5 K7-Encoded Viral Inhibitor of Apoptosis (vIAP)

In common with other gammaherpesviruses, HHV-8 specifies a homologue of cellular Bcl-2 proteins. However, HHV-8 encodes an additional BH-like domain-containing antiapoptotic protein that is a homologue of cellular survivin. This

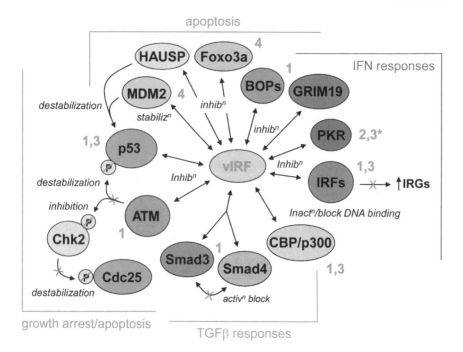

Fig. 2 Summary of inhibitory interactions of vIRFs with cellular proteins. The particular vIRFs interacting with each target are indicated by the adjacent numbering (*red lettering*), and the effect of each interaction is indicated (*italics*). The activities of vIRF interactions are grouped into four general, overlapping biological categories, as indicated by the brackets and color coding. The vIRF:protein interactions and their significance are discussed fully in the text. (*Functional inhibition; no physical interaction detected)

HHV-8 protein is referred to as K7 (corresponding to the encoding ORF) or viral inhibitor of apoptosis (vIAP). K7/vIAP is a membrane-associated protein that contains a putative mitochondrial-targeting signal and localizes to mitochondria, ER, and possibly other membranes as well (Feng et al. 2002; Wang et al. 2002). It has been reported that K7/vIAP binds to cellular Bcl-2 via its C-terminal BH2-like domain and to activated (proteolytically cleaved) caspase 3, bridging the two proteins and functionally inhibiting caspase 3 proteolytic activity (Wang et al. 2002). Interaction with and inhibition of terminal caspase 3 in the apoptotic cascade is analogous to the activities of cellular IAPs, which include survivin, XIAP, and cIAPs 1 and 2. However, the interaction between K7/vIAP and Bcl-2 is a property not reported for its cellular counterparts. The functional and biological significance of this interaction remains to be determined. Nonetheless, K7/vIAP is able to inhibit pro-apoptotic signaling in transfected cells treated with agents such as Fas antibody and TNF-α, indicating its potential to act as a promoter of lytic replication via its pro-survival activity during lytic cycle-induced stress (Wang et al. 2002). K7/vIAP also interacts with calcium-modulating cyclophilin ligand

(CAML), which regulates intracellular calcium ion concentrations (Bram and Crabtree 1994; Feng et al. 2002). The functional significance of this interaction is evident from the ability of wild-type but not CAML binding-refractory K7/vIAP to inhibit chemically-induced mitochondrial depolarization (i.e., apoptotic triggering) in transfected cells (Feng et al. 2002). Thus, in addition to its inhibitory binding to caspase 3, K7/vIAP appears to mediate apoptotic inhibition via CAML interaction. K7/vIAP also interacts with the cellular protein PLIC1 (protein-linking integrin-associated protein and cytoskeleton-1, also called ubiquilin), which associates with ubiquitin-conjugated proteins to inhibit their proteasomal degradation (Feng et al. 2004; Kleijnen et al. 2000; Wu et al. 1999). K7/vIAP appears to antagonize PLIC1 activity, thereby destabilizing ubiquitinated proteins, as demonstrated for p53 and NF-κB-inhibitory IκB (Feng et al. 2004). Together, the inhibitory interactions between K7/vIAP and cellular proteins PLIC1, caspase 3/Bcl-2, and CAML may promote cell survival during lytic replication and consequently enhance the efficiency of virus production.

4 Terminal Membrane Proteins

4.1 K1/Variable ITAM-Containing Protein (VIP)

The K1 ORF of HHV-8 is located at the left end of the genome and is collinear with other gammaherpesvirus genes encoding signaling membrane proteins. These include saimiri transformation-associated protein (STP) of herpesvirus saimiri (HVS), latency membrane protein-1 (LMP-1) of EBV, and the K1-homologous R1 receptor of rhesus rhadinovirus (Albrecht et al. 1992; Damania et al. 1999; Kaye et al. 1993; Lagunoff and Ganem 1997; Murthy et al. 1989). The K1 protein is a type I transmembrane signaling protein containing a functional immunoreceptor tyrosine-based activation motif (ITAM) in its cytoplasmic C-tail (Lagunoff et al. 2001; Lee et al. 1998a). Sequencing of K1 in different HHV-8 isolates identified an unusual degree of amino acid sequence variability in the extracellular regions of the encoded proteins (Nicholas et al. 1998; Zong et al. 1999), hence the naming of the K1 protein as variable ITAM-containing protein (VIP). While the functional significance of this variability has not been established, the K1 locus has served as a basis of epidemiological studies of HHV-8 strain distribution and infectivity (Hayward and Zong 2007; Mbulaiteye et al. 2006; Whitby et al. 2004). Based initially on the genomic position of K1 and subsequently on the constitutive signaling and transforming properties of K1/VIP, the protein was implicated as a potential contributor to HHV-8 pathogenesis. K1/VIP, like RRV R1, was able to substitute functionally for the positionally equivalent ORF1/STP of HVS in in vivo tumorigenesis assays and to promote cell growth and transformation in isolation (Lee et al. 1998b; Prakash et al. 2002). K1/VIP activation of the AKT pathway and consequent activation of mTOR (associated with cell growth) and the inactivation of pro-apoptotic GSK3, Bad, and forkhead transcription factors have been implicated in these activities (Tomlinson and Damania 2004; Wang et al. 2006). However, as K1 appears to be expressed primarily or exclusively during lytic replication (Jenner et al. 2001; Lagunoff and Ganem 1997; Nakamura et al. 2003; Paulose-Murphy et al. 2001), its potential role in KS, PEL, and MCD may be restricted to paracrine effects of K1/VIP-induced cellular cytokines (see below) rather than direct effects suggested by initial functional analyses. While immunodetection of K1/VIP in KS and MCD tissues has been reported, this has not been associated with latently infected cells (Lee et al. 2003a; Wang et al. 2006). It should be noted, however, that in situ detection of K1 transcripts in some KS cells lacking detectable lytic marker (major capsid protein) mRNA expression suggests the possibility that K1/VIP may be expressed in at least some latently infected KS cells (Wang et al. 2006).

K1/VIP recruits and activates Src-family kinases, PI3K, and PLCγ to mediate signal transduction via several pathways by ligand-independent, constitutive signaling (Lagunoff et al. 2001; Lee et al. 2005, 1998a; Samaniego et al. 2001; Tomlinson and Damania 2004). It has been suggested that K1/VIP may contribute to HHV-8-associated disease, especially KS, by induction of cellular cytokines. K1/VIP-induced cytokines include pro-inflammatory IL-1β, IL-6, and GM-CSF and angiogenic factors VEGF, CXCL8, and bFGF (Lee et al. 2005; Prakash et al. 2002; Samaniego et al. 2001; Wang et al. 2006). Contributions of the HHV-8 receptor to pathogenesis via cellular cytokine induction could theoretically occur during lytic replication or during latency. In KS, PEL, and MCD, small proportions of cells support lytic reactivation, enabling lytically expressed proteins, like K1, to exert paracrine influence on surrounding latently infected and uninfected cells (Aoki et al. 2003; Aoki and Tosato 2003; Ensoli et al. 2001).

4.2 K15-Encoded Membrane Protein

The K15-encoded protein in its full-length form is a twelve-transmembrane domain-containing signaling receptor. Like K1, K15 may play a role in pathogenesis via cytokine dysregulation, and it could conceivably contribute to malignant disease through pro-survival signaling during latency. Transcripts from the K15 locus are expressed predominantly during the lytic cycle, but some K15 products have been detected in resting (latent) PEL cultures (Choi et al. 2000; Glenn et al. 1999; Sharp et al. 2002). The issue is complex because the K15 primary transcript contains eight exons and can be differentially spliced; the resulting mRNAs and encoded proteins may be expressed differently based on cell type and whether the virus is in the latent or productive phase. All forms of K15 contain C-terminal protein sequences with functional signaling motifs (see below), but the protein isoforms differ in their complement of transmembrane domains. K15 transcripts and a 23-kDa protein isoform have been detected during latency in PEL cells, but K15 mRNA levels are induced considerably upon lytic reactivation (Choi et al. 2000; Glenn et al. 1999; Sharp et al. 2002; Tsai et al. 2009).

Full-length K15 protein has been detected in HHV-8 bacmid-containing HEK293 cells, though only after lytic induction with butyrate treatment (Brinkmann et al. 2007). However, the full-length protein has not been observed in cells naturally infected with HHV-8. The ability of immediate early, lytic trigger protein RTA to activate transcription from the K15 promoter is consistent with predominant lytic expression of K15 (Wong and Damania 2006). Nonetheless, uncertainty remains regarding the expression characteristics of K15 transcripts and proteins and whether K15 receptor signaling could contribute to latency and HHV-8 neoplasia in an autocrine manner.

Signaling motifs in the cytoplasmic tail of all K15 isoforms and in both M and P allelic protein types (Poole et al. 1999) include two SH2- and single SH3- and TRAF-binding sites (Brinkmann et al. 2003; Choi et al. 2000; Glenn et al. 1999). SH2 binding-mediated interactions with Src-family kinases occur via the Y_{481}EEV motif, which is the primary site of K15 phosphorylation (Brinkmann et al. 2003; Choi et al. 2000). This, together with the SH3-binding sequence (PPLP), leads to inhibition of B-cell receptor (BCR) signaling. PPLP-motif interactions with intersectin 2 (endocytic adaptor protein) and with Src kinases (such as Lyn and Hck) are important for this activity (Lim et al. 2007; Pietrek et al. 2010). BCR inhibition by the K15 receptor parallels that of the collinearly encoded LMP-2 of EBV, and each is likely to promote latency by inhibiting lytic cycle reactivation promoted by BCR signaling. The Y_{481}EEV motif has been implicated in the activation of NF-κB and mitogen-activated protein kinases (MAPKs) ERK and JNK, which occurs after Y_{481} phosphorylation (Brinkmann and Schulz 2006; Choi et al. 2000; Pietrek et al. 2010). Interaction of the K15 receptor with TRAFs 1, 2, and 3 is likely to contribute to NF-κB and MAPK signaling (Brinkmann and Schulz 2006; Glenn et al. 1999). The second tyrosine-containing motif, Y_{432}ASI, is not detectably phosphorylated, and its significance is uncertain. However, its interaction with apoptotic regulatory protein HAX-1 and ER and mitochondrial colocalization of HAX-1 with K15 suggest that the viral receptor may function to promote cell survival via this motif (Sharp et al. 2002).

Examination of the downstream effects of K15 signaling has provided insight into possible functions of the receptor in HHV-8 biology and its potential contributions to viral pathogenesis. In addition to suspected antiapoptotic activity via interaction with HAX-1, K15 can induce the expression of several antiapoptotic genes, including *A20*, *Bcl-2A1*, *Birc2*, and *Birc3* (Brinkmann et al. 2007). Induction of these genes may help promote cell survival during the lytic cycle, further enhancing productive replication. If K15 is expressed during latency, its pro-survival signaling could contribute to prolonged latent cell viability and maintenance of latency pools in vivo and to viral pathogenesis. On the other hand, the observed induction of cellular cytokines in K15-transduced cells suggests a mechanism by which K15 can affect surrounding cells (latently infected and uninfected) by paracrine signaling from lytically infected cells. K15-expressing latently infected cells could exert similar effects on the microenvironment. Cytokines induced by K15 receptor signaling include IL-6, CCL2, CXCL3, and CXCL8; each of these possesses angiogenic activity and has been implicated in KS pathogenesis (Brinkmann et al. 2007; Caselli et al. 2007; Ensoli and Sturzl 1998). It is intriguing that K15 also induces expression of genes representing downstream targets of angiogenic VEGF signaling. This would clearly implicate K15 as an autocrine contributor to pathogenesis should the receptor be expressed during latency, but such activity of K15 could also contribute to productive replication. Angiogenic targets of K15 include *Dscr-1* and *Cox-2* (Brinkmann et al. 2007). It is notable that Cox-2 has been reported to be induced during *de novo* and subsequent latent infection of endothelial cells by HHV-8, and Cox-2 is important for the production of several inflammatory and angiogenic factors (Sharma-Walia et al. 2010). Thus, pro-survival and paracrine-mediated pro-angiogenic roles of K15 in HHV-8 lytic replication and pathogenesis seem likely, and there is potential for autocrine activity via pro-survival signaling contributing to latency and neoplasia if the receptor is latently expressed.

5   Summary

The discovery and study of HHV-8 has provided the opportunity to identify unique virus-specified activities encoded by proteins either not previously known among viruses or those not previously investigated or characterized in depth in other viral systems. HHV-8 has also provided a model for the identification and character-ization of viral miRNAs, a new area of research that has provided unique and important insights into viral manipulation of host cell processes as part of normal virus biology and potentially in viral pathogenesis. The properties of the charac-terized protein and miRNA players in these processes have been described in detail, and several key points emerge. First, the notion that only autocrine, latent viral activities are relevant to virus-associated neoplasia needs revision, certainly in the case of KS and possibly for PEL and MCD. Paracrine factors (viral and/or cellular) produced during lytic replication can contribute to pro-proliferative, pro-survival, and other functions of pathogenic relevance. The latent and lytic viral proteins implicated in HHV-8 pathogenesis and their likely autocrine and para-crine contributions to disease are summarized in Fig. 3. Secondly, "lytic" and "latent" classifications for viral products are not as distinct as once thought. For example, vIL-6, vIRF-1, vIRF-3, K1, and K15 are clearly expressed maximally during productive replication, but there is evidence for their expression during latency as well in some cell types. Furthermore, it is notable that latently expressed vIL-6 and vIRF-3 are of demonstrable importance for PEL cell growth and via-bility. A third key point is that the virally encoded chemokines vCCL-1 and vCCL-2, while secreted during productive replication and thought to function to promote virus production via paracrine effects on the microenvironment (most notably to evade host immune responses), can also act directly on the cells in which they are produced to enhance virus production via antiapoptotic signaling. Such direct pro-replication activity has also been demonstrated for vGPCR. For the v-chemokines and vGPCR, it is possible that induced cellular cytokines may serve similar and/or

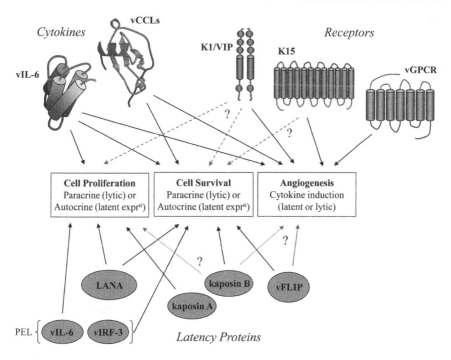

Fig. 3 Overview of potential contributions of HHV-8 proteins to virus-associated neoplasia. The general activities of relevance to HHV-8 malignant pathogenesis are indicated. Both latent and lytic proteins have the potential to contribute to disease, via autocrine and paracrine activities, respectively. Latent expression of signaling receptors encoded by ORFs K1 and K15 has so far not been demonstrated conclusively. Therefore, whether the autocrine activities detected experimentally (dotted lines) reflect direct roles of the receptors in HHV-8 pathogenesis remains uncertain (?). However, contributions of K1 and K15 receptors could contribute to neoplasia via induction of mitogenic, pro-survival, and angiogenic secreted cellular factors. This is also true of the viral chemokine receptor, vGPCR. The viral cytokines (lytic) are secreted and can act by both autocrine and paracrine mechanisms to influence cell growth and survival. These activities can promote virus replication (by autocrine signaling) in addition to contributing to viral pathogenesis (via paracrine effects). Latent expression of vIL-6 in PEL is likely to contribute to pathogenesis, mainly via intracrine signaling. vIRF-3 is also expressed during latency in PEL and like vIL-6 promotes PEL cell viability. The latency proteins have the potential to contribute to HHV-8-associated malignancies by direct autocrine effects on cell proliferation and survival by mechanisms typical of oncogenes and tumor suppressors (see the text). Kaposin B and vFLIP have the potential to function as promoters of cell proliferation and/or angiogenesis, but these functions have not been demonstrated directly (?)

additional activities to promote virus replication in an autocrine fashion as well as having broader effects on the host microenvironment. Finally, several of the HHV-8 proteins have multiple interactions with a broad range of host factors, a point summarized in Table 2 and exemplified by LANA and vIRF-1. Thus, viral proteins can have extraordinarily multifaceted activities via numerous protein interactions, and detailed characterization of these interactions and their functional

Table 2 Virus–host protein interactions and their activities

Protein	Class	Target(s)	Activity/Function	References
LANA	Latent, replication	p53	Pro-survival	Friborg et al. (1999)
		Angiogenin/ annexin A2	Pro-survival	Paudel et al. (2012)
		pRb	Pro-mitogenic	Radkov et al. (2000)
		GSK3β	Pro-mitogenic	Fujimuro et al. (2003)
		c-Myc	Pro-mitogenic	Liu et al. (2007b)
		Histones H2A/B	Viral genome-chromosome tethering	Barbera et al. (2006)
		DNMT3a	Transcriptional repression	Shamay et al. (2006)
		SUV39H1	Transcriptional repression	Sakakibara et al. (2004)
		mSin3/SAP30/CIR	Transcriptional repression	Krithivas et al. (2000)
K13/vFLIP	Latent, signaling	IKKα/β	NF-κB activation/survival	Matta et al. (2007)
		TRAF2/IKKγ?	NF-κB/Jnk activation	Guasparri et al. (2006)
		Procaspase-8	Inhibition of caspase activation/survival	Belanger et al. (2001)
Kaposin A	Latent, signaling	Cytohesin-1, Septin 4 variant	GTPase-mediated signal transduction, activation of several signaling kinases; roles in survival/proliferation?	Kliche et al. (2001), Lin et al. (2007)
Kaposin B	Latent, mRNA reg.	MK2 kinase	MK2 activation, stabilization of ARE-containing mRNAs (e.g. IL-6, PROX1); potential pro-survival and/or mitogenic activity, endothelial reprogramming	McCormick and Ganem (2005)
vIL-6	Ligand, cytokine	gp130/gp80	STAT/MAPK activation; proliferation/survival, pro-inflammatory/angiogenic	Chow et al. (2001), Boulanger et al. (2004)
		VKORC1v2	Proliferation/survival (PEL), unknown mechanism	Chen et al. (2012)

(continued)

Table 2 (continued)

Protein	Class	Target(s)	Activity/Function	References
vCCL-1	Ligand, chemokine	CCR8	Agonist, Th2 polarization; pro-survival, promotes virus replication; pro-angiogenic	Dairaghi et al. (1999), Choi and Nicholas (2008)
vCCL-2	Ligand, chemokine	CCR3, CCR8, CCR5?	Agonist, Th2 polarization; pro-survival, promotes virus replication; pro-angiogenic	Choi and Nicholas (2008),Boshoff et al. (1997), Nakano et al. (2003)
		CCR5, CCR2, CCR10, CXCR4, CX₃CR1, XCR1	Chemokine antagonist; pro-survival, promotes virus replication	Chen et al. (1998), Boshoff et al. (1997), Kledal et al. (1997), Luttichau et al. (2001), (2007)
vCCL-3	Ligand, chemokine	CCR4, XCR1	Agonist, Th2 polarization; pro-angiogenic	Stine et al. (2000), Luttichau et al. (2007)
vGPCR	Signaling receptor	Gα (i, q, 12/13), SHP2	Activation of various pro-survival and mitogenic signaling pathways; promotes virus replication; pro-angiogenic	Couty et al. (2001), Liu et al. (2004), Philpott et al. (2011), Shepard et al. (2001), Sandford et al. (2009)
vIRF-1	Innate response reg.	p53, ATM, p300/CBP, GRIM19, Smads 3/4, BOPs (incl. Bim & Bid), IRF1, IRF3	Pro-survival; IFN signaling inhibition	Burysek et al. (1999a), Choi et al. (2012), Li et al. (2000), Lin et al. (2001), Nakamura et al. (2001), Seo et al. (2001), (2002), (2005), Shin et al. (2006)
vIRF-2	Innate response reg.	PKR, IRF1, IRF2, p65, p300	IFN signaling inhibition	Burysek and Pitha (2001), Burysek et al. (1999b)
vIRF-3	Innate response reg.	p53, IRF3, IRF5, IRF7, 14-3-3, HIF-1α, Foxo3a	Pro-survival; IFN signaling inhibition; angiogenic signaling	Joo et al. 2007, Munoz-Fontela et al. (2007), Rivas et al. (2001), Shin et al. (2008), Weis et al. (2008)

(continued)

Table 2 (continued)

Protein	Class	Target(s)	Activity/Function	References
vIRF-4	Innate response reg.	MDM2	Pro-survival (via p53 destabilization)	Lee et al. (2009b)
		HAUSP	Pro-survival (via p53 destabilization)	Lee et al. (2011)
		CSL	Unknown	Heinzelmann et al. (2010)
K7/vIAP	Apoptosis regulation	Bcl-2/caspase 3	Pro-survival	Wang et al. (2002)
		CAML	Pro-survival	Feng et al. (2002)
		PLIC	Pro-survival	Feng et al. (2004)
K1/VIP	Apoptosis regulation	Src kinases, PI3K, SHP2, PLCγ	Pro-survival/ mitogenic; pro-angiogenic/ inflammatory	Lee et al. (1998a), (2005)
K15	Signaling receptor	Src kinases, TRAFs, intersectin 2, HAX-1	Pro-survival/ mitogenic; pro-angiogenic	Brinkmann et al. (2003), Lim et al. (2007), Pietrek et al. (2010), Sharp et al. (2002)

effects is important for understanding their individual and combined contributions to virus biology and pathogenesis. Such characterization can potentially provide the basis for the development of new antiviral and therapeutic drugs designed to interfere with specific virus–host interactions of critical importance for viral replication or pathogenesis. This chapter has attempted to overview the various novel HHV-8–host interactions and related activities that contribute to these processes and that could perhaps be targeted in this way. The interactions described also illustrate the breadth and complexity of virus–host interactions and suggest that similar activities and mechanisms may be operative in other viral systems.

References

Abend JR, Uldrick T, Ziegelbauer JM (2010) Regulation of tumor necrosis factor-like weak inducer of apoptosis receptor protein (TWEAKR) expression by Kaposi's sarcoma-associated herpesvirus microRNA prevents TWEAK-induced apoptosis and inflammatory cytokine expression. J Virol 84:12139–12151

Albrecht JC, Nicholas J, Cameron KR, Newman C, Fleckenstein B, Honess RW (1992) Herpesvirus saimiri has a gene specifying a homologue of the cellular membrane glycoprotein CD59. Virology 190:527–530

Alexander L, Denekamp L, Knapp A, Auerbach MR, Damania B, Desrosiers RC (2000) The primary sequence of rhesus monkey rhadinovirus isolate 26–95: sequence similarities to

Kaposi's sarcoma-associated herpesvirus and rhesus monkey rhadinovirus isolate 17577. J Virol 74:3388–3398

Ambrosini G, Adida C, Altieri DC (1997) A novel anti-apoptosis gene, survivin, expressed in cancer and lymphoma. Nat Med 3:917–921

An FQ, Compitello N, Horwitz E, Sramkoski M, Knudsen ES, Renne R (2005) The latency-associated nuclear antigen of Kaposi's sarcoma-associated herpesvirus modulates cellular gene expression and protects lymphoid cells from p16 INK4A-induced cell cycle arrest. J Biol Chem 280:3862–3874

An J, Sun Y, Sun R, Rettig MB (2003) Kaposi's sarcoma-associated herpesvirus encoded vFLIP induces cellular IL-6 expression: the role of the NF-kappaB and JNK/AP1 pathways. Oncogene 22:3371–3385

Angell JE, Lindner DJ, Shapiro PS, Hofmann ER, Kalvakolanu DV (2000) Identification of GRIM-19, a novel cell death-regulatory gene induced by the interferon-beta and retinoic acid combination, using a genetic approach. J Biol Chem 275:33416–33426

Aoki Y, Feldman GM, Tosato G (2003) Inhibition of STAT3 signaling induces apoptosis and decreases survivin expression in primary effusion lymphoma. Blood 101:1535–1542

Aoki Y, Jaffe ES, Chang Y, Jones K, Teruya-Feldstein J, Moore PS, Tosato G (1999) Angiogenesis and hematopoiesis induced by Kaposi's sarcoma-associated herpesvirus-encoded interleukin-6. Blood 93:4034–4043

Aoki Y, Narazaki M, Kishimoto T, Tosato G (2001) Receptor engagement by viral interleukin-6 encoded by Kaposi sarcoma-associated herpesvirus. Blood 98:3042–3049

Aoki Y, Tosato G (1999) Role of vascular endothelial growth factor/vascular permeability factor in the pathogenesis of Kaposi's sarcoma-associated herpesvirus-infected primary effusion lymphomas. Blood 94:4247–4254

Aoki Y, Tosato G (2003) Targeted inhibition of angiogenic factors in AIDS-related disorders. Curr Drug Targets Infect Disord 3:115–128

Areste C, Mutocheluh M, Blackbourn DJ (2009) Identification of caspase-mediated decay of interferon regulatory factor-3, exploited by a Kaposi sarcoma-associated herpesvirus immunoregulatory protein. J Biol Chem 284:23272–23285

Arvanitakis L, Mesri EA, Nador RG, Said JW, Asch AS, Knowles DM, Cesarman E (1996) Establishment and characterization of a primary effusion (body cavity-based) lymphoma cell line (BC-3) harboring kaposi's sarcoma-associated herpesvirus (KSHV/HHV-8) in the absence of Epstein-Barr virus. Blood 88:2648–2654

Bais C, Santomasso B, Coso O, Arvanitakis L, Raaka EG, Gutkind JS, Asch AS, Cesarman E, Gershengorn MC, Mesri EA (1998) G-protein-coupled receptor of Kaposi's sarcoma-associated herpesvirus is a viral oncogene and angiogenesis activator. Nature 391:86–89

Barbera AJ, Chodaparambil JV, Kelley-Clarke B, Joukov V, Walter JC, Luger K, Kaye KM (2006) The nucleosomal surface as a docking station for Kaposi's sarcoma herpesvirus LANA. Science 311:856–861

Bartel DP (2004) MicroRNAs: genomics, biogenesis, mechanism, and function. Cell 116:281–297

Bartoli M, Gu X, Tsai NT, Venema RC, Brooks SE, Marrero MB, Caldwell RB (2000) Vascular endothelial growth factor activates STAT proteins in aortic endothelial cells. J Biol Chem 275:33189–33192

Belanger C, Gravel A, Tomoiu A, Janelle ME, Gosselin J, Tremblay MJ, Flamand L (2001) Human herpesvirus 8 viral FLICE-inhibitory protein inhibits Fas-mediated apoptosis through binding and prevention of procaspase-8 maturation. J Hum Virol 4:62–73

Bellare P, Ganem D (2009) Regulation of KSHV lytic switch protein expression by a virus-encoded microRNA: an evolutionary adaptation that fine-tunes lytic reactivation. Cell Host Microbe 6:570–575

Bertin J, Armstrong RC, Ottilie S, Martin DA, Wang Y, Banks S, Wang GH, Senkevich TG, Alnemri ES, Moss B, Lenardo MJ, Tomaselli KJ, Cohen JI (1997) Death effector domain-containing herpesvirus and poxvirus proteins inhibit both Fas- and TNFR1-induced apoptosis. Proc Natl Acad Sci U S A 94:1172–1176

Boshoff C, Endo Y, Collins PD, Takeuchi Y, Reeves JD, Schweickart VL, Siani MA, Sasaki T, Williams TJ, Gray PW, Moore PS, Chang Y, Weiss RA (1997) Angiogenic and HIV-inhibitory functions of KSHV-encoded chemokines. Science 278:290–294

Boulanger MJ, Chow DC, Brevnova EE, Garcia KC (2003) Hexameric structure and assembly of the interleukin-6/IL-6 alpha-receptor/gp130 complex. Science 300:2101–2104

Boulanger MJ, Cow DC, Brevnova, EE, Martick M. Sandford G, Nicholas J, Garcia KC (2004) Molecular mechanisms for viral mimicry of a human cytokine: activation of gp130 by HHV-8 interleukin 6. J Mol Biol 335:641–654

Bram RJ, Crabtree GR (1994) Calcium signalling in T cells stimulated by a cyclophilin B-binding protein. Nature 371:355–358

Brinkmann MM, Glenn M, Rainbow L, Kieser A, Henke-Gendo C, Schulz TF (2003) Activation of mitogen-activated protein kinase and NF-kappaB pathways by a Kaposi's sarcoma-associated herpesvirus K15 membrane protein. J Virol 77:9346–9358

Brinkmann MM, Pietrek M, Dittrich-Breiholz O, Kracht M, Schulz TF (2007) Modulation of host gene expression by the K15 protein of Kaposi's sarcoma-associated herpesvirus. J Virol 81:42–58

Brinkmann MM, Schulz TF (2006) Regulation of intracellular signalling by the terminal membrane proteins of members of the Gammaherpesvirinae. J Gen Virol 87:1047–1074

Brown HJ, Song MJ, Deng H, Wu TT, Cheng G, Sun R (2003) NF-kappaB inhibits gammaherpesvirus lytic replication. J Virol 77:8532–8540

Bubman D, Guasparri I, Cesarman E (2007) Deregulation of c-Myc in primary effusion lymphoma by Kaposi's sarcoma herpesvirus latency-associated nuclear antigen. Oncogene 26:4979–4986

Burger R, Neipel F, Fleckenstein B, Savino R, Ciliberto G, Kalden JR, Gramatzki M (1998) Human herpesvirus type 8 interleukin-6 homologue is functionally active on human myeloma cells. Blood 91:1858–1863

Burger R, Wendler J, Antoni K, Helm G, Kalden JR, Gramatzki M (1994) Interleukin-6 production in B-cell neoplasias and Castleman's disease: evidence for an additional paracrine loop. Ann Hematol 69:25–31

Burysek L, Pitha PM (2001) Latently expressed human herpesvirus 8-encoded interferon regulatory factor 2 inhibits double-stranded RNA-activated protein kinase. J Virol 75:2345–2352

Burysek L, Yeow WS, Lubyova B, Kellum M, Schafer SL, Huang YQ, Pitha PM (1999a) Functional analysis of human herpesvirus 8-encoded viral interferon regulatory factor 1 and its association with cellular interferon regulatory factors and p300. J Virol 73:7334–7342

Burysek L, Yeow WS, Pitha PM (1999b) Unique properties of a second human herpesvirus 8-encoded interferon regulatory factor (vIRF-2). J Hum Virol 2:19–32

Cai X, Lu S, Zhang Z, Gonzalez CM, Damania B, Cullen BR (2005) Kaposi's sarcoma-associated herpesvirus expresses an array of viral microRNAs in latently infected cells. Proc Natl Acad Sci U S A 102:5570–5575

Cannon M (2007) The KSHV and other human herpesviral G protein-coupled receptors. Curr Top Microbiol Immunol 312:137–156

Cannon M, Philpott NJ, Cesarman E (2003) The Kaposi's sarcoma-associated herpesvirus G protein-coupled receptor has broad signaling effects in primary effusion lymphoma cells. J Virol 77:57–67

Carbone A, Cilia AM, Gloghini A, Capello D, Perin T, Bontempo D, Canzonieri V, Tirelli U, Volpe R, Gaidano G (2000) Primary effusion lymphoma cell lines harbouring human herpesvirus type-8. Leuk Lymphoma 36:447–456

Caselli E, Fiorentini S, Amici C, Di Luca D, Caruso A, Santoro MG (2007) Human herpesvirus 8 acute infection of endothelial cells induces monocyte chemoattractant protein 1-dependent capillary-like structure formation: role of the IKK/NF-kappaB pathway. Blood 109:2718–2726

Chandriani S, Ganem D (2010) Array-based transcript profiling and limiting-dilution reverse transcription-PCR analysis identify additional latent genes in Kaposi's sarcoma-associated herpesvirus. J Virol 84:5565–5573

Chang Y, Moore PS (1996) Kaposi's Sarcoma (KS)-associated herpesvirus and its role in KS. Infect Agents Dis 5:215–222

Chaudhary PM, Jasmin A, Eby MT, Hood L (1999) Modulation of the NF-kappa B pathway by virally encoded death effector domains-containing proteins. Oncogene 18:5738–5746

Chen D, Cousins E, Sandford G, Nicholas J (2012) Human herpesvirus 8 viral interleukin-6 interacts with splice variant 2 of vitamin K epoxide reductase complex subunit 1. J Virol 86:1577–1588

Chen D, Nicholas J (2006) Structural requirements for gp80 independence of human herpesvirus 8 interleukin-6 (vIL-6) and evidence for gp80 stabilization of gp130 signaling complexes induced by vIL-6. J Virol 80:9811–9821

Chen D, Sandford G, Nicholas J (2009a) Intracellular signaling mechanisms and activities of human herpesvirus 8 interleukin-6. J Virol 83:722–733

Chen S, Bacon KB, Li L, Garcia GE, Xia Y, Lo D, Thompson DA, Siani MA, Yamamoto T, Harrison JK, Feng L (1998) In vivo inhibition of CC and CX3C chemokine-induced leukocyte infiltration and attenuation of glomerulonephritis in Wistar-Kyoto (WKY) rats by vMIP-II. J Exp Med 188:193–198

Chen W, Hilton IB, Staudt MR, Burd CE, Dittmer DP (2010) Distinct p53, p53:LANA, and LANA complexes in Kaposi's Sarcoma–associated Herpesvirus Lymphomas. J Virol 84:3898–3908

Chen X, Cheng L, Jia X, Zeng Y, Yao S, Lv Z, Qin D, Fang X, Lei Y, Lu C (2009b) Human immunodeficiency virus type 1 Tat accelerates Kaposi sarcoma-associated herpesvirus Kaposin A-mediated tumorigenesis of transformed fibroblasts in vitro as well as in nude and immunocompetent mice. Neoplasia 11:1272–1284

Choi JK, Lee BS, Shim SN, Li M, Jung JU (2000) Identification of the novel K15 gene at the rightmost end of the Kaposi's sarcoma-associated herpesvirus genome. J Virol 74:436–446

Choi YB, Nicholas J (2008) Autocrine and paracrine promotion of cell survival and virus replication by human herpesvirus 8 chemokines. J Virol 82:6501–6513

Choi YB, Nicholas J (2010) Bim nuclear translocation and inactivation by viral interferon regulatory factor. PLoS Pathog 6:e1001031

Choi YB, Sandford G, Nicholas J (2012) Human herpesvirus 8 interferon regulatory factor-mediated BH3-only protein Inhibition via Bid BH3-B mimicry. PLoS Pathog 8(6):e1002748

Chow D, He X, Snow AL, Rose-John S, Garcia KC (2001) Structure of an extracellular gp130 cytokine receptor signaling complex. Science 291:2150–2155

Chugh P, Matta H, Schamus S, Zachariah S, Kumar A, Richardson JA, Smith AL, Chaudhary PM (2005) Constitutive NF-kappaB activation, normal Fas-induced apoptosis, and increased incidence of lymphoma in human herpes virus 8 K13 transgenic mice. Proc Natl Acad Sci U S A 102:12885–12890

Couty JP, Geras-Raaka E, Weksler BB, Gershengorn MC (2001) Kaposi's sarcoma-associated herpesvirus G protein-coupled receptor signals through multiple pathways in endothelial cells. J Biol Chem 276:33805–33811

Crump MP, Elisseeva E, Gong J, Clark-Lewis I, Sykes BD (2001) Structure/function of human herpesvirus-8 MIP-II (1–71) and the antagonist N-terminal segment (1–10). FEBS Lett 489:171–175

Cunningham C, Barnard S, Blackbourn DJ, Davison AJ (2003) Transcription mapping of human herpesvirus 8 genes encoding viral interferon regulatory factors. J Gen Virol 84:1471–1483

Dairaghi DJ, Fan RA, McMaster BE, Hanley MR, Schall TJ (1999) HHV8-encoded vMIP-I selectively engages chemokine receptor CCR8. Agonist and antagonist profiles of viral chemokines. J Biol Chem 274:21569–21574

Damania B, Li M, Choi JK, Alexander L, Jung JU, Desrosiers RC (1999) Identification of the R1 oncogene and its protein product from the rhadinovirus of rhesus monkeys. J Virol 73:5123–5131

Dittmer DP (2003) Transcription profile of Kaposi's sarcoma-associated herpesvirus in primary Kaposi's sarcoma lesions as determined by real-time PCR arrays. Cancer Res 63:2010–2015

Djerbi M, Screpanti V, Catrina AI, Bogen B, Biberfeld P, Grandien A (1999) The inhibitor of death receptor signaling, FLICE-inhibitory protein defines a new class of tumor progression factors. J Exp Med 190:1025–1032

Dolken L, Malterer G, Erhard F, Kothe S, Friedel CC, Suffert G, Marcinowski L, Motsch N, Barth S, Beitzinger M, Lieber D, Bailer SM, Hoffmann R, Ruzsics Z, Kremmer E, Pfeffer S, Zimmer R, Koszinowski UH, Grasser F, Meister G, Haas J (2010) Systematic analysis of viral and cellular microRNA targets in cells latently infected with human gamma-herpesviruses by RISC immunoprecipitation assay. Cell Host Microbe 7:324–334

Ensoli B, Sgadari C, Barillari G, Sirianni MC, Sturzl M, Monini P (2001) Biology of Kaposi's sarcoma. Eur J Cancer 37:1251–1269

Ensoli B, Sturzl M (1998) Kaposi's sarcoma: a result of the interplay among inflammatory cytokines, angiogenic factors and viral agents. Cytokine Growth Factor Rev 9:63–83

Esteban M, Garcia MA, Domingo-Gil E, Arroyo J, Nombela C, Rivas C (2003) The latency protein LANA2 from Kaposi's sarcoma-associated herpesvirus inhibits apoptosis induced by dsRNA-activated protein kinase but not RNase L activation. J Gen Virol 84:1463–1470

Feng P, Park J, Lee BS, Lee SH, Bram RJ, Jung JU (2002) Kaposi's sarcoma-associated herpesvirus mitochondrial K7 protein targets a cellular calcium-modulating cyclophilin ligand to modulate intracellular calcium concentration and inhibit apoptosis. J Virol 76:11491–11504

Feng P, Scott CW, Cho NH, Nakamura H, Chung YH, Monteiro MJ, Jung JU (2004) Kaposi's sarcoma-associated herpesvirus K7 protein targets a ubiquitin-like/ubiquitin-associated domain-containing protein to promote protein degradation. Mol Cell Biol 24:3938–3948

Field N, Low W, Daniels M, Howell S, Daviet L, Boshoff C, Collins M (2003) KSHV vFLIP binds to IKK-gamma to activate IKK. J Cell Sci 116:3721–3728

Fielding CA, McLoughlin RM, Colmont CS, Kovaleva M, Harris DA, Rose-John S, Topley N, Jones SA (2005) Viral IL-6 blocks neutrophil infiltration during acute inflammation. J Immunol 175:4024–4029

Friborg J Jr, Kong W, Hottiger MO, Nabel GJ (1999) p53 inhibition by the LANA protein of KSHV protects against cell death. Nature 402:889–894

Fujimuro M, Hayward SD (2003) The latency-associated nuclear antigen of Kaposi's sarcoma-associated herpesvirus manipulates the activity of glycogen synthase kinase-3beta. J Virol 77:8019–8030

Fujimuro M, Wu FY, ApRhys C, Kajumbula H, Young DB, Hayward GS, Hayward SD (2003) A novel viral mechanism for dysregulation of beta-catenin in Kaposi's sarcoma-associated herpesvirus latency. Nat Med 9:300–306

Fuld S, Cunningham C, Klucher K, Davison AJ, Blackbourn DJ (2006) Inhibition of interferon signaling by the Kaposi's sarcoma-associated herpesvirus full-length viral interferon regulatory factor 2 protein. J Virol 80:3092–3097

Gaidano G, Pastore C, Gloghini A, Volpe G, Capello D, Polito P, Vaccher E, Tirelli U, Saglio G, Carbone A (1997) Human herpesvirus type-8 (HHV-8) in haematopoietic neoplasia. Leuk Lymphoma 24:257–266

Glas R, Franksson L, Une C, Eloranta ML, Ohlen C, Orn A, Karre K (2000) Recruitment and activation of natural killer (NK) cells in vivo determined by the target cell phenotype. An adaptive component of NK cell-mediated responses. J Exp Med 191:129–138

Glenn M, Rainbow L, Aurade F, Davison A, Schulz TF (1999) Identification of a spliced gene from Kaposi's sarcoma-associated herpesvirus encoding a protein with similarities to latent membrane proteins 1 and 2A of Epstein-Barr virus. J Virol 73:6953–6963

Godfrey A, Anderson J, Papanastasiou A, Takeuchi Y, Boshoff C (2005) Inhibiting primary effusion lymphoma by lentiviral vectors encoding short hairpin RNA. Blood 105:2510–2518

Gottwein E, Corcoran DL, Mukherjee N, Skalsky RL, Hafner M, Nusbaum JD, Shamulailatpam P, Love CL, Dave SS, Tuschl T, Ohler U, Cullen BR (2011) Viral microRNA targetome of KSHV-infected primary effusion lymphoma cell lines. Cell Host Microbe 10:515–526

Gottwein E, Cullen BR (2010) A human herpesvirus microRNA inhibits p21 expression and attenuates p21-mediated cell cycle arrest. J Virol 84:5229–5237

Gottwein E, Mukherjee N, Sachse C, Frenzel C, Majoros WH, Chi JT, Braich R, Manoharan M, Soutschek J, Ohler U, Cullen BR (2007) A viral microRNA functions as an orthologue of cellular miR-155. Nature 450:1096–1099

Grossmann C, Ganem D (2008) Effects of NFkappaB activation on KSHV latency and lytic reactivation are complex and context-dependent. Virology 375:94–102

Grossmann C, Podgrabinska S, Skobe M, Ganem D (2006) Activation of NF-kappaB by the latent vFLIP gene of Kaposi's sarcoma-associated herpesvirus is required for the spindle shape of virus-infected endothelial cells and contributes to their proinflammatory phenotype. J Virol 80:7179–7185

Guasparri I, Keller SA, Cesarman E (2004) KSHV vFLIP is essential for the survival of infected lymphoma cells. J Exp Med 199:993–1003

Guasparri I, Wu H, Cesarman E (2006) The KSHV oncoprotein vFLIP contains a TRAF-interacting motif and requires TRAF2 and TRAF3 for signalling. EMBO Rep 7:114–119

Haddad L, El Hajj H, Abou-Merhi R, Kfoury Y, Mahieux R, El-Sabban M, Bazarbachi A (2008) KSHV-transformed primary effusion lymphoma cells induce a VEGF-dependent angiogenesis and establish functional gap junctions with endothelial cells. Leukemia 22:826–834

Hansen A, Henderson S, Lagos D, Nikitenko L, Coulter E, Roberts S, Gratrix F, Plaisance K, Renne R, Bower M, Kellam P, Boshoff C (2010) KSHV-encoded miRNAs target MAF to induce endothelial cell reprogramming. Genes Dev 24:195–205

Hayward GS, Zong JC (2007) Modern evolutionary history of the human KSHV genome. Curr Top Microbiol Immunol 312:1–42

Heinrich PC, Behrmann I, Haan S, Hermanns HM, Muller-Newen G, Schaper F (2003) Principles of interleukin (IL)-6-type cytokine signalling and its regulation. Biochem J 374:1–20

Heinrich PC, Behrmann I, Muller-Newen G, Schaper F, Graeve L (1998) Interleukin-6-type cytokine signalling through the gp130/Jak/STAT pathway. Biochem J 334(Pt 2):297–314

Heinzelmann K, Scholz BA, Nowak A, Fossum E, Kremmer E, Haas J, Frank R, Kempkes B (2010) Kaposi's sarcoma-associated herpesvirus viral interferon regulatory factor 4 (vIRF4/K10) is a novel interaction partner of CSL/CBF1, the major downstream effector of Notch signaling. J Virol 84:12255–12264

Hu S, Vincenz C, Buller M, Dixit VM (1997) A novel family of viral death effector domain-containing molecules that inhibit both CD-95- and tumor necrosis factor receptor-1-induced apoptosis. J Biol Chem 272:9621–9624

Ishiyama T, Nakamura S, Akimoto Y, Koike M, Tomoyasu S, Tsuruoka N, Murata Y, Sato T, Wakabayashi Y, Chiba S (1994) Immunodeficiency and IL-6 production by peripheral blood monocytes in multicentric Castleman's disease. Br J Haematol 86:483–489

Jenner RG, Alba MM, Boshoff C, Kellam P (2001) Kaposi's sarcoma-associated herpesvirus latent and lytic gene expression as revealed by DNA arrays. J Virol 75:891–902

Jones KD, Aoki Y, Chang Y, Moore PS, Yarchoan R, Tosato G (1999) Involvement of interleukin-10 (IL-10) and viral IL-6 in the spontaneous growth of Kaposi's sarcoma herpesvirus-associated infected primary effusion lymphoma cells. Blood 94:2871–2879

Joo CH, Shin YC, Gack M, Wu L, Levy D, Jung JU (2007) Inhibition of interferon regulatory factor 7 (IRF7)-mediated interferon signal transduction by the Kaposi's sarcoma-associated herpesvirus viral IRF homolog vIRF3. J Virol 81:8282–8292

Karim R, Tse G, Putti T, Scolyer R, Lee S (2004) The significance of the Wnt pathway in the pathology of human cancers. Pathology 36:120–128

Kawa K (2000) Epstein-Barr virus–associated diseases in humans. Int J Hematol 71:108–117

Kaye KM, Izumi KM, Kieff E (1993) Epstein-Barr virus latent membrane protein 1 is essential for B-lymphocyte growth transformation. Proc Natl Acad Sci U S A 90:9150–9154

Keller SA, Schattner EJ, Cesarman E (2000) Inhibition of NF-kappaB induces apoptosis of KSHV-infected primary effusion lymphoma cells. Blood 96:2537–2542

Kishimoto T, Akira S, Narazaki M, Taga T (1995) Interleukin-6 family of cytokines and gp130. Blood 86:1243–1254

Kledal TN, Rosenkilde MM, Coulin F, Simmons G, Johnsen AH, Alouani S, Power CA, Luttichau HR, Gerstoft J, Clapham PR, Clark-Lewis I, Wells TN, Schwartz TW (1997) A broad-spectrum chemokine antagonist encoded by Kaposi's sarcoma-associated herpesvirus. Science 277:1656–1659

Kleijnen MF, Shih AH, Zhou P, Kumar S, Soccio RE, Kedersha NL, Gill G, Howley PM (2000) The hPLIC proteins may provide a link between the ubiquitination machinery and the proteasome. Mol Cell 6:409–419

Kliche S, Nagel W, Kremmer E, Atzler C, Ege A, Knorr T, Koszinowski U, Kolanus W, Haas J (2001) Signaling by human herpesvirus 8 kaposin A through direct membrane recruitment of cytohesin-1. Mol Cell 7:833–843

Kovaleva M, Bussmeyer I, Rabe B, Grotzinger J, Sudarman E, Eichler J, Conrad U, Rose-John S, Scheller J (2006) Abrogation of viral interleukin-6 (vIL-6)-induced signaling by intracellular retention and neutralization of vIL-6 with an anti-vIL-6 single-chain antibody selected by phage display. J Virol 80:8510–8520

Krithivas A, Young DB, Liao G, Greene D, Hayward SD (2000) Human herpesvirus 8 LANA interacts with proteins of the mSin3 corepressor complex and negatively regulates Epstein-Barr virus gene expression in dually infected PEL cells. J Virol 74:9637–9645

Kuwana T, Bouchier-Hayes L, Chipuk JE, Bonzon C, Sullivan BA, Green DR, Newmeyer DD (2005) BH3 domains of BH3-only proteins differentially regulate Bax-mediated mitochondrial membrane permeabilization both directly and indirectly. Mol Cell 17:525–535

Lagunoff M, Ganem D (1997) The structure and coding organization of the genomic termini of Kaposi's sarcoma-associated herpesvirus. Virology 236:147–154

Lagunoff M, Lukac DM, Ganem D (2001) Immunoreceptor tyrosine-based activation motif-dependent signaling by Kaposi's sarcoma-associated herpesvirus K1 protein: effects on lytic viral replication. J Virol 75:5891–5898

Lee BS, Connole M, Tang Z, Harris NL, Jung JU (2003a) Structural analysis of the Kaposi's sarcoma-associated herpesvirus K1 protein. J Virol 77:8072–8086

Lee BS, Lee SH, Feng P, Chang H, Cho NH, Jung JU (2005) Characterization of the Kaposi's sarcoma-associated herpesvirus K1 signalosome. J Virol 79:12173–12184

Lee H, Guo J, Li M, Choi JK, DeMaria M, Rosenzweig M, Jung JU (1998a) Identification of an immunoreceptor tyrosine-based activation motif of K1 transforming protein of Kaposi's sarcoma-associated herpesvirus. Mol Cell Biol 18:5219–5228

Lee H, Veazey R, Williams K, Li M, Guo J, Neipel F, Fleckenstein B, Lackner A, Desrosiers RC, Jung JU (1998b) Deregulation of cell growth by the K1 gene of Kaposi's sarcoma-associated herpesvirus. Nat Med 4:435–440

Lee HR, Choi WC, Lee S, Hwang J, Hwang E, Guchhait K, Haas J, Toth Z, Jeon YH, Oh TK, Kim MH, Jung JU (2011) Bilateral inhibition of HAUSP deubiquitinase by a viral interferon regulatory factor protein. Nat Struct Mol Biol 18:1336–1344

Lee HR, Kim MH, Lee JS, Liang C, Jung JU (2009a) Viral interferon regulatory factors. J Interferon Cytokine Res 29:621–627

Lee HR, Toth Z, Shin YC, Lee JS, Chang H, Gu W, Oh TK, Kim MH, Jung JU (2009b) Kaposi's sarcoma-associated herpesvirus viral interferon regulatory factor 4 targets MDM2 to deregulate the p53 tumor suppressor pathway. J Virol 83:6739–6747

Lee Y, Ahn C, Han J, Choi H, Kim J, Yim J, Lee J, Provost P, Radmark O, Kim S, Kim VN (2003b) The nuclear RNase III Drosha initiates microRNA processing. Nature 425:415–419

Lei X, Bai Z, Ye F, Xie J, Kim CG, Huang Y, Gao SJ (2010) Regulation of NF-kappaB inhibitor IkappaBalpha and viral replication by a KSHV microRNA. Nat Cell Biol 12:193–199

Li H, Komatsu T, Dezube BJ, Kaye KM (2002) The Kaposi's sarcoma-associated herpesvirus K12 transcript from a primary effusion lymphoma contains complex repeat elements, is spliced, and initiates from a novel promoter. J Virol 76:11880–11888

Li M, Damania B, Alvarez X, Ogryzko V, Ozato K, Jung JU (2000) Inhibition of p300 histone acetyltransferase by viral interferon regulatory factor. Mol Cell Biol 20:8254–8263

Liang D, Lin X, Lan K (2012) Looking at Kaposi's Sarcoma-Associated Herpesvirus-Host Interactions from a microRNA Viewpoint. Front Microbiol 2:271

Lim CS, Seet BT, Ingham RJ, Gish G, Matskova L, Winberg G, Ernberg I, Pawson T (2007) The K15 protein of Kaposi's sarcoma-associated herpesvirus recruits the endocytic regulator intersectin 2 through a selective SH3 domain interaction. Biochemistry 46:9874–9885

Lin CW, Tu PF, Hsiao NW, Chang CY, Wan L, Lin YT, Chang HW (2007) Identification of a novel septin 4 protein binding to human herpesvirus 8 kaposin A protein using a phage display cDNA library. J Virol Methods 143:65–72

Lin R, Genin P, Mamane Y, Sgarbanti M, Battistini A, Harrington WJ Jr, Barber GN, Hiscott J (2001) HHV-8 encoded vIRF-1 represses the interferon antiviral response by blocking IRF-3 recruitment of the CBP/p300 coactivators. Oncogene 20:800–811

Lin X, Liang D, He Z, Deng Q, Robertson ES, Lan K (2011) miR-K12-7-5p encoded by Kaposi's sarcoma-associated herpesvirus stabilizes the latent state by targeting viral ORF50/RTA. PLoS One 6:e16224

Lindner SE, Sugden B (2007) The plasmid replicon of Epstein-Barr virus: mechanistic insights into efficient, licensed, extrachromosomal replication in human cells. Plasmid 58:1–12

Liu C, Okruzhnov Y, Li H, Nicholas J (2001) Human herpesvirus 8 (HHV-8)-encoded cytokines induce expression of and autocrine signaling by vascular endothelial growth factor (VEGF) in HHV-8-infected primary-effusion lymphoma cell lines and mediate VEGF-independent antiapoptotic effects. J Virol 75:10933–10940

Liu C, Sandford G, Fei G, Nicholas J (2004) Galpha protein selectivity determinant specified by a viral chemokine receptor-conserved region in the C tail of the human herpesvirus 8 g protein-coupled receptor. J Virol 78:2460–2471

Liu J, Martin H, Shamay M, Woodard C, Tang QQ, Hayward SD (2007a) Kaposi's sarcoma-associated herpesvirus LANA protein downregulates nuclear glycogen synthase kinase 3 activity and consequently blocks differentiation. J Virol 81:4722–4731

Liu J, Martin HJ, Liao G, Hayward SD (2007b) The Kaposi's sarcoma-associated herpesvirus LANA protein stabilizes and activates c-Myc. J Virol 81:10451–10459

Liu L, Eby MT, Rathore N, Sinha SK, Kumar A, Chaudhary PM (2002) The human herpes virus 8-encoded viral FLICE inhibitory protein physically associates with and persistently activates the Ikappa B kinase complex. J Biol Chem 277:13745–13751

Liu Y, Sun R, Lin X, Liang D, Deng Q, Lan K (2012) Kaposi's sarcoma-associated herpesvirus-encoded microRNA miR-K12-11 attenuates transforming growth factor beta signaling through suppression of SMAD5. J Virol 86:1372–1381

Louahed J, Struyf S, Demoulin JB, Parmentier M, Van Snick J, Van Damme J, Renauld JC (2003) CCR8-dependent activation of the RAS/MAPK pathway mediates anti-apoptotic activity of I-309/ CCL1 and vMIP-I. Eur J Immunol 33:494–501

Lu CC, Li Z, Chu CY, Feng J, Sun R, Rana TM (2010a) MicroRNAs encoded by Kaposi's sarcoma-associated herpesvirus regulate viral life cycle. EMBO Rep 11:784–790

Lu F, Stedman W, Yousef M, Renne R, Lieberman PM (2010b) Epigenetic regulation of Kaposi's sarcoma-associated herpesvirus latency by virus-encoded microRNAs that target Rta and the cellular Rbl2-DNMT pathway. J Virol 84:2697–2706

Lubyova B, Kellum MJ, Frisancho JA, Pitha PM (2007) Stimulation of c-Myc transcriptional activity by vIRF-3 of Kaposi sarcoma-associated herpesvirus. J Biol Chem 282:31944–31953

Lubyova B, Pitha PM (2000) Characterization of a novel human herpesvirus 8-encoded protein, vIRF-3, that shows homology to viral and cellular interferon regulatory factors. J Virol 74:8194–8201

Lund E, Guttinger S, Calado A, Dahlberg JE, Kutay U (2004) Nuclear export of microRNA precursors. Science 303:95–98

Luttichau HR (2008) The herpesvirus 8 encoded chemokines vCCL2 (vMIP-II) and vCCL3 (vMIP-III) target the human but not the murine lymphotactin receptor. Virol J 5:50

Luttichau HR, Johnsen AH, Jurlander J, Rosenkilde MM, Schwartz TW (2007) Kaposi sarcoma-associated herpes virus targets the lymphotactin receptor with both a broad spectrum antagonist vCCL2 and a highly selective and potent agonist vCCL3. J Biol Chem 282:17794–17805

Luttichau HR, Lewis IC, Gerstoft J, Schwartz TW (2001) The herpesvirus 8-encoded chemokine vMIP-II, but not the poxvirus-encoded chemokine MC148, inhibits the CCR10 receptor. Eur J Immunol 31:1217–1220

Ma X, Kalakonda S, Srinivasula SM, Reddy SP, Platanias LC, Kalvakolanu DV (2007) GRIM-19 associates with the serine protease HtrA2 for promoting cell death. Oncogene 26:4842–4849

Mandel-Gutfreund Y, Kosti I, Larisch S (2011) ARTS, the unusual septin: structural and functional aspects. Biol Chem 392:783–790

Marcos-Villar L, Lopitz-Otsoa F, Gallego P, Munoz-Fontela C, Gonzalez-Santamaria J, Campagna M, Shou-Jiang G, Rodriguez MS, Rivas C (2009) Kaposi's sarcoma-associated herpesvirus protein LANA2 disrupts PML oncogenic domains and inhibits PML-mediated transcriptional repression of the survivin gene. J Virol 83:8849–8858

Matta H, Chaudhary PM (2004) Activation of alternative NF-kappa B pathway by human herpes virus 8-encoded Fas-associated death domain-like IL-1 beta-converting enzyme inhibitory protein (vFLIP). Proc Natl Acad Sci U S A 101:9399–9404

Matta H, Mazzacurati L, Schamus S, Yang T, Sun Q, Chaudhary PM (2007) Kaposi's sarcoma-associated herpesvirus (KSHV) oncoprotein K13 bypasses TRAFs and directly interacts with the IkappaB kinase complex to selectively activate NF-kappaB without JNK activation. J Biol Chem 282:24858–24865

Mbulaiteye S, Marshall V, Bagni RK, Wang CD, Mbisa G, Bakaki PM, Owor AM, Ndugwa CM, Engels EA, Katongole-Mbidde E, Biggar RJ, Whitby D (2006) Molecular evidence for mother-to-child transmission of Kaposi sarcoma-associated herpesvirus in Uganda and K1 gene evolution within the host. J Infect Dis 193:1250–1257

McCormick C, Ganem D (2005) The kaposin B protein of KSHV activates the p38/MK2 pathway and stabilizes cytokine mRNAs. Science 307:739–741

McCormick C, Ganem D (2006) Phosphorylation and function of the kaposin B direct repeats of Kaposi's sarcoma-associated herpesvirus. J Virol 80:6165–6170

Mesri EA, Cesarman E, Boshoff C (2010) Kaposi's sarcoma and its associated herpesvirus. Nat Rev Cancer 10:707–719

Miles SA, Rezai AR, Salazar-Gonzalez JF, Vander Meyden M, Stevens RH, Logan DM, Mitsuyasu RT, Taga T, Hirano T, Kishimoto T et al (1990) AIDS Kaposi sarcoma-derived cells produce and respond to interleukin 6. Proc Natl Acad Sci U S A 87:4068–4072

Montaner S, Sodhi A, Molinolo A, Bugge TH, Sawai ET, He Y, Li Y, Ray PE, Gutkind JS (2003) Endothelial infection with KSHV genes in vivo reveals that vGPCR initiates Kaposi's sarcomagenesis and can promote the tumorigenic potential of viral latent genes. Cancer Cell 3:23–36

Montaner S, Sodhi A, Servitja JM, Ramsdell AK, Barac A, Sawai ET, Gutkind JS (2004) The small GTPase Rac1 links the Kaposi sarcoma-associated herpesvirus vGPCR to cytokine secretion and paracrine neoplasia. Blood 104:2903–2911

Moore PS, Boshoff C, Weiss RA, Chang Y (1996) Molecular mimicry of human cytokine and cytokine response pathway genes by KSHV. Science 274:1739–1744

Mori Y, Nishimoto N, Ohno M, Inagi R, Dhepakson P, Amou K, Yoshizaki K, Yamanishi K (2000) Human herpesvirus 8-encoded interleukin-6 homologue (viral IL-6) induces endogenous human IL-6 secretion. J Med Virol 61:332–335

Munoz-Fontela C, Collado M, Rodriguez E, Garcia MA, Alvarez-Barrientos A, Arroyo J, Nombela C, Rivas C (2005) Identification of a nuclear export signal in the KSHV latent protein LANA2 mediating its export from the nucleus. Exp Cell Res 311:96–105

Munoz-Fontela C, Marcos-Villar L, Gallego P, Arroyo J, Da Costa M, Pomeranz KM, Lam EW, Rivas C (2007) Latent protein LANA2 from Kaposi's sarcoma-associated herpesvirus interacts with 14-3-3 proteins and inhibits FOXO3a transcription factor. J Virol 81:1511–1516

Muralidhar S, Pumfery AM, Hassani M, Sadaie MR, Kishishita M, Brady JN, Doniger J, Medveczky P, Rosenthal LJ (1998) Identification of kaposin (open reading frame K12) as a human herpesvirus 8 (Kaposi's sarcoma-associated herpesvirus) transforming gene. J Virol 72:4980–4988

Muralidhar S, Veytsmann G, Chandran B, Ablashi D, Doniger J, Rosenthal LJ (2000) Characterization of the human herpesvirus 8 (Kaposi's sarcoma-associated herpesvirus) oncogene, kaposin (ORF K12). J Clin Virol 16:203–213

Murthy SC, Trimble JJ, Desrosiers RC (1989) Deletion mutants of herpesvirus saimiri define an open reading frame necessary for transformation. J Virol 63:3307–3314

Nachmani D, Stern-Ginossar N, Sarid R, Mandelboim O (2009) Diverse herpesvirus microRNAs target the stress-induced immune ligand MICB to escape recognition by natural killer cells. Cell Host Microbe 5:376–385

Nakamura H, Li M, Zarycki J, Jung JU (2001) Inhibition of p53 tumor suppressor by viral interferon regulatory factor. J Virol 75:7572–7582

Nakamura H, Lu M, Gwack Y, Souvlis J, Zeichner SL, Jung JU (2003) Global changes in Kaposi's sarcoma-associated virus gene expression patterns following expression of a tetracycline-inducible Rta transactivator. J Virol 77:4205–4220

Nakano K, Isegawa Y, Zou P, Tadagaki K, Inagi R, Yamanishi K (2003) Kaposi's sarcoma-associated herpesvirus (KSHV)-encoded vMIP-I and vMIP-II induce signal transduction and chemotaxis in monocytic cells. Arch Virol 148:871–890

Neipel F, Albrecht JC, Ensser A, Huang YQ, Li JJ, Friedman-Kien AE, Fleckenstein B (1997a) Human herpesvirus 8 encodes a homolog of interleukin-6. J Virol 71:839–842

Neipel F, Albrecht JC, Fleckenstein B (1997b) Cell-homologous genes in the Kaposi's sarcoma-associated rhadinovirus human herpesvirus 8: determinants of its pathogenicity? J Virol 71:4187–4192

Nicholas J (2005) Human gammaherpesvirus cytokines and chemokine receptors. J Interferon Cytokine Res 25:373–383

Nicholas J, Ruvolo VR, Burns WH, Sandford G, Wan X, Ciufo D, Hendrickson SB, Guo HG, Hayward GS, Reitz MS (1997) Kaposi's sarcoma-associated human herpesvirus-8 encodes homologues of macrophage inflammatory protein-1 and interleukin-6. Nat Med 3:287–292

Nicholas J, Zong JC, Alcendor DJ, Ciufo DM, Poole LJ, Sarisky RT, Chiou CJ, Zhang X, Wan X, Guo HG, Reitz MS, Hayward GS (1998) Novel organizational features, captured cellular genes, and strain variability within the genome of KSHV/HHV8. J Natl Cancer Inst Monogr 23:79–88

Offermann MK (2007) Kaposi sarcoma herpesvirus-encoded interferon regulator factors. Curr Top Microbiol Immunol 312:185–209

Okano M (2000) Haematological associations of Epstein-Barr virus infection. Baillieres Best Pract Res Clin Haematol 13:199–214

Pati S, Cavrois M, Guo HG, Foulke JS Jr, Kim J, Feldman RA, Reitz M (2001) Activation of NF-kappaB by the human herpesvirus 8 chemokine receptor ORF74: evidence for a paracrine model of Kaposi's sarcoma pathogenesis. J Virol 75:8660–8673

Paudel N, Sadagopan S, Balasubramanian S, Chandran B (2012) Kaposi's Sarcoma-Associated Herpesvirus Latency-Associated Nuclear Antigen and Angiogenin Interact with Common Host Proteins, Including Annexin A2, Which Is Essential for Survival of Latently Infected Cells. J Virol 86:1589–1607

Paulose-Murphy M, Ha NK, Xiang C, Chen Y, Gillim L, Yarchoan R, Meltzer P, Bittner M, Trent J, Zeichner S (2001) Transcription program of human herpesvirus 8 (kaposi's sarcoma-associated herpesvirus). J Virol 75:4843–4853

Pearce M, Matsumura S, Wilson AC (2005) Transcripts encoding K12, v-FLIP, v-cyclin, and the microRNA cluster of Kaposi's sarcoma-associated herpesvirus originate from a common promoter. J Virol 79:14457–14464

Petre CE, Sin SH, Dittmer DP (2007) Functional p53 signaling in Kaposi's sarcoma-associated herpesvirus lymphomas: implications for therapy. J Virol 81:1912–1922

Pfeffer S, Sewer A, Lagos-Quintana M, Sheridan R, Sander C, Grasser FA, van Dyk LF, Ho CK, Shuman S, Chien M, Russo JJ, Ju J, Randall G, Lindenbach BD, Rice CM, Simon V, Ho DD, Zavolan M, Tuschl T (2005) Identification of microRNAs of the herpesvirus family. Nat Methods 2:269–276

Pfeffer S, Zavolan M, Grasser FA, Chien M, Russo JJ, Ju J, John B, Enright AJ, Marks D, Sander C, Tuschl T (2004) Identification of virus-encoded microRNAs. Science 304:734–736

Philpott N, Bakken T, Pennell C, Chen L, Wu J, Cannon M (2011) The Kaposi's sarcoma-associated herpesvirus G protein-coupled receptor contains an immunoreceptor tyrosine-based inhibitory motif that activates Shp2. J Virol 85:1140–1144

Pietrek M, Brinkmann MM, Glowacka I, Enlund A, Havemeier A, Dittrich-Breiholz O, Kracht M, Lewitzky M, Saksela K, Feller SM, Schulz TF (2010) Role of the Kaposi's sarcoma-associated herpesvirus K15 SH3 binding site in inflammatory signaling and B-cell activation. J Virol 84:8231–8240

Platt G, Carbone A, Mittnacht S (2002) p16INK4a loss and sensitivity in KSHV associated primary effusion lymphoma. Oncogene 21:1823–1831

Poole LJ, Zong JC, Ciufo DM, Alcendor DJ, Cannon JS, Ambinder R, Orenstein JM, Reitz MS, Hayward GS (1999) Comparison of genetic variability at multiple loci across the genomes of the major subtypes of Kaposi's sarcoma-associated herpesvirus reveals evidence for recombination and for two distinct types of open reading frame K15 alleles at the right-hand end. J Virol 73:6646–6660

Pozharskaya VP, Weakland LL, Zimring JC, Krug LT, Unger ER, Neisch A, Joshi H, Inoue N, Offermann MK (2004) Short duration of elevated vIRF-1 expression during lytic replication of human herpesvirus 8 limits its ability to block antiviral responses induced by alpha interferon in BCBL-1 cells. J Virol 78:6621–6635

Prakash O, Tang ZY, Peng X, Coleman R, Gill J, Farr G, Samaniego F (2002) Tumorigenesis and aberrant signaling in transgenic mice expressing the human herpesvirus-8 K1 gene. J Natl Cancer Inst 94:926–935

Pyakurel P, Pak F, Mwakigonja AR, Kaaya E, Heiden T, Biberfeld P (2006) Lymphatic and vascular origin of Kaposi's sarcoma spindle cells during tumor development. Int J Cancer 119:1262–1267

Qin Z, Freitas E, Sullivan R, Mohan S, Bacelieri R, Branch D, Romano M, Kearney P, Oates J, Plaisance K, Renne R, Kaleeba J, Parsons C (2010) Upregulation of xCT by KSHV-encoded microRNAs facilitates KSHV dissemination and persistence in an environment of oxidative stress. PLoS Pathog 6:e1000742

Radkov SA, Kellam P, Boshoff C (2000) The latent nuclear antigen of Kaposi sarcoma-associated herpesvirus targets the retinoblastoma-E2F pathway and with the oncogene Hras transforms primary rat cells. Nat Med 6:1121–1127

Rivas C, Thlick AE, Parravicini C, Moore PS, Chang Y (2001) Kaposi's sarcoma-associated herpesvirus LANA2 is a B-cell-specific latent viral protein that inhibits p53. J Virol 75:429–438

Rodriguez A, Vigorito E, Clare S, Warren MV, Couttet P, Soond DR, van Dongen S, Grocock RJ, Das PP, Miska EA, Vetrie D, Okkenhaug K, Enright AJ, Dougan G, Turner M, Bradley A (2007) Requirement of bic/microRNA-155 for normal immune function. Science 316:608–611

Rosenkilde MM, Waldhoer M, Luttichau HR, Schwartz TW (2001) Virally encoded 7TM receptors. Oncogene 20:1582–1593

Roth WK (1991) HIV-associated Kaposi's sarcoma: new developments in epidemiology and molecular pathology. J Cancer Res Clin Oncol 117:186–191

Russo JJ, Bohenzky RA, Chien MC, Chen J, Yan M, Maddalena D, Parry JP, Peruzzi D, Edelman IS, Chang Y, Moore PS (1996) Nucleotide sequence of the Kaposi sarcoma-associated herpesvirus (HHV8). Proc Natl Acad Sci U S A 93:14862–14867

Sadagopan S, Sharma-Walia N, Veettil MV, Bottero V, Levine R, Vart RJ, Chandran B (2009) Kaposi's sarcoma-associated herpesvirus upregulates angiogenin during infection of human dermal microvascular endothelial cells, which induces 45S rRNA synthesis, antiapoptosis, cell proliferation, migration, and angiogenesis. J Virol 83:3342–3364

Sadagopan S, Valiya Veettil M, Paudel N, Bottero V, Chandran B (2011) Kaposi's sarcoma-associated herpesvirus-induced angiogenin plays roles in latency via the phospholipase C gamma pathway: blocking angiogenin inhibits latent gene expression and induces the lytic cycle. J Virol 85:2666–2685

Sadagopan S, Veettil MV, Chakraborty S, Sharma-Walia N, Paudel N, Bottero V, Chandran B (2012) Angiogenin functionally interacts with p53 and regulates p53-mediated apoptosis and cell survival. Oncogene 31:4835–4847

Sadler R, Wu L, Forghani B, Renne R, Zhong W, Herndier B, Ganem D (1999) A complex translational program generates multiple novel proteins from the latently expressed kaposin (K12) locus of Kaposi's sarcoma-associated herpesvirus. J Virol 73:5722–5730

Sakakibara S, Ueda K, Nishimura K, Do E, Ohsaki E, Okuno T, Yamanishi K (2004) Accumulation of heterochromatin components on the terminal repeat sequence of Kaposi's sarcoma-associated herpesvirus mediated by the latency-associated nuclear antigen. J Virol 78:7299–7310

Samaniego F, Pati S, Karp JE, Prakash O, Bose D (2001) Human herpesvirus 8 K1-associated nuclear factor-kappa B-dependent promoter activity: role in Kaposi's sarcoma inflammation? J Natl Cancer Inst Monogr 28:15–23

Samols MA, Hu J, Skalsky RL, Renne R (2005) Cloning and identification of a microRNA cluster within the latency-associated region of Kaposi's sarcoma-associated herpesvirus. J Virol 79:9301–9305

Samols MA, Skalsky RL, Maldonado AM, Riva A, Lopez MC, Baker HV, Renne R (2007) Identification of cellular genes targeted by KSHV-encoded microRNAs. PLoS Pathogens 3:e65

Sandford G, Choi YB, Nicholas J (2009) Role of ORF74-encoded viral G protein-coupled receptor in human herpesvirus 8 lytic replication. J Virol 83:13009–13014

Schmidt K, Wies E, Neipel F (2011) Kaposi's sarcoma-associated herpesvirus viral interferon regulatory factor 3 inhibits gamma interferon and major histocompatibility complex class II expression. J Virol 85:4530–4537

Schwarz DS, Hutvagner G, Du T, Xu Z, Aronin N, Zamore PD (2003) Asymmetry in the assembly of the RNAi enzyme complex. Cell 115:199–208

Schwarz M, Murphy PM (2001) Kaposi's sarcoma-associated herpesvirus G protein-coupled receptor constitutively activates NF-kappa B and induces proinflammatory cytokine and chemokine production via a C-terminal signaling determinant. J Immunol 167:505–513

Searles RP, Bergquam EP, Axthelm MK, Wong SW (1999) Sequence and genomic analysis of a Rhesus macaque rhadinovirus with similarity to Kaposi's sarcoma-associated herpesvirus/human herpesvirus 8. J Virol 73:3040–3053

Sears R, Nuckolls F, Haura E, Taya Y, Tamai K, Nevins JR (2000) Multiple Ras-dependent phosphorylation pathways regulate Myc protein stability. Genes Dev 14:2501–2514

Seo T, Lee D, Lee B, Chung JH, Choe J (2000) Viral interferon regulatory factor 1 of Kaposi's sarcoma-associated herpesvirus (human herpesvirus 8) binds to, and inhibits transactivation of, CREB-binding protein. Biochem Biophys Res Commun 270:23–27

Seo T, Lee D, Shim YS, Angell JE, Chidambaram NV, Kalvakolanu DV, Choe J (2002) Viral interferon regulatory factor 1 of Kaposi's sarcoma-associated herpesvirus interacts with a cell death regulator, GRIM19, and inhibits interferon/retinoic acid-induced cell death. J Virol 76:8797–8807

Seo T, Park J, Choe J (2005) Kaposi's sarcoma-associated herpesvirus viral IFN regulatory factor 1 inhibits transforming growth factor-beta signaling. Cancer Res 65:1738–1747

Seo T, Park J, Lee D, Hwang SG, Choe J (2001) Viral interferon regulatory factor 1 of Kaposi's sarcoma-associated herpesvirus binds to p53 and represses p53-dependent transcription and apoptosis. J Virol 75:6193–6198

Shamay M, Krithivas A, Zhang J, Hayward SD (2006) Recruitment of the de novo DNA methyltransferase Dnmt3a by Kaposi's sarcoma-associated herpesvirus LANA. Proc Natl Acad Sci U S A 103:14554–14559

Sharma-Walia N, Paul AG, Bottero V, Sadagopan S, Veettil MV, Kerur N, Chandran B (2010) Kaposi's sarcoma associated herpes virus (KSHV) induced COX-2: a key factor in latency, inflammation, angiogenesis, cell survival and invasion. PLoS Pathog 6:e1000777

Sharp TV, Wang HW, Koumi A, Hollyman D, Endo Y, Ye H, Du MQ, Boshoff C (2002) K15 protein of Kaposi's sarcoma-associated herpesvirus is latently expressed and binds to HAX-1, a protein with antiapoptotic function. J Virol 76:802–816

Shepard LW, Yang M, Xie P, Browning DD, Voyno-Yasenetskaya T, Kozasa T, Ye RD (2001) Constitutive activation of NF-kappa B and secretion of interleukin-8 induced by the G protein-coupled receptor of Kaposi's sarcoma-associated herpesvirus involve G alpha(13) and RhoA. J Biol Chem 276:45979–45987

Shim WS, Ho IA, Wong PE (2007) Angiopoietin: a TIE(d) balance in tumor angiogenesis. Mol Cancer Res 5:655–665

Shin YC, Joo CH, Gack MU, Lee HR, Jung JU (2008) Kaposi's sarcoma-associated herpesvirus viral IFN regulatory factor 3 stabilizes hypoxia-inducible factor-1 alpha to induce vascular endothelial growth factor expression. Cancer Res 68:1751–1759

Shin YC, Nakamura H, Liang X, Feng P, Chang H, Kowalik TF, Jung JU (2006) Inhibition of the ATM/p53 signal transduction pathway by Kaposi's sarcoma-associated herpesvirus interferon regulatory factor 1. J Virol 80:2257–2266

Skalsky RL, Samols MA, Plaisance KB, Boss IW, Riva A, Lopez MC, Baker HV, Renne R (2007) Kaposi's sarcoma-associated herpesvirus encodes an ortholog of miR-155. J Virol 81:12836–12845

Staskus KA, Zhong W, Gebhard K, Herndier B, Wang H, Renne R, Beneke J, Pudney J, Anderson DJ, Ganem D, Haase AT (1997) Kaposi's sarcoma-associated herpesvirus gene expression in endothelial (spindle) tumor cells. J Virol 71:715–719

Stern-Ginossar N, Elefant N, Zimmermann A, Wolf DG, Saleh N, Biton M, Horwitz E, Prokocimer Z, Prichard M, Hahn G, Goldman-Wohl D, Greenfield C, Yagel S, Hengel H, Altuvia Y, Margalit H, Mandelboim O (2007) Host immune system gene targeting by a viral miRNA. Science 317:376–381

Stern-Ginossar N, Gur C, Biton M, Horwitz E, Elboim M, Stanietsky N, Mandelboim M, Mandelboim O (2008) Human microRNAs regulate stress-induced immune responses mediated by the receptor NKG2D. Nat Immunol 9:1065–1073

Stine JT, Wood C, Hill M, Epp A, Raport CJ, Schweickart VL, Endo Y, Sasaki T, Simmons G, Boshoff C, Clapham P, Chang Y, Moore P, Gray PW, Chantry D (2000) KSHV-encoded CC chemokine vMIP-III is a CCR4 agonist, stimulates angiogenesis, and selectively chemoattracts TH2 cells. Blood 95:1151–1157

Sturzl M, Blasig C, Schreier A, Neipel F, Hohenadl C, Cornali E, Ascherl G, Esser S, Brockmeyer NH, Ekman M, Kaaya EE, Tschachler E, Biberfeld P (1997) Expression of HHV-8 latency-associated T0.7 RNA in spindle cells and endothelial cells of AIDS-associated, classical and African Kaposi's sarcoma. Int J Cancer 72:68–71

Suffert G, Malterer G, Hausser J, Viiliainen J, Fender A, Contrant M, Ivacevic T, Benes V, Gros F, Voinnet O, Zavolan M, Ojala PM, Haas JG, Pfeffer S (2011) Kaposi's sarcoma herpesvirus microRNAs target caspase 3 and regulate apoptosis. PLoS Pathog 7:e1002405

Sun Q, Matta H, Lu G, Chaudhary PM (2006) Induction of IL-8 expression by human herpesvirus 8 encoded vFLIP K13 via NF-kappaB activation. Oncogene 25:2717–2726

Sun R, Lin SF, Staskus K, Gradoville L, Grogan E, Haase A, Miller G (1999) Kinetics of Kaposi's sarcoma-associated herpesvirus gene expression. J Virol 73:2232–2242

Thomas M, Augustin HG (2009) The role of the Angiopoietins in vascular morphogenesis. Angiogenesis 12:125–137

Thome M, Schneider P, Hofmann K, Fickenscher H, Meinl E, Neipel F, Mattmann C, Burns K, Bodmer JL, Schroter M, Scaffidi C, Krammer PH, Peter ME, Tschopp J (1997) Viral FLICE-inhibitory proteins (FLIPs) prevent apoptosis induced by death receptors. Nature 386:517–521

Tomkowicz B, Singh SP, Cartas M, Srinivasan A (2002) Human herpesvirus-8 encoded Kaposin: subcellular localization using immunofluorescence and biochemical approaches. DNA Cell Biol 21:151–162

Tomkowicz B, Singh SP, Lai D, Singh A, Mahalingham S, Joseph J, Srivastava S, Srinivasan A (2005) Mutational analysis reveals an essential role for the LXXLL motif in the transformation function of the human herpesvirus-8 oncoprotein, kaposin. DNA Cell Biol 24:10–20

Tomlinson CC, Damania B (2004) The K1 protein of Kaposi's sarcoma-associated herpesvirus activates the Akt signaling pathway. J Virol 78:1918–1927

Tsai YH, Wu MF, Wu YH, Chang SJ, Lin SF, Sharp TV, Wang HW (2009) The M type K15 protein of Kaposi's sarcoma-associated herpesvirus regulates microRNA expression via its SH2-binding motif to induce cell migration and invasion. J Virol 83:622–632

Umbach JL, Cullen BR (2010) In-depth analysis of Kaposi's sarcoma-associated herpesvirus microRNA expression provides insights into the mammalian microRNA-processing machinery. J Virol 84:695–703

Verma SC, Lan K, Robertson E (2007) Structure and function of latency-associated nuclear antigen. Curr Top Microbiol Immunol 312:101–136

Verschuren EW, Jones N, Evan GI (2004) The cell cycle and how it is steered by Kaposi's sarcoma-associated herpesvirus cyclin. J Gen Virol 85:1347–1361

Verzijl D, Fitzsimons CP, Van Dijk M, Stewart JP, Timmerman H, Smit MJ, Leurs R (2004) Differential activation of murine herpesvirus 68- and Kaposi's sarcoma-associated herpesvirus-encoded ORF74 G protein-coupled receptors by human and murine chemokines. J Virol 78:3343–3351

Wang HW, Sharp TV, Koumi A, Koentges G, Boshoff C (2002) Characterization of an anti-apoptotic glycoprotein encoded by Kaposi's sarcoma-associated herpesvirus which resembles a spliced variant of human survivin. EMBO J 21:2602–2615

Wang HW, Trotter MW, Lagos D, Bourboulia D, Henderson S, Makinen T, Elliman S, Flanagan AM, Alitalo K, Boshoff C (2004) Kaposi sarcoma herpesvirus-induced cellular reprogramming contributes to the lymphatic endothelial gene expression in Kaposi sarcoma. Nat Genet 36:687–693

Wang L, Dittmer DP, Tomlinson CC, Fakhari FD, Damania B (2006) Immortalization of primary endothelial cells by the K1 protein of Kaposi's sarcoma-associated herpesvirus. Cancer Res 66:3658–3666

Weber KS, Grone HJ, Rocken M, Klier C, Gu S, Wank R, Proudfoot AE, Nelson PJ, Weber C (2001) Selective recruitment of Th2-type cells and evasion from a cytotoxic immune response mediated by viral macrophage inhibitory protein-II. Eur J Immunol 31:2458–2466

Whitby D, Marshall VA, Bagni RK, Wang CD, Gamache CJ, Guzman JR, Kron M, Ebbesen P, Biggar RJ (2004) Genotypic characterization of Kaposi's sarcoma-associated herpesvirus in asymptomatic infected subjects from isolated populations. J Gen Virol 85:155–163

Wies E, Hahn AS, Schmidt K, Viebahn C, Rohland N, Lux A, Schellhorn T, Holzer A, Jung JU, Neipel F (2009) The Kaposi's Sarcoma-associated Herpesvirus-encoded vIRF-3 Inhibits Cellular IRF-5. J Biol Chem 284:8525–8538

Wies E, Mori Y, Hahn A, Kremmer E, Sturzl M, Fleckenstein B, Neipel F (2008) The viral interferon-regulatory factor-3 is required for the survival of KSHV-infected primary effusion lymphoma cells. Blood 111:320–327

Willis SN, Adams JM (2005) Life in the balance: how BH3-only proteins induce apoptosis. Curr Opin Cell Biol 17:617–625

Wong EL, Damania B (2006) Transcriptional regulation of the Kaposi's sarcoma-associated herpesvirus K15 gene. J Virol 80:1385–1392

Wu AL, Wang J, Zheleznyak A, Brown EJ (1999) Ubiquitin-related proteins regulate interaction of vimentin intermediate filaments with the plasma membrane. Mol Cell 4:619–625

Yang TY, Chen SC, Leach MW, Manfra D, Homey B, Wiekowski M, Sullivan L, Jenh CH, Narula SK, Chensue SW, Lira SA (2000) Transgenic expression of the chemokine receptor encoded by human herpesvirus 8 induces an angioproliferative disease resembling Kaposi's sarcoma. J Exp Med 191:445–454

Yoo J, Kang J, Lee HN, Aguilar B, Kafka D, Lee S, Choi I, Lee J, Ramu S, Haas J, Koh CJ, Hong YK (2010) Kaposin-B enhances the PROX1 mRNA stability during lymphatic reprogramming of vascular endothelial cells by Kaposi's sarcoma herpes virus. PLoS Pathog 6:e1001046

Yoshizaki K, Matsuda T, Nishimoto N, Kuritani T, Taeho L, Aozasa K, Nakahata T, Kawai H, Tagoh H, Komori T et al (1989) Pathogenic significance of interleukin-6 (IL-6/BSF-2) in Castleman's disease. Blood 74:1360–1367

Young LS, Murray PG (2003) Epstein-Barr virus and oncogenesis: from latent genes to tumours. Oncogene 22:5108–5121

Zeng Y, Yi R, Cullen BR (2003) MicroRNAs and small interfering RNAs can inhibit mRNA expression by similar mechanisms. Proc Natl Acad Sci U S A 100:9779–9784

Zhang YJ, Bonaparte RS, Patel D, Stein DA, Iversen PL (2008) Blockade of viral interleukin-6 expression of Kaposi's sarcoma-associated herpesvirus. Mol Cancer Ther 7:712–720

Zhao J, Punj V, Matta H, Mazzacurati L, Schamus S, Yang Y, Yang T, Hong Y, Chaudhary PM (2007) K13 blocks KSHV lytic replication and deregulates vIL6 and hIL6 expression: a model of lytic replication induced clonal selection in viral oncogenesis. PLoS One 2:e1067

Zhong W, Wang H, Herndier B, Ganem D (1996) Restricted expression of Kaposi sarcoma-associated herpesvirus (human herpesvirus 8) genes in Kaposi sarcoma. Proc Natl Acad Sci U S A 93:6641–6646

Ziech D, Franco R, Pappa A, Malamou-Mitsi V, Georgakila S, Georgakilas AG, Panayiotidis MI (2010) The role of epigenetics in environmental and occupational carcinogenesis. Chem Biol Interact 188:340–349

Ziegelbauer JM, Sullivan CS, Ganem D (2009) Tandem array-based expression screens identify host mRNA targets of virus-encoded microRNAs. Nat Genet 41:130–134

Zimring JC, Goodbourn S, Offermann MK (1998) Human herpesvirus 8 encodes an interferon regulatory factor (IRF) homolog that represses IRF-1-mediated transcription. J Virol 72:701–707

Zong JC, Ciufo DM, Alcendor DJ, Wan X, Nicholas J, Browning PJ, Rady PL, Tyring SK, Orenstein JM, Rabkin CS, Su IJ, Powell KF, Croxson M, Foreman KE, Nickoloff BJ, Alkan S, Hayward GS (1999) High-level variability in the ORF-K1 membrane protein gene at the left end of the Kaposi's sarcoma-associated herpesvirus genome defines four major virus subtypes and multiple variants or clades in different human populations. J Virol 73:4156–4170

Anti-Viral Treatment and Cancer Control

Wei-Liang Shih, Chi-Tai Fang and Pei-Jer Chen

Abstract

Hepatitis B virus (HBV), hepatitis C virus (HCV), human papillomavirus (HPV), and Epstein–Barr virus (EBV) contribute to about 10–15 % global burden of human cancers. Conventional chemotherapy or molecular target therapies have been used to treat virus-associated cancers. However, a more proactive approach would be the use of antiviral treatment to suppress or eliminate viral infections to prevent the occurrence of cancer in the first place. Antiviral treatments against chronic HBV and HCV infections have achieved this goal, with significant reduction in the incidence of hepatocellular carcinoma in treated patients. Antiviral treatments for EBV, Kaposi's sarcoma-associated herpesvirus (KSHV), and human T-cell lymphotropic virus type 1 (HTLV-1) had limited success in treating refractory EBV-associated lymphoma and post-transplant lymphoproliferative disorder, KSHV-associated Kaposi's sarcoma in AIDS patients, and HTLV-1-associated acute, chronic, and smoldering subtypes of adult T-cell lymphoma, respectively. Therapeutic HPV vaccine and RNA-interference-based therapies for treating HPV-associated cervical cancers also showed some encouraging results. Taken together, antiviral therapies have yielded promising results in cancer prevention and treatment. More large-scale studies are necessary to confirm the efficacy of antiviral therapy. Further investigation for more effective and convenient antiviral regimens warrants more attention.

P.-J. Chen (✉)
Institute of Clinical Medicine, National Taiwan University, Taipei, Taiwan
e-mail: peijerchen@ntu.edu.tw

W.-L. Shih · C.-T. Fang
Institute of Epidemiology and Preventive Medicine, National Taiwan University, Taipei, Taiwan

M. H. Chang and K.-T. Jeang (eds.), *Viruses and Human Cancer*,
Recent Results in Cancer Research 193, DOI: 10.1007/978-3-642-38965-8_14,
© Springer-Verlag Berlin Heidelberg 2014

Contents

1   Introduction

Hepatitis B virus (HBV), hepatitis C virus (HCV), human papilloma virus (HPV), Epstein–Barr virus (EBV), human T-cell lymphotropic virus type 1 (HTLV-1), Kaposi's sarcoma-associated herpesvirus (KSHV) and Merkel cell polyomavirus (MCV) are the seven viruses that are currently known to cause chronic infection and to be associated with specific cancers in human (Moore and Chang 2010). In total, viruses contributed to the development of 10–15 % human cancer cases worldwide. Although only a small proportion of infected individuals actually develop cancers, the clinical prognosis is usually very poor.

Viral factors have long been proposed to play important roles in carcinogenesis. Advances in molecular technologies now allow rapid and accurate quantification of viral load, a marker of virus replication activity in human body. Accumulated data show that viral load and cancer risk often parallel for virus-associated cancers, such as HBV and HCV for hepatocellular carcinoma (HCC), and EBV for nasopharyngeal carcinoma (NPC). In light of the importance of virus replication activity in the carcinogenesis of virus-associated cancers, antiviral therapy that can suppress or eliminate viruses could be one important strategy for cancer prevention. Currently, antiviral therapy has been applied on these seven viruses-associated cancers (Table 1), and clinical benefit has been proved for antiviral therapy for HBV and HCV.

In this chapter, we reviewed the current evidences on the correlation between viral load and clinical outcomes (e.g., cancer risk and survival) and the current understandings on the effect of antiviral treatments for cancer-associated viruses.

Table 1 Human cancer viruses, associated cancers, and their specific antiviral therapies

Virus[a]	Cancer type[b]	Antiviral therapy[c]	
HBV	HCC	Interferons	(Interferon-α and pegylated interferon-α2a)
		Nucleos(t)ide analogues	(Lamivudine, adefovir, entecavir, telbivudine, and tenofovir)
HCV	HCC	Peglyated interferon plus ribavirin	
		Direct antiviral agents (protease inhibitor)	(Boceprevir and telaprevir)
KSHV	AIDS-KS	Antiviral herpesvirus drug	(Ganciclovir and valganciclovir)
		HAART, PI based	
		HAART, NNRTI based	
	HIV-negative KS	HIV protease inhibitor	(Indinavir)
HTLV-1	ATL (acute, chronic, and smoldering forms)	Interferon-α plus zidovudine/zalcitabine alone or combined with chemotherapy	
EBV	PTLD, NPC, HL	Immunotherapy	(EBV-specific cytotoxic T cells)
	EBV-associate lymphoma	Virus directed	(Lytic replication inducer plus acyclovir/ganciclovir)
HPV	Cervical cancer	Therapeutic HPV vaccine (focus on HPV-16 and HPV-18)	
		RNA-interference-based therapy	(Antisense oligonucleotides, ribozymes, and siRNAs)
MCV	MCC	Interferon	(Interferon-α and interferon-β)

[a] *HBV* hepatitis B virus, *HCV* hepatitis C virus, *EBV* Epstein–Barr virus, *HPV* human papillomavirus, *HTLV-1* human T-cell lymphotropic virus type 1, *KSHV* Kaposi's associated sarcoma virus, *MCV* Merkel cell polyomavirus

[b] *HCC* hepatocellular carcinoma, *AIDS-KS* AIDS-Kaposi's sarcoma, *ATL* adult T-cell lymphoma, *PTLD* post-transplant lymphoproliferative disorder, *NPC* nasopharyngeal carcinoma, *HL* Hodgkin's lymphoma, *MCC* Merkel cell carcinoma

[c] *HAART* highly active antiretroviral treatment, *PI* protease inhibitor, NNRTI: non-nucleoside reverse transcriptase inhibitor

2 Hepatitis B Virus and Hepatitis C Virus

Chronic HBV and HCV infections are major etiological factors in 75–80 % of HCC cases worldwide. Universal HBV vaccination on newborns in Taiwan since 1985 has lead to a 70 % reduction in HBV-related HCC in young-aged children

(Chang et al. 2009). Nevertheless, millions of individuals who had already chronically infected by HBV are still under the risk of developing HCC. Unfortunately, no effective vaccine is available for preventing HCV infection now. The proportion of HCV-related HCC has progressively increased, especially in the developed countries (El-Serag 2012). For these people persistently infected by either HBV or HCV, antiviral therapies' target on HBV and HCV can be an effective strategy to reduce risk of HCC.

2.1 Anti-HBV Therapies

HBV replication is the key force to drive the progression of HBV-related diseases (Liaw 2006). HBeAg is a well-known HBV replication marker (Chen et al. 2009; Fang et al. 2003). Epidemiological studies also consistently show that the elevated HBV viral load, which represented a higher HBV replication activity, is associated with high HCC risk, worse progression, and poor survival (Chen et al. 2009). Reducing HBV DNA to undetectable level or induction of HBeAg seroconversion has been the main therapeutic endpoints (Feld et al. 2009).

Conventional interferon-α, pegylated interferon-α2a, and nucleos(t)ide analogues (NUCs), including lamivudine, adefovir, entecavir, telbivudine, and tenofovir, are currently approved treatments for chronic HBV infection. Treatment with newer NUCs, including entecavir, telbivudine, and tenofovir, could suppress HBV DNA on average by 6.2–6.9 and 4.6–5.2 $\log_{10}$ IU/mL in HBeAg-positive (Chang et al. 2006; Lai et al. 2007; Marcellin et al. 2008) and HBeAg-negative patients (Lai et al. 2006, 2007; Marcellin et al. 2008), respectively. Undetectable levels of HBV DNA can be achieved in 60–80 % of HBeAg-positive patients (Chang et al. 2006; Dienstag 2009; Lai et al. 2007; Marcellin et al. 2008) and 88–95 % of HBeAg-negative patients (Dienstag 2009; Lai et al. 2006, 2007; Marcellin et al. 2008). Treatment with lamivudine and adefovir yields less reduction in HBV DNA and lower proportion of undetectable HBV DNA in both HBeAg-positive and HBeAg-negative patients. In addition, HBeAg seroconversion was observed in 12–23 % of NUC-treated patients (Chang et al. 2006; Dienstag 2009; Lai et al. 2007; Marcellin et al. 2008). After one-year course of NUC treatment, sustained virological response was maintained in relatively small proportion of initial responders (Dienstag 2009). Thus, NUCs were usually used as a long-term therapy. No optimal duration of NUC therapy is available now, but extending treatment for at least more than 6 months after HBeAg seroconversion to consolidate the sustained response is currently an acceptable practice.

Although NUCs are highly potent and safe, emergence of drug resistance after prolonged use of NUCs is a major concern. It has been reported that patients with lamivudine resistance had higher HCC risk than NUC-naïve patients (Papatheodoridis et al. 2010). Lamivudine resistance would accumulate rapidly to 15–25 % after 12 months and to 60–65 % after 4 years of treatment (Papatheodoridis et al. 2008). Resistance to adefovir and telbivudine can also reach 25–30 %

Table 2 Meta-analyses of nucleos(t)ide analogues and interferon-based therapy for HBV and HCV infections

Study	Design	No. of subjects	Therapy	Endpoint	Main findings
HBV					
Mommeja-Marin et al. (2003)	26 prospective studies	$N = 3,428$ (2,524 HBeAg positive)	IFN vs. CN & NUCs vs. CN	Histological response Biochemical response Serological response	Treatment-induced HBV DNA reduction correlated with histological, biochemical, and serological responses
Sung et al. (2008)	12 studies	$N = 2,742$ 1,292 treated 1,450 untreated	IFN vs. CN	HBV-related HCC	$RR = 0.66$ (0.48–0.89)
	5 studies	$N = 2,289$ 1,267 treated 1,022 untreated	NUCs vs. CN	HBV-related HCC	$RR = 0.22$ (0.10–0.50)
Yang et al. (2009)	11 clinical trials	$N = 2,082$ 1,006 treated 1,076 untreated	IFN vs. CN	HBV-related HCC incidence	$RR = 0.59$ (0.43–0.81)
	5 clinical trials	$N = 935$ 516 treated 419 untreated	IFN vs. CN	HBV-related cirrhosis	$RR = 0.65$ (0.47–0.91)
Wong et al. (2010)	11 studies	$N = 2,122$ 975 treated 1,147 untreated	IFN vs. CN	Overall hepatic events	$RR = 0.55$ (0.43–0.70)
				Liver-related mortality	$RR = 0.63$ (0.42–0.96)
Papatheodoridis et al. (2010)	21 studies	$N = 4,415$ 3,881 treated 534 untreated	NUCs vs. CN	HCC incidence	HCC incidence: treated: 2.8 % untreated: 6.4 %
Zhang et al. (2011)	2 RCTs	$N = 1,58$ 95 treated 63 untreated	Non-maintenance IFN treated vs. CN	HBV-related HCC incidence	$RR = 0.23$ (0.05–1.04)

(continued)

Table 2 (continued)

Study	Design	No. of subjects	Therapy	Endpoint	Main findings
HCV					
Cammá et al. (2001)	14 studies	HCV-related cirrhosis $N = 3,109$	IFN treated vs. non-treated	HCC incidence	$OR = 0.28$ (0.22–0.36)
Papatheodoridis et al. (2001)	11 studies	HCV-related cirrhosis $N = 2178$	CN vs. IFN	HCC incidence	$OR = 3.02$ (2.35–3.89)
	5 studies	HCV-related cirrhosis $N = 683$	Non-SVR vs. SVR	HCC incidence	$OR = 3.65$ (1.71–7.78)
Zhang et al. (2011)	4 RCTs	HCV patients $N = 378$	Non-maintenance IFN treated vs. non-treated	HCV-related HCC incidence	$RR = 0.39$ (0.26–0.59)
	2 RCTs	HCV patients $N = 223$	Non-maintenance IFN treated vs. non-treated	HCV-SVR	$RR = 0.30$ (0.04–2.15)
	2 RCTs	Initial non-responders of IFN therapy $N = 1,101$	Maintenance IFN treated vs. non-treated	HCV-related HCC incidence	$RR = 0.96$ (0.59–1.56)
Singal et al. (2010)	20 studies	HCV-related cirrhosis $N = 4,700$	Treated vs. non-treated	HCC incidence	$RR = 0.43$ (0.33–0.56)
	14 studies		SVR vs. non-SVR	HCC incidence	$RR = 0.35$ (0.26–0.46)
Qu et al (2012)	8 RCTs	HCV-related cirrhosis $N = 1505$	Treated vs. non-treated	HCC incidence	$OR = 0.29$ (0.10–0.80)
	3 RCTs	HCV-related cirrhosis $N = 1155$	Maintenance IFN therapy treated vs. non-treated	HCC incidence	$OR = 0.54$ (0.32–0.90)

CN control group (placebo or no treatment), *RR* relative risk, *OR* odds ratio, *SVR* sustained virological response

IFN interferon, *NUCs* nucleos(t)ide analogues

RCTs randomized controlled trials

after long-term treatment. Only treatment with entecavir and tenofovir showed negligible resistance. Thus, due to better resistance profile, excellent safety profile, and superior efficacy, entecavir and tenofovir have now been suggested as the first-line therapy (Dienstag 2009).

In contrast to NUCs, interferon-based therapy was less used for HBV infection due to side effects and poor tolerability. Nevertheless, recent studies have provided more supportive evidence for its role in anti-HBV therapy. Treatment of HBeAg-positive patients with pegylated interferon for 48–56 weeks achieved undetectable HBV DNA in 10–25 % of patients and <2–4.5 $\log_{10}$ copies/mL mean reduction in HBV DNA (Janssen et al. 2005; Lau et al. 2005). HBeAg loss and HBeAg seroconversion were durable in 81 and 70 % of initial responders with or without concomitant lamivudine therapy, respectively, for a mean of 3 years of follow-up after treatment (Buster et al. 2008). The proportion of HBV DNA undetectability in HBeAg-negative patients was 63 %, and average reduction of viral load was 4.1 $\log_{10}$ copies/mL after 48 weeks of therapy of pegylated interferon (Marcellin et al. 2004). After 24 weeks of follow-up (week 72), rate of HBV DNA undetectability decreased to 19 % and average reduction in viral load became 2.3 $\log_{10}$ copies/mL (Marcellin et al. 2004). A durable suppression of HBV DNA to undetectable level was observed in 46 % of HBeAg-negative initial responders at 3 years of follow-up after the end of treatment (Marcellin et al. 2009). Although pegylated interferon resulted in virological response in a small proportion of patients, it seemed to have higher probability to achieve sustained off-therapy response. Due to this advantage and the fixed duration of treatment, pegylated interferon still has a therapeutic role in selected patients.

The long-term effect of anti-HBV therapy with NUCs and interferon on survival and incidence of HCC has also been investigated. For interferon-based therapy, two studies reported that sustained virological responders showed significantly better survival and lower risk of developing HCC (Niederau et al. 1996; van Zonneveld et al. 2004). Compared with non-treated patients, interferon-treated patients had lower HCC incidence with *RR* of 0.23–0.66 (Table 2). For NUCs, two early, large randomized control trials of treatment for chronic HBV-infected patients with advanced liver diseases showed that lamivudine could reduce risk of disease progression including developing HCC in treated patients and in patients with persistent viral suppression (Di Marco et al. 2004; Liaw et al. 2004). Several meta-analyses consistently concluded that NUC treatment was associated with a lower HCC incidence (Table 2).

2.2   Anti-HCV Therapies

For HCV, there is still no vaccine available. Fortunately, effective anti-HCV therapies are available to provide good clinical outcomes for HCV patients. Interferon-based treatments by using pegylated interferon with ribavirin are the current standard of care for HCV patients. This regimen resulted in sustained

virological response (SVR) in about 50 % of HCV patients with genotype 1 and
80 % of HCV patients with genotypes 2 and 3 (Munir et al. 2010). The
achievement of SVR is durable (Hofmann and Zeuzem 2011) and highly associated with good overall clinical outcome, including decreased risk of HCC
(Table 2) and liver-related deaths (Masuzaki et al. 2010). Nevertheless, non-SVR
patients did not show risk reduction in disease progression, and maintenance of
anti-HCV treatment provided no further benefits on clinical outcomes in these
patients (Table 2).

High viral load and genotype 1 were used to predict a lower response rate to
anti-HCV therapies in chronic hepatitis C patients. But the percentage of non-
responders has decreased in recent years because of the advance in therapy.
Pegylated interferon combined with ribavirin increased SVR to about 40 % in
HCV genotype 1b patients with high viral load (Masuzaki et al. 2010). In addition,
these non-responders now could be identified by many clinical markers (e.g.,
obesity, IL-28 polymorphisms) before treatment to allow treatment optimization
(Masarone and Persico 2011). Furthermore, upcoming new drugs (boceprevir and
telaprevir, two HCV non-structural 3/4A protease inhibitors) may provide satisfactory HCV RNA suppression in patients with genotype 1 infection, which did
not respond to interferon-based regimens (Hofmann and Zeuzem 2011). We are
now at the dawn of a new era of more effective and friendlier anti-HCV therapies.

3   HBV–HCV Coinfection

Treatment of HBV–HCV coinfection is important in endemic area because of its
fairly high prevalence due to shared routes of transmission and increased risk for
liver diseases, including HCC. Although no specific treatment guidelines are
established, individualized therapy according to hepatitis virology, history of
antiviral treatment, and stage and grade of liver disease is recommended (Potthoff
et al. 2010).

Before initiation of therapy, determination of dominant virus infection is
required. Combination of interferon/pegylated interferon and ribavirin are current
choice of treatment for patients with dominant HCV infection. There were
43–69 % of treated patients who showed HCV-SVR at the end of 24 weeks of
therapy with interferon plus ribavirin (Potthoff et al. 2010). After 24 or 48 weeks
of combination therapy in genotype 2/3 HCV and genotype 1 HCV, respectively,
high HCV-SVR rate can be achieved in both genotypes (83 % for genotype 2/3
and 72 % for genotype 1) (Liu et al. 2009). At the end of 24–48 weeks of follow-
up, undetectability of HBV DNA was obtained in 11–18 % for interferon plus
ribavirin and 56 % for pegylated interferon plus ribavirin (Liu et al. 2009; Potthoff
et al. 2010). Interestingly, the combination therapy seemed to have higher chance
of HBsAg clearance, which is the optimal treatment goal. The proportion of
HBsAg clearance was 11–19 % for these two types of regimens (Liu et al. 2009;
Yeh et al. 2011). However, as severe hepatic flares and recurrence of HBV after

therapy have been reported, using this combination therapy with caution and long-term virological monitoring in some coinfected patients are necessary. In HBV-dominant or dually active HBV–HCV-coinfected patients, the optimal therapeutic regimen remains unestablished.

4   HBV–HIV Coinfection and HCV–HIV Coinfection

Coinfection of HBV–HIV and HCV–HIV usually hastens the development of liver diseases, including fibrosis and HCC. Generally speaking, monotherapy of interferons, pegylated interferons, lamivudine, adefovir, entecavir, and telbivudine has yielded much less satisfactory responses in HBV–HIV-coinfected patients (Lacombe and Rockstroh 2012). Nevertheless, tenofovir monotherapy offered some benefits, such as effective virological suppression, good resistance profile, and histological improvement. However, interruption use must be avoided because of viral breakthrough-induced deleterious consequences.

According to the guidelines proposed by the European AIDS Clinical Society (Lacombe and Rockstroh 2012; Soriano et al. 2010), HBV–HIV-coinfected patients who do not need HIV treatment (>500 CD4 + cells per uL) are suggested to receive early combination therapy of tenofovir and lamivudine/emtricitabine or to receive NUC therapy (either adefovir or telbivudine) in "early add-on" strategy when their HBV DNA level was more than 2,000 IU/mL. In this group of patients who have favorable response to interferon therapy, a 48-week course of interferon is suggested. When both HBV and HIV treatments must be applied, the appropriate option for treatment depends on prior use of lamivudine. In lamivudine-naïve patients, combination of tenofovir plus lamivudine/emtricitabine is the therapeutic option. And for lamivudine-treated patients, adding tenofovir, entecavir, or other HIV nucleos(t)ide reverse transcriptase inhibitors is recommended. Besides, lamivudine-containing HAART has been shown to be able to suppress HBV replication in HBeAg-negative HBV–HIV-coinfected patients (Fang et al. 2003). Although the above guidelines have been proposed, more solid evidence is still needed to make evidence-based therapeutic decisions.

In regard to treatment of HCV–HIV-coinfected patients, SVR rates could be achieved to 27–50 % by using treatment of pegylated interferon with ribavirin, which is the current standard therapy. Higher SVR (44–73 %) was observed in genotypes 2 and 3. In contrast, only 17–35 % genotype 1 and genotype 4 patients reached SVR (Lacombe and Rockstroh 2012).

Regimens for HCV–HIV-coinfected patients would be different according to HCV genotype and virological response. For non-HCV genotype 1 patients, in patients with rapid virological response (undetectable HCV viral load at week 4), 24 and 48 weeks of pegylated interferon/ribavirin combination therapy are recommended for genotype 2/3 and genotype 4, respectively. If the virological response is not achieved at week 12, treatment should be stopped. Patients with virological response at week 12 and undetectable HCV RNA at week 24 are recommended to receive 48- and 72-week courses of treatment for genotype 2/3 and genotype 4, respectively (Lacombe and Rockstroh 2012; Soriano et al. 2010). For HCV genotype 1 patients, longer duration of treatment may provide benefits under the conditions of early virological response at week 12 and undetectable virus at week 24. However, patients with higher fibrosis stages are recommended to have therapy of interferon/ribavirin combination plus HCV protease inhibitors. Recent pilot trials showed that this triple therapy in coinfected patients increased the SVR rate when compared with patients receiving combined treatment (Ingiliz and Rockstroh 2012). In addition, based on the results from clinical studies, HAART may offer positive impact on control of prognosis of liver damage in HCV–HIV-coinfected patients, and most of first-line HAART has good fitness in these patients (Jones and Nunez 2011).

5   Kaposi's Sarcoma-Associated Herpesvirus

Kaposi's sarcoma-associated herpesvirus (KSHV), also known as human herpesvirus 8 (HHV-8), is a necessary factor for Kaposi's sarcoma (KS), the most common malignancy in HIV patients who became immunocompromised. Elevated KSHV viral load was observed more frequently in KS patients than in asymptomatic KSHV-infected patients and was associated with a higher risk of AIDS-KS (Gantt and Casper 2011; Sunil et al. 2010). There is still no effective vaccine for KSHV. Nevertheless, several studies showed that anti-herpes virus drugs, such as ganciclovir and valganciclovir, not only reduced KSHV viral load but also prevented AIDS-KS (Gantt and Casper 2011). Though these small studies still need further confirmation (Gantt and Casper 2011), it will be interesting to examine the effects of these promising agents in combination with DNA synthesis blockers or lytic replication inducer in future clinical trials.

The replication of KSHV strongly depends on HIV-induced immunodeficiency (Mesri et al. 2010). Early use of highly active antiretroviral treatment (HAART) can restore host immunity and decrease the incidence and mortality of AIDS-KS in HAART-treated patients (Bower et al. 2006; Mesri et al. 2010). Compared with that in pre-HAART era, KS incidence in HAART era dropped by sixfold (Sunil et al. 2010). Even in resource-limited regions, early use of HAART can result in reduced KS incidence (Casper 2011). In the individual patient level, HAART is also significantly associated with control of KSHV viremia (Bourboulia et al. 2004). However, for late presenters who had already developed KS at the time of initial diagnosis, HAART alone induces complete remission in only half of these patients (Nguyen et al. 2008). For AIDS-KS patients, who did not reach complete remission with HAART alone, co-administration of HAART and chemotherapy could improve the response rate to 81.5 % (Bower et al. 2006). HAART can be classified to protease inhibitor (PI)-based and non-nucleoside reverse transcriptase inhibitor (NNRTI)-based regimens. No difference in KS incidence rate between these two types of regimens was observed in small observational studies (Gantt and Casper 2011), but PI-based HAART seemed to have better efficacy because more complete remission and relapse were reported when switching therapy from PI-based to NNRTI-based HAART (Gantt and Casper 2011). Nevertheless, whether PI-based HAART does have a therapeutic advantage for KSHV or not, this superiority still requires more convincing evidence to support, and controlled trials with large sample size and optimized HAART regimens are urgently needed.

For KS in HIV-negative patients, a 28-patient trial study showed that treatment with individual ART (indinavir) induced tumor regression and stabilization of disease progression in some patients (Monini et al. 2009). In addition to KS, KSHV was also linked to primary effusion lymphoma (PEL) and multicentric Castleman's disease (MCD) (Chang et al. 2006), which are rare malignancies with very poor prognosis. Recently, few studies proved that treatment with HAART alone or with other therapy (e.g., monoclonal antibodies) for PEL and MCD may prolong resolution of symptoms (Sunil et al. 2010).

6 Human T-Cell Lymphotropic Virus Type 1

Human T-cell lymphotropic virus type 1 (HTLV-1) is the etiological agent of adult T-cell lymphoma (ATL) (Poiesz et al. 1980). Although only a small proportion (<6 %) of 15–20 million estimated infected people worldwide developed ATL after a 10- to 40-year latent period (Proietti et al. 2005), the prognosis for ATL patients is very poor. Median survivals of acute and lymphomatous ATLs are less than one year (Goncalves et al. 2010).

Compared with asymptomatic carriers, ATL patients had significantly higher level of HTLV-1 proviral DNA and antibody titer (Manns et al. 1999). The quantity of HTLV-1 proviral DNA is a predictive factor for the development of ATL and is correlated with clinical outcomes (Etoh et al. 1999; Iwanaga et al. 2010; Okayama et al. 2004). Based on the accumulated data, antiviral therapy now is one of the treatment options for ATL (Tsukasaki et al. 2009). Treatment for ATL with antiviral drugs was investigated in the early 1980s in Japan, but encouraging improvement was obtained from two trials that combined interferon-α and zidovudine (AZT) in treating ATL patients reported in 1995 (Gill et al. 1995; Hermine et al. 1995). These two studies described impressive high response rate (more than 50 %) and mild toxic effects, and the survival time was prolonged to more than one year. After that, many small studies (all less than 30 patients) using interferon-α and AZT/zalcitabine were performed in France (Hermine et al. 2002), United Kingdom (Matutes et al. 2001), Martinique (French West Indies) (Besson et al. 2002), and United States (Ratner et al. 2009). Overall, these studies showed consistent efficacy of antiviral therapy. Recently, a worldwide meta-analysis with 254 ATL patients treated with interferon-α and AZT combination and/or with chemotherapy provided further evidence that combination of interferon-α and AZT treatment resulted in high response and significantly prolonged survival (Bazarbachi et al. 2010). However, the survival advantage was limited to acute, chronic, and smoldering subtypes. For lymphomatous ATL, chemotherapy seemed to be more effective than antiviral therapy. Besides, treatment with combination of interferon-α/AZT and arsenic trioxide also resulted in promising outcome on 7 patients with relapsed/refractory acute or lymphomatous ATL (Hermine et al. 2004). Another study using 10 chronic ATL patients showed 100 % response to the treatment (Kchour et al. 2009). Although these were preliminary observations, the feasibility of this regimen has been considered.

Improvement of treatment response and survival that resulted from antiviral drugs was an important advance in treating ATL. Studies also showed a reduction in HTLV-1 proviral DNA load in ATL patients after antiviral therapy, and the decrease was correlated with the response of treatment (Kchour et al. 2007, 2009). However, all these results came from studies with small sample size.

7   Epstein–Barr Virus

As the first identified carcinogenic virus, EBV has been linked to many malignancies, including Burkitt's lymphoma, Hodgkin's lymphoma (HL), immune-suppressed B-cell lymphoma, post-transplant lymphoproliferative disorder (PTLD), gastric carcinoma, and NPC (Kutok and Wang 2006). Although there is still no standard assay (De Paoli et al. 2007), quantification of EBV viral load has been used in the management of PTLD and NPC patients. Because of different and complex etiopathological and clinical features, the clinical significance of EBV viral load varies across all types of EBV-associated malignancies (Kimura et al. 2008).

Pharmacological therapy, immunotherapy, and virus-directed approaches are existing antiviral therapies for EBV-associated diseases. For EBV-associated malignancies, the maintenance of latent replication makes pharmacological therapy with nucleoside analogues produce no effect. Immunotherapy, which uses adoptive EBV-specific cytotoxic T cells (EBV-CTLs) to kill infected cells, has been used to treat PTLD (Rooney et al. 1995), NPC (Chua et al. 2001), and HL (Bollard et al. 2004). Application of such therapy in PTLD provided some beneficial consequences including EBV DNA reduction in 2–3 log scale and stabilization of viral load (Gustafsson et al. 2000; Rooney et al. 1995), prevention of development for PTLD (Heslop et al. 2010), and prolonged activity of infused CTLs against viral reactivation for up to 9 years (Heslop et al. 2010). But, results from treatment of NPC and HL patients were only partially satisfied. For NPC, 15–19 % of patients with metastatic and locally advanced diseases remained in poor survival outcome under the standard care of chemoradiotherapy. Immunotherapy was considered because of the expression of EBNA1, LMB1, and LMB2 EBV antigens, which were the targets of EBV-CTLs, in NPC. Focussing mainly on refractory and relapse NPC patients, the encouraging immunotherapeutic results came from achievement of complete remission (Louis et al. 2010; Straathof et al. 2005), lack of significant toxicity, and in some patients, the control of disease progression (Comoli et al. 2005). For relapsed HL patients, complete/partial remission, decrease in EBV DNA in PBMC, and stable disease progression were observed (Bollard et al. 2004). Current data on immunotherapy were limited by the small number of patients and the short-lasting effects. Therefore, all kinds of immunotherapy for EBV-associated malignancies are still under investigation (Merlo et al. 2011).

Virus-directed approaches, which use the EBV viral genome in tumor cells as the target, could be divided into two groups. One group includes designs of inhibition of EBV oncoprotein expression, inducing loss of EBV episome, and production of cellular toxins in EBV-infected tumors. The other approach combined the induction of EBV lytic replication and then followed by the use of acyclovir and ganciclovir (Ghosh et al. 2012; Israel and Kenney 2003). Phase I/II clinical trials for EBV-associated lymphoma found that combination of arginine butyrate (one kind of HDAC inhibitor) and ganciclovir obtained noticeable clinical response in some refractory patients (Faller et al. 2001; Perrine et al. 2007). For lytic replication inducer, many chemical agents such as phorbol esters, DNA methyltransferase inhibitor (e.g., 5-azacytidine), and HDAC inhibitors have been developed and evaluated in *in vitro* and animal studies, but optimal agent remains undetermined today. Taken together, studies of antiviral therapies on EBV-associated malignancies provided some encouraging results to pave the way for future large and long-term studies.

8   Human Papillomavirus

Of more than 100 identified human papillomavirus (HPV) types, 15 high-risk HPV types have been linked to about 12 cancers (Bharti et al. 2009; Woodman et al. 2007). The causative role of HPV-16 and HPV-18 in the cervical cancer is the most well documented. Viral load of HPV in general did not correlate well with disease severity, duration of infection, clearance of disease, or prognosis of disease (Boulet et al. 2008; Woodman et al. 2007), probably due to the extremely large variation in oncogenic potential among different HPV genotypes (Josefsson et al. 2000). For the highest risk genotype HPV-16, nevertheless, high viral load appears to be associated with prevalent cervical cancer precursors and may be used to predict the development of incident disease and persistence of infection (Gravitt et al. 2007; Josefsson et al. 2000; Xi et al. 2011).

Several antiviral strategies for treating HPV infection and HPV-induced cervical cancer are under development, among which therapeutic HPV vaccines and RNA-interference-based therapies are two possible approaches. Besides, a recent *in vitro* study demonstrated that two synthetic polyamides could cause loss of HPV episomal DNA, and their therapeutic utility is under assessment (Edwards et al. 2011).

Therapeutic HPV vaccines against HPV-16 and HPV-18, including proteins-/peptides-based vaccines and DNA vaccines, are currently under testing (Bermúdez-Humarán and Langella 2010; Bharti et al. 2009). In animal model, vaccines against

E6 and E7 early HPV proteins showed regressive effect of disease progression (Gomez-Gutierrez et al. 2007; Zwaveling et al. 2002). In Phase I/II clinical trials, anti-HPV CTL responses and HPV clearance had been observed in some but not all patients with advanced and early stage lesions (Bharti et al. 2009; Brinkman et al. 2007). Similar results were observed for vaccines using autologous dendritic cells pulsed with HPV antigens. In a Phase II trial involving 127 patients, DNA vaccine induces sustained cytotoxic T-lymphocyte immune responses in women with high-grade lesions, with partial or even complete regression in some of the patients (Brinkman et al. 2007). These encouraging results had demonstrated the potential of therapeutic HPV vaccines.

RNA interference therapies used antisense oligonucleotides, ribozymes, and small-interfering RNAs (siRNAs) to inhibit the expression of HPV E6 and E7 genes at post-transcriptional level. Several studies proved that RNAi therapies induced apoptosis, reduction in growth rate, and cell death in cervical cancer cell and mouse model (Benitez-Hess et al. 2011; Jonson et al. 2008; Sima et al. 2007; Zhou et al. 2012) and also could increase sensitivity of cancer cells to some chemotherapy, such as cisplatin (Koivusalo et al. 2005; Putral et al. 2005). Nevertheless, most of these beneficial results came from pre-clinical studies and Phase I trials, so that practical implication and the long-term potential as treatment are still unclear. Furthermore, problems in using antisense molecules or ribozymes, including target selection, delivery efficiency, and short-term maintenance, still need to be overcome.

9   Merkel Cell Polyomavirus

Merkel cell polyomavirus (MCV) is a polyomavirus newly discovered in 2008 (Feng et al. 2008). MCV can be detected in 25–64 % healthy individuals (Carter et al. 2009; Kean et al. 2009; Tolstov et al. 2009). MCV can also be found in many normal tissues such as esophagus, liver, colon, lung, and bladder (Loyo et al. 2010). MCV is a causal factor of Merkel cell carcinoma (MCC), which is a rare but aggressive and lethal neuroendocrine skin cancer (Feng et al. 2008). MCV had a significantly higher MCV viral load than other tissues (Loyo et al. 2010). Given the high association of MCV with MCC, antiviral therapy may be a new approach to treat MCC beyond the current surgical excision, lymph node surgery, radiation therapy, and chemotherapy (Schrama et al. 2012).

Studies on antiviral therapy against MCV for MCC are still limited. An *in vitro* study showed that interferon-α may have anti-tumor effect on MCC (Krasagakis et al. 2008). Interferon-α has been used to treat two MCV-positive MCC patients, without clinical response (Biver-Dalle et al. 2011). Favorable response was reported, however, in a 62-year-old woman with MCV-positive MCC with interferon-β at a dose of 3,000,000 IU per day (Nakajima et al. 2009). The *in vitro* and *in vitro* data also suggested that interferon-α and interferon-β may have strong anti-tumor effect on MCC, especially for MCV-positive MCC (Willmes et al. 2012). Due to limited data, much study is needed to evaluate the effect of antiviral therapy on MCC.

10 Conclusion and Future Perspective

The tremendous advances in antiviral therapy against oncogenic viruses, particularly HBV and HCV, have revolutionized the concepts and practice of cancer prevention. Primary prevention of HCC in chronically HBV- and/or HCV-infected patients is now a realistic goal, along with significant life and cost savings (Kim et al. 2007; Toy et al. 2008). The widespread use of HAART also dramatically reduced the incidence of KS in HIV-infected patients. For other oncogenic viruses, there is still a lack of convenient, effective antiviral pharmacological agents that can be used for primary prevention. Nevertheless, antiviral therapy can effectively prolong patient's survival in some types of HTLV-1-induced ATL and has become one of the treatment options. Further studies on experimental antiviral therapy and technology may yield promising results in adjuvant therapy for EBV-, HPV-, and MCV-associated malignancies.

Furthermore, to facilitate the development of new preventive or therapeutic approaches and agents, future researches are needed to continually improve our understanding in biological mechanism of viral-related carcinogenesis and the relationship between dynamics of virus and natural history of diseases. These efforts could add to our arsenal against these viruses and yield a more favorable/satisfactory and clinical outcomes in the future.

References

Bazarbachi A, Plumelle Y, Carlos Ramos J, Tortevoye P, Otrock Z, Taylor G, Gessain A, Harrington W, Panelatti G, Hermine O (2010) Meta-analysis on the use of zidovudine and interferon-alfa in adult T-cell leukemia/lymphoma showing improved survival in the leukemic subtypes. J Clin Oncol 28:4177–4183

Benitez-Hess ML, Reyes-Gutierrez P, Alvarez-Salas LM (2011) Inhibition of human papillomavirus expression using DNAzymes. Methods Mol Biol 764:317–335

Bermúdez-Humarán LG, Langella P (2010) Perspectives for the development of human papillomavirus vaccines and immunotherapy. Expert Rev Vaccines 9:35–44

Besson C, Panelatti G, Delaunay C, Gonin C, Brebion A, Hermine O, Plumelle Y (2002) Treatment of adult T-cell leukemia-lymphoma by CHOP followed by therapy with antinucleosides, alpha interferon and oral etoposide. Leuk Lymphoma 43:2275–2279

Bharti AC, Shukla S, Mahata S, Hedau S, Das BC (2009) Anti-human papillomavirus therapeutics: facts & future. Indian J Med Res 130:296–310

Biver-Dalle C, Nguyen T, Touze A, Saccomani C, Penz S, Cunat-Peultier S, Riou-Gotta MO, Humbert P, Coursaget P, Aubin F (2011) Use of interferon-alpha in two patients with Merkel cell carcinoma positive for Merkel cell polyomavirus. Acta Oncol 50:479–480

Bollard CM, Aguilar L, Straathof KC, Gahn B, Huls MH, Rousseau A, Sixbey J, Gresik MV, Carrum G, Hudson M, Dilloo D, Gee A, Brenner MK, Rooney CM, Heslop HE (2004) Cytotoxic T lymphocyte therapy for Epstein-Barr virus[+] Hodgkin's disease. J Exp Med 200:1623–1633

Boulet GA, Horvath CA, Berghmans S, Bogers J (2008) Human papillomavirus in cervical cancer screening: important role as biomarker. Cancer Epidemiol Biomarkers Prev 17:810–817

Bourboulia D, Aldam D, Lagos D, Allen E, Williams I, Cornforth D, Copas A (2004) Short- and long-term effects of highly active antiretroviral therapy on Kaposi sarcoma-associated herpesvirus immune responses and viraemia. AIDS 18:485–493

Bower M, Palmieri C, Dhillon T (2006) AIDS-related malignancies: changing epidemiology and the impact of highly active antiretroviral therapy. Curr Opin Infect Dis 19:14–19

Brinkman JA, Hughes SH, Stone P, Caffrey AS, Muderspach LI, Roman LD, Weber JS, Kast WM (2007) Therapeutic vaccination for HPV induced cervical cancers. Dis Markers 23:337–352

Buster EH, Flink HJ, Cakaloglu Y, Simon K, Trojan J, Tabak F, So TM, Feinman SV, Mach T, Akarca US, Schutten M, Tielemans W, van Vuuren AJ, Hansen BE, Janssen HL (2008) Sustained HBeAg and HBsAg loss after long-term follow-up of HBeAg-positive patients treated with peginterferon alpha-2b. Gastroenterology 135:459–467

Cammá C, Giunta M, Andreone P, Craxi A (2001) Interferon and prevention of hepatocellular carcinoma in viral cirrhosis: an evidence-based approach. J Hepatol 34:593–602

Carter JJ, Paulson KG, Wipf GC, Miranda D, Madeleine MM, Johnson LG, Lemos BD, Lee S, Warcola AH, Iyer JG, Nghiem P, Galloway DA (2009) Association of Merkel cell polyomavirus-specific antibodies with Merkel cell carcinoma. J Natl Cancer Inst 101:1510–1522

Casper C (2011) The increasing burden of HIV-associated malignancies in resource-limited regions. Annu Rev Med 62:157–170

Chang TT, Gish RG, de Man R, Gadano A, Sollano J, Chao YC, Lok AS, Hamn KH, Goodman Z, Zho J, Cross A, DeHertogh D, Wilber R, Colonno R, Apelian D, Group tBAS (2006) A comparison of entecavir and lamivudine for HBeAg-positive chronic hepatitis B. N Engl J Med 354: 1001–1010

Chang MH, You SL, Chen CJ, Liu CJ, Lee CM, Lin SM, Chu HC, Wu TC, Yang SS, Kuo HS, Chen DS (2009) Decreased incidence of hepatocellular carcinoma in hepatitis B vaccinees: a 20-year follow-up study. J Natl Cancer Inst 101:1348–1355

Chen CJ, Yang HI, Iloeje UH (2009) Hepatitis B virus DNA levels and outcomes in chronic hepatitis B. Hepatology 49:S72–S84

Chua D, Huang J, Zheng B, Lau SY, Luk W, Kwong DL, Sham JS, Moss D, Yuen KY, Im SW, Ng MH (2001) Adoptive transfer of autologous Epstein-Barr virus-specific cytotoxic T cells for nasopharyngeal carcinoma. Int J Cancer 94:73–80

Comoli P, Pedrazzoli P, Maccario R, Basso S, Carminati O, Labirio M, Schiavo R, Secondino S, Frasson C, Perotti C, Moroni M, Locatelli F, Siena S (2005) Cell therapy of stage IV nasopharyngeal carcinoma with autologous Epstein-Barr virus-targeted cytotoxic T lympho-cytes. J Clin Oncol 23:8942–8949

De Paoli P, Pratesi C, Bortolin MT (2007) The Epstein Barr virus DNA levels as a tumor marker in EBV-associated cancers. J Cancer Res Clin Oncol 133:809–815

Di Marco V, Marzano A, Lampertico P, Andreone P, Santantonio T, Almasio PL, Rizzetto M, Craxi A (2004) Clinical outcome of HBeAg-negative chronic hepatitis B in relation to virological response to lamivudine. Hepatology 40:883–891

Dienstag JL (2009) Benefits and risks of nucleoside analog therapy for hepatitis B. Hepatology 49:S112–S121

Edwards TG, Koeller KJ, Slomczynska U, Fok K, Helmus M, Bashkin JK, Fisher C (2011) HPV episome levels are potently decreased by pyrrole-imidazole polyamides. Antiviral Res 91:177–186

El-Serag HB (2012) Epidemiology of viral hepatitis and hepatocellular carcinoma. Gastroen-terology 142:1264–1273.e1

Etoh KI, Yamaguchi K, Tokudome S, Watanabe T, Okayama A, Stuver S, Mueller N, Takatsuki K, Matsuoka M (1999) Rapid quantification of HTLV-1 provirus load: detection of monoclonal proliferation of HTLV-1-infected cells among blood donors. Int J Cancer 81:859–864

Faller DV, Mentzer SJ, Perrine SP (2001) Induction of the Epstein-Barr virus thymidine kinase gene with concomitant nucleoside antivirals as a therapeutic strategy for Epstein-Barr virus-associated malignancies. Curr Opin Oncol 13:360–367

Fang CT, Chen PJ, Chen MY, Hung CC, Chang SC, Chang AL, Chen DS (2003) Dynamics of plasma hepatitis B virus levels after highly active antiretroviral therapy in patients with HIV infection. J Hepatol 39:1028–1035

Feld JJ, Wong DK, Heathcote EJ (2009) Endpoints of therapy in chronic hepatitis B. Hepatology 49:S96–S102

Feng H, Shuda M, Chang Y, Moore PS (2008) Clonal integration of a polyomavirus in human Merkel cell carcinoma. Science 319:1096–1100

Gantt S, Casper C (2011) Human herpesvirus 8-associated neoplasms: the roles of viral replication and antiviral treatment. Curr Opin Infect Dis 24:295–301

Ghosh SK, Perrine SP, Faller DV (2012) Advances in virus-directed therapeutics against Epstein-Barr virus-associated malignancies. Adv Virol 2012:509296

Gill PS, Harrington W Jr, Kaplan MH, Ribeiro RC, Bennett JM, Liebman HA, Bernstein-Singer M, Espina BM, Cabral L, Allen S, Kornblau S, Pike MC, Levine AM (1995) Treatment of adult T-cell leukemia-lymphoma with a combination of interferon alfa and zidovudine. N Engl J Med 332:1744–1748

Gomez-Gutierrez JG, Elpek KG, Montes de Oca-Luna R, Shirwan H, Sam Zhou H, McMasters KM (2007) Vaccination with an adenoviral vector expressing calreticulin-human papillomavirus 16 E7 fusion protein eradicates E7 expressing established tumors in mice. Cancer Immunol Immunother 56:997–1007

Goncalves DU, Proietti FA, Ribas JG, Araujo MG, Pinheiro SR, Guedes AC, Carneiro-Proietti AB (2010) Epidemiology, treatment, and prevention of human T-cell leukemia virus type 1-associated diseases. Clin Microbiol Rev 23:577–589

Gravitt PE, Kovacic MB, Herrero R, Schiffman M, Bratti C, Hildesheim A, Morales J, Alfaro M, Sherman ME, Wacholder S, Rodriguez AC, Burk RD (2007) High load for most high risk human papillomavirus genotypes is associated with prevalent cervical cancer precursors but only HPV16 load predicts the development of incident disease. Int J Cancer 121:2787–2793

Gustafsson Å, Levitsky V, Zou JZ, Frisan T, Dalianis T, Ljungman P, Ringden O, Winiarski J, Ernberg I, Masucci MG (2000) Epstein-Barr virus (EBV) load in bone marrow transplant recipients at risk to develop posttransplant lymphoproliferative disease: prophylactic infusion of EBV-specific cytotoxic T cells. Blood 95:807–814

Hermine O, Bouscary D, Gessain A, Turlure P, Leblond V, Franck N, Buzyn-Veil A, Rio B, Macintyre E, Dreyfus F, Bazarbachi A (1995) Treatment of adult T-cell leukemia-lymphoma with zidovudine and interferon alfa. N Engl J Med 332:1749–1751

Hermine O, Allard I, Lévy V, Arnulf B, Gessain A, Bazarbachi A (2002) A prospective phase II clinical trial with the use of zidovudine and interferon-alpha in the acute and lymphoma forms of adult T-cell leukemia/lymphoma. Hematol J 3:276–282

Hermine O, Dombret H, Poupon J, Arnulf B, Lefrere F, Rousselot P, Damaj G, Delarue R, Fermand JP, Brouet JC, Degos L, Varet B, de The H, Bazarbachi A (2004) Phase II trial of arsenic trioxide and alpha interferon in patients with relapsed/refractory adult T-cell leukemia/lymphoma. Hematol J 5:130–134

Heslop HE, Slobod KS, Pule MA, Hale GA, Rousseau A, Smith CA, Bollard CM, Liu H, Wu MF, Rochester RJ, Amrolia PJ, Hurwitz JL, Brenner MK, Rooney CM (2010) Long-term outcome of EBV-specific T-cell infusions to prevent or treat EBV-related lymphoproliferative disease in transplant recipients. Blood 115:925–935

Hofmann WP, Zeuzem S (2011) A new standard of care for the treatment of chronic HCV infection. Nat Rev Gastroenterol Hepatol 8:257–264

Ingiliz P, Rockstroh JK (2012) HIV-HCV co-infection facing HCV protease inhibitor licensing: implications for clinicians. Liver Int 32:1194–1199

Israel BF, Kenney SC (2003) Virally targeted therapies for EBV-associated malignancies. Oncogene 22:5122–5130

Iwanaga M, Watanabe T, Utsunomiya A, Okayama A, Uchimaru K, Koh KR, Ogata M, Kikuchi H, Sagara Y, Uozumi K, Mochizuki M, Tsukasaki K, Saburi Y, Yamamura M, Tanaka J, Moriuchi Y, Hino S, Kamihira S, Yamaguchi K (2010) Human T-cell leukemia virus type I (HTLV-1) proviral load and disease progression in asymptomatic HTLV-1 carriers: a nationwide prospective study in Japan. Blood 116:1211–1219

Janssen HLA, van Zonneveld M, Senturk H, Zeuzem S, Akarca US, Cakaloglu Y, Simon C, So TMK, Gerken G, de Man RA, Niesters HGM, Zondervan P, Hansen B, Schalm SW (2005) Pegylated interferon alfa-2b alone or in combination with lamivudine for HBeAg-positive chronic hepatitis B: a randomised trial. Lancet 365:123–129

Jones M, Nunez M (2011) HIV and hepatitis C co-infection: the role of HAART in HIV/hepatitis C virus management. Curr Opin HIV AIDS 6:546–552

Jonson AL, Rogers LM, Ramakrishnan S, Downs LS Jr (2008) Gene silencing with siRNA targeting E6/E7 as a therapeutic intervention in a mouse model of cervical cancer. Gynecol Oncol 111:356–364

Josefsson AM, Magnusson PKE, Ylitalo N, Sørensen P, Qwarforth-Tubbin P, Andersen PK, Melbye M, Adami H-O, Gyllensten UB (2000) Viral load of human papilloma virus 16 as a determinant for development of cervical carcinoma in situ: a nested case-control study. Lancet 355:2189–2193

Kchour G, Makhoul NJ, Mahmoudi M, Kooshyar MM, Shirdel A, Rastin M, Rafatpanah H, Tarhini M, Zalloua PA, Hermine O, Farid R, Bazarbachi A (2007) Zidovudine and interferon-alpha treatment induces a high response rate and reduces HTLV-1 proviral load and VEGF plasma levels in patients with adult T-cell leukemia from North East Iran. Leuk Lymphoma 48:330–336

Kchour G, Tarhini M, Kooshyar MM, El Hajj H, Wattel E, Mahmoudi M, Hatoum H, Rahimi H, Maleki M, Rafatpanah H, Rezaee SA, Yazdi MT, Shirdel A, de The H, Hermine O, Farid R, Bazarbachi A (2009) Phase 2 study of the efficacy and safety of the combination of arsenic trioxide, interferon alpha, and zidovudine in newly diagnosed chronic adult T-cell leukemia/lymphoma (ATL). Blood 113:6528–6532

Kean JM, Rao S, Wang M, Garcea RL (2009) Seroepidemiology of human polyomaviruses. PLoS Pathog 5:e1000363

Kim WR, Benson JT, Hindman A, Brosgart C, Fortner-Burton C (2007) Decline in the need for liver transplantation for end stage liver disease secondary to hepatitis B in the US [Abstract]. Hepatology 46(Suppl):238A

Kimura H, Ito Y, Suzuki R, Nishiyama Y (2008) Measuring Epstein-Barr virus (EBV) load: the significance and application for each EBV-associated disease. Rev Med Virol 18:305–319

Koivusalo R, Krausz E, Helenius H, Hietanen S (2005) Chemotherapy compounds in cervical cancer cells primed by reconstitution of p53 function after short interfering RNA-mediated degradation of human papillomavirus 18 E6 mRNA: opposite effect of siRNA in combination with different drugs. Mol Pharmacol 68:372–382

Krasagakis K, Kruger-Krasagakis S, Tzanakakis GN, Darivianaki K, Stathopoulos EN, Tosca AD (2008) Interferon-alpha inhibits proliferation and induces apoptosis of merkel cell carcinoma in vitro. Cancer Invest 26:562–568

Kutok JL, Wang F (2006) Spectrum of Epstein-Barr virus-associated diseases. Annu Rev Pathol Mech Dis 1:375–404

Lacombe K, Rockstroh J (2012) HIV and viral hepatitis coinfections: advances and challenges. Gut 61(Suppl 1):i47–i58

Lai CL, Shouval D, Lok AS, Chang TT, Cheinquer H, Goodman Z, DeHertogh D, Wilber R, Zink RC, Cross A, Colonno R, Fernandes L, Group tBAS (2006) Entecavir versus lamivudine for patients with HBeAg-negative chronic hepatitis B. N Engl J Med 354:1011–1020

Lai CL, Gane E, Liaw YF, Hsu CW, Thongsawat S, Wang Y, Chen Y, Heathcote EJ, Rasenack J, Bzowej B, Naoumov NV, Di Bisceglie AM, Zeuzem S, Moon YM, Goodman Z, Chao G,

Constance BF, Brown NA, Group tGS (2007) Telbivudine versus lamivudine in patients with chronic hepatitis B. N Engl J Med 357:2576–2588

Lau GKK, Piratvisuth T, Luo KX, Marcellin P, Thongsawat S, Cooksley G, Gane E, Fried MW, Chow WC, Paik SW, Chang WY, Berg T, Flisiak R, McCloud P, Pluck N, Group tPA-aH-PCHBS (2005) Peginterferon alfa-2a, lamivudine, and combination for HBeAg-positive chronic hepatitis B. N Engl J Med 352:2682–2695

Liaw YF (2006) Hepatitis B virus replication and liver disease progression: the impact of antiviral therapy. Antivir Ther 11:669–679

Liaw YF, Sung JJ, Chow WC, Farrell G, Lee CZ, Yuen H, Tanwandee T, Tao QM, Shue K, Keene ON, Dixon JS, Gray DF, Sabbat J, Group CALMS (2004) Lamivudine for patients with chronic hepatitis B and advanced liver disease. N Engl J Med 351:1521–1531

Liu CJ, Chuang WL, Lee CM, Yu ML, Lu SN, Wu SS, Liao LY, Chen CL, Kuo HT, Chao YC, Tung SY, Yang SS, Kao JH, Liu CH, Su WW, Lin CL, Jeng YM, Chen PJ, Chen DS (2009) Peginterferon alfa-2a plus ribavirin for the treatment of dual chronic infection with hepatitis B and C viruses. Gastroenterology 136:496–504.e3

Louis CU, Straathof K, Bollard CM, Ennamuri S, Gerken C, Lopez TT, Huls MH, Sheehan A, Wu MF, Liu H, Gee A, Brenner MK, Rooney CM, Heslop HE, Gottschalk S (2010) Adoptive transfer of EBV-specific T cells results in sustained clinical responses in patients with locoregional nasopharyngeal carcinoma. J Immunother 33:983–990

Loyo M, Guerrero-Preston R, Brait M, Hoque MO, Chuang A, Kim MS, Sharma R, Liegeois NJ, Koch WM, Califano JA, Westra WH, Sidransky D (2010) Quantitative detection of Merkel cell virus in human tissues and possible mode of transmission. Int J Cancer 126:2991–2996

Manns A, Miley WJ, Wilks RJ, Morgan OSC, Hanchard B, Wharfe G, Caranston B, Maloney W, Welles SL, Blattner WA, Waters D (1999) Quantitative proviral DNA and antibody levels in the natural history of HTLV-1 infection. J Infect Dis 180:1487–1493

Marcellin P, Lau GKK, Bonino F, Farci P, Hadziyannis S, Jin R, Lu ZM, Piratvisuth T, Germanidis G, Yurdaydin C, Diago M, Gurel S, Lai MY, Button P, Pluck N, Group tPA-aH-NCHBS (2004) Peginterferon alfa-2a alone, lamivudine alone, and the two in combination in patients with HBeAg-negative chronic hepatitis B. N Engl J Med 351:1206–1217

Marcellin P, Heathcote EJ, Buti M, Gane E, de Man RA, Krastev Z, Germanidis G, Lee SS, Flisiak R, Kaita K, Manns M, Kotzev I, Tchernev K, Buggisch P, Weilert F, Kurdas OO, Shiffman ML, Trinh H, Washington MK, Sorbel J, Anderson J, Snow-Lampart A, Mondou E, Quinn J, Rousseau F (2008) Tenofovir disoproxil fumarate versus adefovir dipivoxil for chronic hepatitis B. N Engl J Med 359:2442–2455

Marcellin P, Bonino F, Lau GK, Farci P, Yurdaydin C, Piratvisuth T, Jin R, Gurel S, Lu ZM, Wu J, Popescu M, Hadziyannis S (2009) Sustained response of hepatitis B e antigen-negative patients 3 years after treatment with peginterferon alpha-2a. Gastroenterology 136:2169–2179.e1–4

Masarone M, Persico M (2011) Antiviral therapy: why does it fail in HCV-related chronic hepatitis? Expert Rev Anti Infect 9:535–543

Masuzaki R, Yoshida H, Omata M (2010) Interferon reduces the risk of hepatocellular carcinoma in hepatitis C virus-related chronic hepatitis/liver cirrhosis. Oncology 78(Suppl 1):17–23

Matutes E, Taylor GP, Cavenagh J, Pagliuca A, Bareford D, Domingo A, Hamblin M, Kelsey S, Mir N, Reilly JT (2001) Interferon α and zidovudine therapy in adult T-cell leukaemia lymphoma: response and outcome in 15 patients. Br J Haematol 113:779–784

Merlo A, Turrini R, Dolcetti R, Zanovello P, Rosato A (2011) Immunotherapy for EBV-associated malignancies. Int J Hematol 93:281–293

Mesri EA, Cesarman E, Boshoff C (2010) Kaposi's sarcoma and its associated herpesvirus. Nat Rev Cancer 10:707–719

Mommeja-Marin H, Mondou E, Blum MR, Rousseau F (2003) Serum HBV DNA as a marker of efficacy during therapy for chronic HBV infection: analysis and review of the literature. Hepatology 37:1309–1319

Monini P, Sgadari C, Garosso MG, Bellino S, Biagio AD, Toschi E, Bacigalupo S, Sabbatucci M, Cencioni G, Salvi E, Leone P, Ensoli B, Sarcoma tCAoKs (2009) Clinical course of classic Kaposi's sarcoma in HIV-negative patients treated with the HIV protease inhibitor indinavir. AIDS 23:534–538

Moore PS, Chang Y (2010) Why do viruses cause cancer? Highlights of the first century of human tumour virology. Nat Rev Cancer 10:878–889

Munir S, Saleem S, Idrees M, Tariq A, Butt S, Rauff B, Hussain A, Badar S, Naudhani M, Fatima Z, Ali M, Ali L, Akram M, Aftab M, Khubaib B, Awan Z (2010) Hepatitis C treatment: current and future perspectives. Virol J 7:296

Nakajima H, Takaishi M, Yamamoto M, Kamijima R, Kodama H, Tarutani M, Sano S (2009) Screening of the specific polyoma virus as diagnostic and prognostic tools for Merkel cell carcinoma. J Dermatol Sci 56:211–213

Nguyen HQ, Magaret AS, Kitahata MM, van Rompaey SE, Wald A, Casper C (2008) Persistent Kaposi sarcoma in era of highly active antiretroviral therapy: characterizing the predictors of clinical response. AIDS 22:937–945

Niederau C, Heintges T, Lange S, Goldmann G, Niederau CM, Mohr L, Häussinger D (1996) Long-term follow-up of HBeAg-positive patients treated with interferon alfa for chronic hepatitis B. N Engl J Med 334:1422–1427

Okayama A, Stuver S, Matsuoka M, Ishizaki J, Tanaka G, Kubuki Y, Mueller N, Hsieh CC, Tachibana N, Tsubouchi H (2004) Role of HTLV-1 proviral DNA load and clonality in the development of adult T-cell leukemia/lymphoma in asymptomatic carriers. Int J Cancer 110:621–625

Papatheodoridis GV, Papatheodoridis VC, Hadziyannis SJ (2001) Effect of interferon therapy on the development of hepatocellular carcinoma in patients with hepatitis C virus-related cirrhosis: a meta-analysis. Aliment Pharmacol Ther 15:689–698

Papatheodoridis GV, Manolakopoulos S, Dusheiko G, Archimandritis AJ (2008) Therapeutic strategies in the management of patients with chronic hepatitis B virus infection. Lancet Infect Dis 8:167–178

Papatheodoridis GV, Lampertico P, Manolakopoulos S, Lok A (2010) Incidence of hepatocellular carcinoma in chronic hepatitis B patients receiving nucleos(t)ide therapy: a systematic review. J Hepatol 53:348–356

Perrine SP, Hermine O, Small T, Suarez F, O'Reilly R, Boulad F, Fingeroth J, Askin M, Levy A, Mentzer SJ, Di Nicola M, Gianni AM, Klein C, Horwitz S, Faller DV (2007) A phase 1/2 trial of arginine butyrate and ganciclovir in patients with Epstein-Barr virus-associated lymphoid malignancies. Blood 109:2571–2578

Poiesz BJ, Ruscetti FW, Gazdar AF, Bunn PA, Minna JD, Gallo RC (1980) Detection and isolation of type C retrovirus particles from fresh and cultured lymphocytes of a patient with cutaneous T-cell lymphoma. Proc Natl Acad Sci U S A 77:7514–7519

Potthoff A, Manns MP, Wedemeyer H (2010) Treatment of HBV/HCV coinfection. Expert Opin Pharmacother 11:919–928

Proietti FA, Carneiro-Proietti AB, Catalan-Soares BC, Murphy EL (2005) Global epidemiology of HTLV-I infection and associated diseases. Oncogene 24:6058–6068

Putral LN, Bywater MJ, Gu W, Saunders NA, Gabrielli BG, Leggatt GR, McMillan NA (2005) RNA interference against human papillomavirus oncogenes in cervical cancer cells results in increased sensitivity to cisplatin. Mol Pharmacol 68:1311–1319

Qu LS, Chen H, Kuai XL, Xu ZF, Jin F, Zhou GX (2012) Effects of interferon therapy on development of hepatocellular carcinoma in patients with hepatitis C-related cirrhosis: A meta-analysis of randomized controlled trials. Hepatol Res 42:782–789

Ratner L, Harrington W, Feng X, Grant C, Jacobson S, Noy A, Sparano J, Lee J, Ambinder R, Campbell N, Lairmore M (2009) Human T cell leukemia virus reactivation with progression of adult T-cell leukemia-lymphoma. PLoS ONE 4:e4420

Rooney CM, Smith CA, Ng CYC, Loftin S, Li C, Krance RA, Brenner MK, Heslop HE (1995) Use of gene-modified virus-specific T lymphocytes to control Epstein-Barr-virus-related lymphoproliferation. Lancet 345:9–13

Schrama D, Ugurel S, Becker JC (2012) Merkel cell carcinoma: recent insights and new treatment options. Curr Opin Oncol 24:141–149

Sima N, Wang S, Wang W, Kong D, Xu Q, Tian X, Luo A, Zhou J, Xu G, Meng L, Lu Y, Ma D (2007) Antisense targeting human papillomavirus type 16 E6 and E7 genes contributes to apoptosis and senescence in SiHa cervical carcinoma cells. Gynecol Oncol 106:299–304

Singal AK, Singh A, Jaganmohan S, Guturu P, Mummadi R, Kuo YF, Sood GK (2010) Antiviral therapy reduces risk of hepatocellular carcinoma in patients with hepatitis C virus-related cirrhosis. Clin Gastroenterol Hepatol 8:192–199

Soriano V, Vispo E, Labarga P, Medrano J, Barreiro P (2010) Viral hepatitis and HIV co-infection. Antiviral Res 85:303–315

Straathof KCM, Bollard CM, Popat U, Huls MH, Lopez T, Morriss MC, Gresik MV, Gee AP, Russell HV, Brenner MK, Rooney CM, Heslop HE (2005) Treatment of nasopharyngeal carcinoma with Epstein-Barr virus-specific T lymphocytes. Blood 105:1898–1904

Sung JJ, Tsoi KK, Wong VW, Li KC, Chan HL (2008) Meta-analysis: treatment of hepatitis B infection reduces risk of hepatocellular carcinoma. Aliment Pharmacol Ther 28:1067–1077

Sunil M, Reid E, Lechowicz MJ (2010) Update on HHV-8-associated malignancies. Curr Infect Dis Rep 12:147–154

Tolstov YL, Pastrana DV, Feng H, Becker JC, Jenkins FJ, Moschos S, Chang Y, Buck CB, Moore PS (2009) Human Merkel cell polyomavirus infection II. MCV is a common human infection that can be detected by conformational capsid epitope immunoassays. Int J Cancer 125:1250–1256

Toy M, Veldhuijzen IK, De Man RA, Richardus J, Schalm SW (2008) The potential impact of long-term nucleoside therapy on the mortality and morbidity of high viremic chronic hepatitis B [Abstract]. Hepatology 48(Suppl):717A

Tsukasaki K, Hermine O, Bazarbachi A, Ratner L, Ramos JC, Harrington W Jr, O'Mahony D, Janik JE, Bittencourt AL, Taylor GP, Yamaguchi K, Utsunomiya A, Tobinai K, Watanabe T (2009) Definition, prognostic factors, treatment, and response criteria of adult T-cell leukemia-lymphoma: a proposal from an international consensus meeting. J Clin Oncol 27:453–459

van Zonneveld M, Honkoop P, Hansen BE, Niesters HG, Darwish Murad S, de Man RA, Schalm SW, Janssen HL (2004) Long-term follow-up of alpha-interferon treatment of patients with chronic hepatitis B. Hepatology 39:804–810

Willmes C, Adam C, Alb M, Volkert L, Houben R, Becker JC, Schrama D (2012) Type I and II IFNs inhibit Merkel cell carcinoma via modulation of the Merkel cell polyomavirus T antigens. Cancer Res 72:2120–2128

Wong GL, Yiu KK, Wong VW, Tsoi KK, Chan HL (2010) Meta-analysis: reduction in hepatic events following interferon-alfa therapy of chronic hepatitis B. Aliment Pharmacol Ther 32:1059–1068

Woodman CB, Collins SI, Young LS (2007) The natural history of cervical HPV infection: unresolved issues. Nat Rev Cancer 7:11–22

Xi LF, Hughes JP, Castle PE, Edelstein ZR, Wang C, Galloway DA, Koutsky LA, Kiviat NB, Schiffman M (2011) Viral load in the natural history of human papillomavirus type 16 infection: a nested case-control study. J Infect Dis 203:1425–1433

Yang YF, Zhao W, Zhong YD, Xia HM, Shen L, Zhang N (2009) Interferon therapy in chronic hepatitis B reduces progression to cirrhosis and hepatocellular carcinoma: a meta-analysis. J Viral Hepat 16:265–271

Yeh ML, Hung CH, Huang JF, Liu CJ, Lee CM, Dai CY, Wang JH, Lin ZY, Lu SN, Hu TH, Yu ML, Kao JH, Chuang WL, Chen PJ, Chen DS (2011) Long-term effect of interferon plus ribavirin on hepatitis B surface antigen seroclearance in patients dually infected with hepatitis B and C viruses. PLoS ONE 6:e20752

Zhang CH, Xu GL, Jia WD, Li JS, Ma JL, Ge YS (2011) Effects of interferon treatment on development and progression of hepatocellular carcinoma in patients with chronic virus infection: a meta-analysis of randomized controlled trials. Int J Cancer 129:1254–1264

Zhou J, Peng C, Li B, Wang F, Zhou C, Hong D, Ye F, Cheng X, Lu W, Xie X (2012) Transcriptional gene silencing of HPV16 E6/E7 induces growth inhibition via apoptosis in vitro and in vivo. Gynecol Oncol 124:296–302

Zwaveling S, Ferreira Mota SC, Nouta J, Johnson M, Lipford GB, Offringa R, van der Burg SH, Melief CJM (2002) Established human papillomavirus type 16-expressing tumors are effectively eradicated following vaccination with long peptides. J Immunol 169:350–358

Lightning Source UK Ltd.
Milton Keynes UK
UKOW07n1019081214

242770UK00002B/18/P